MW00846266

Springer Series in Materials Science

Volume 228

The Springer Series in Materials Science covers the complete spectrum of materials physics, including fundamental principles, physical properties, materials theory and design. Recognizing the increasing importance of materials science in future device technologies, the book titles in this series reflect the state-of-the-art in understanding and controlling the structure and properties of all important classes of materials.

More information about this series at http://www.springer.com/series/856

Jan Seidel
Editor

Topological Structures in Ferroic Materials

Domain Walls, Vortices and Skyrmions

Springer

Editor
Jan Seidel
School of Materials Science and Engineering
University of New South Wales
Sydney, NSW
Australia

ISSN 0933-033X ISSN 2196-2812 (electronic)
Springer Series in Materials Science
ISBN 978-3-319-25299-5 ISBN 978-3-319-25301-5 (eBook)
DOI 10.1007/978-3-319-25301-5

Library of Congress Control Number: 2015958337

Printed on acid-free paper

This Springer imprint is published by SpringerNature
The registered company is Springer International Publishing AG Switzerland

Preface

Topological defects play important roles in nature. They are found in fields as diverse as cosmology, particle physics, superfluidity, liquid crystals, and metallurgy, manifesting themselves as screw/edge dislocations in liquid crystals, magnetic flux tubes in superconductors, and vortices in superfluids, for example. They can also be found in ferroic materials, i.e., materials with a spontaneous reversible ordering, such as magnetic materials, ferroelectrics, and ferroelastic materials, which have been studied for a long time and are used widely in sensors, actuators, information technology, and other smart materials applications. Ferroic phases can arise in more than one distinct orientation of the order parameter, thus spatial variations in the orientations of the order parameter are accommodated through the formation of discrete domain structures. Adjacent domains are separated by naturally occurring planar topological defects called domain walls. Over the last few years they have been intensively investigated with respect to their inherent functional behavior in various studies involving ceramics, thin films, and single crystalline material. The fact that the electronic conductivity of domain walls in ferroelectrics and multiferroics can be utilized for nanoscale functional elements and new concepts involving magnetic domain walls for memory and spintronic applications have sparked wide interest and have led to the finding of unique properties associated with such topological structures. The understanding of these phenomena has progressed to a point where we can say that the physical properties of topological structures such as domain walls can be completely different from those of the parent bulk material phase.

Although domain walls are the commonly understood concept of ferroic order, they are not the only way in which spatially varying order parameters can be arranged. Alternatively, more complex patterns can develop in which the order parameter changes in different ways as described, e.g., by the topological theory of defects in ordered media by N.D. Mermin. When combined with local defects or singularities, the number of possible geometrical patterns and textures that arise can be manifold and include vortex structures and skyrmions. Which kind of ferroic micro- or nanostructure developments depends critically on the relative magnitudes of various energies associated with exchange, spin–orbit interaction, crystallographic anisotropy, and surfaces and interfaces in ferroic materials. Physical dimensions of the specific material and its morphology are also important in determining the exact nature of the ferroic patterns that develop in equilibrium, which include flux-closure structures and other complex topological patterns such as periodic arrays of magnetic skyrmions that can be observed, e.g., by magnetic force microscopy.

It is only within the last few years that experiments have focused on trying to find and study such topological structures in ferroic materials. It is clear today that these interesting structures can form in ferroelectrics and magnetic systems; however, the study of their basic properties and the exploration of their potential for future applications has only just begun. This book is an effort to capture some of the interesting developments in this rapidly changing field of research. Tuning, manipulating, and exploiting the physical properties of such topological structures provides a new playground for condensed matter and functional materials research. In addition, it offers a novel platform for future nanotechnology.

Sydney Jan Seidel

Contents

Contributors

O. Aktas ETH Zürich, Zurich, Switzerland

Andreas Bauer Physik-Department, Technische Universität München, Garching, Germany

Petr Bednyakov Ceramics Laboratory, Swiss Federal Institute of Technology (EPFL), Lausanne, Switzerland

Shin-Liang Chin Cavendish Laboratory, University of Cambridge, Cambridge, UK

Russell P. Cowburn Cavendish Laboratory, University of Cambridge, Cambridge, UK

Arnaud Crassous Ceramics Laboratory, Swiss Federal Institute of Technology (EPFL), Lausanne, Switzerland

X. Ding State Key Laboratory for Mechanical Behavior of Materials, Xi'an Jiaotong University, Xi'an, People's Republic of China

Eugene A. Eliseev Institute for Problems of Materials Science, National Academy of Sciences of Ukraine, Kyiv, Ukraine

Amalio Fernández-Pacheco Cavendish Laboratory, University of Cambridge, Cambridge, UK

Markus Garst Institut Für Theoretische Physik, Universität Zu Köln, Cologne, Germany

J. Hlinka Institute of Physics, Czech Academy of Science, Praha 8, Czech Republic

Sergei V. Kalinin Oak Ridge National Laboratory, Center for Nanophase Materials Science, Oak Ridge, TN, USA

Mathias Kläui Institute of Physics and Materials Science in Mainz, Johannes Gutenberg University Mainz, Mainz, Germany

Benjamin Krüger Institute of Physics, Johannes Gutenberg University Mainz, Mainz, Germany

Reinoud Lavrijsen Cavendish Laboratory, University of Cambridge, Cambridge, UK; Department of Applied Physics, Center for NanoMaterials, Eindhoven University of Technology, Eindhoven, MB, The Netherlands

JiHyun Lee Cavendish Laboratory, University of Cambridge, Cambridge, UK

Dishant Mahendru Cavendish Laboratory, University of Cambridge, Cambridge, UK

Rhodri Mansell Cavendish Laboratory, University of Cambridge, Cambridge, UK

P. Marton Institute of Physics, Czech Academy of Science, Praha 8, Czech Republic

Masahito Mochizuki Department of Physics and Mathematics, PRESTO, Japan Science and Technology Agency, Aoyama Gakuin University, Tokyo, Japan

Anna N. Morozovska Institute of Physics, National Academy of Sciences of Ukraine, Kyiv, Ukraine

P. Ondrejkovic Institute of Physics, Czech Academy of Science, Praha 8, Czech Republic

Dorothee Petit Cavendish Laboratory, University of Cambridge, Cambridge, UK

Christian Pfleiderer Physik-Department, Technische Universität München, Garching, Germany

E.K.H. Salje Department of Earth Sciences, University of Cambridge, Cambridge, UK

James F. Scott Cavendish Laboratory, Department of Physics, University of Cambridge, Cambridge, UK; Departments of Chemistry and Physics, St. Andrews University, Fife, UK

Tomas Sluka Ceramics Laboratory, Swiss Federal Institute of Technology (EPFL), Lausanne, Switzerland; Department of Quantum Matter Physics, University of Geneva, Geneva, Switzerland

V. Stepkova Institute of Physics, Czech Academy of Science, Praha 8, Czech Republic

Alexander Tagantsev Ceramics Laboratory, Swiss Federal Institute of Technology (EPFL), Lausanne, Switzerland

Alexander Welbourne Cavendish Laboratory, University of Cambridge, Cambridge, UK

Petr Yudin Ceramics Laboratory, Swiss Federal Institute of Technology (EPFL), Lausanne, Switzerland

Chapter 1
Generic Aspects of Skyrmion Lattices in Chiral Magnets

Andreas Bauer and Christian Pfleiderer

Abstract Magnetic skyrmions are topologically non-trivial spin whirls that may not be transformed continuously into topologically trivial states such as ferromagnetic spin alignment. In recent years lattice structures composed of skyrmions have been discovered in certain bulk chiral magnets with non-centrosymmetric crystal structures. The magnetic phase diagrams of these materials share remarkable similarities despite great variations of the characteristic temperature, field, and length scales and regardless whether the underlying electronic state is that of a metal, semiconductor, or insulator.

1.1 Introduction and Outline

In 1961 British nuclear physicist Tony Skyrme proposed a theoretical model in which neutrons and protons arise as topological solitons of pion fields, i.e., fermions are derived from bosonic fields [1–3]. Representing the, perhaps, first example of what is now broadly referred to as fractionalization, the implications of Skyrme's model only began to be fully appreciated two decades later, when Witten and Adkins demonstrated its relevance for real experiments [4]. Since the days of this early work many different variants of Skyrme's original notion have been worked out in entirely different fields of physics. These states and excitations are now rather generously called *skyrmions*. Examples include areas as diverse as particle physics [4–8], the quantum Hall state at half-filling [9–11], Bose-Einstein condensates [12–14], and liquid crystals [15]. However, in recent years skyrmions are probably most actively investigated in the area of solid state magnetism, where certain spin textures are referred to as skyrmions. These magnetic textures display a non-trivial real-space topology, i.e.,

A. Bauer (✉) · C. Pfleiderer
Physik-Department, Technische Universität München, James-Franck-Straße, D-85748 Garching, Germany
e-mail: Andreas.Bauer@frm2.tum.de

C. Pfleiderer
e-mail: Christian.Pfleiderer@frm2.tum.de

J. Seidel (ed.), *Topological Structures in Ferroic Materials*, Springer Series in Materials Science 228, DOI 10.1007/978-3-319-25301-5_1

it is not possible to continuously transform them into conventional (topologically trivial) forms of spin order such as ferromagnetism or antiferromagnetism.

While skyrmions were theoretically predicted to exist in non-centrosymmetric magnetic materials with uniaxial anisotropy as early as 1989 [16, 17], it was despite concerted efforts rather unexpected, when skyrmions in magnetic materials were identified experimentally for the first time in the cubic transition metal compounds MnSi [18] and $Fe_{1-x}Co_xSi$ [19] in the form of a lattice structure . Since then similar topologically non-trivial spin textures have been reported to exist for a rapidly growing number of rather different bulk and thin film systems. The interest driving this search for further materials stabilizing skyrmions is quite diverse, ranging from fundamental questions on the possible break down of Fermi liquid theory [20–22] all the way to new forms of spintronics applications [23]. From a practical point of view the most important implication of the non-trivial topology is their emergent electrodynamics leading to an exceptionally efficient coupling between the spin textures and spin currents [24, 25]. Further, the very detailed understanding of the spin excitations achieved to date suggests strongly that tailored microwave devices may be designed through the combination of different materials [26–30].

A precondition for further advances is a detailed understanding of the magnetic phase diagrams of these compounds. In turn, this chapter provides a review of the most extensively studied class of skyrmion materials to date, namely cubic chiral magnets crystallizing in the space group $P2_13$. We begin in Sect. 1.2 with a brief introduction to the basic properties of this class of compounds focusing on the salient properties of the skyrmion lattice state. This is followed by an introduction to the Ginzburg-Landau model of these materials in Sect. 1.3. The main part of this chapter in Sect. 1.4 is dedicated to an account of the determination of magnetic phase diagrams based on measurements of thermodynamic bulk properties. Despite great variations of the characteristic temperature, field, and length scales between the different materials of interest, the magnetic phase diagrams observed are remarkably similar. This brings us to a summary of the main consequences that arise from the non-trivial topological winding of skyrmions in Sect. 1.5, in particular their emergent electrodynamics. The chapter closes in Sect. 1.6 with a brief account of topologically non-trivial spin structures as recently discovered in other materials.

1.2 Skyrmion Lattice in Cubic Chiral Magnets

The helimagnetism of the materials reviewed in this chapter is homochiral with a modulation wavelength that is large as compared to typical lattice constants. The latter represents an important precondition for the description of the magnetic properties in a continuum model and the characterization of the topological properties. Well-known representatives are the (pseudo-)binary $B20$ transition metal monosilicides and monogermanides MnSi, $Mn_{1-x}Fe_xSi$, $Mn_{1-x}Co_xSi$, $Fe_{1-x}Co_xSi$, FeGe, MnGe, and mixtures thereof, as well as the insulator Cu_2OSeO_3. All of these compounds crystallize in the space group $P2_13$, which lacks inversion symmetry such that two crystalline enantiomers stabilize.

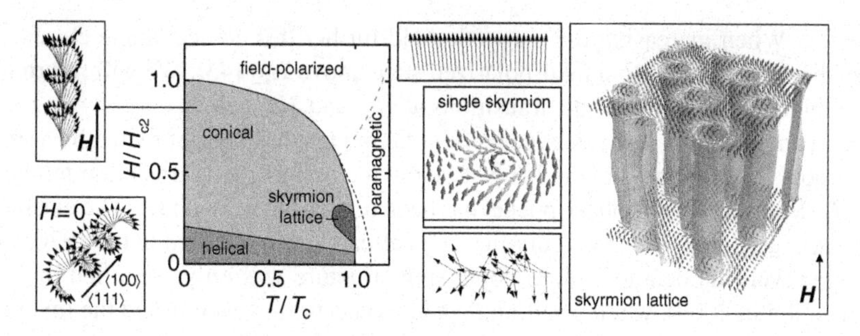

Fig. 1.1 Spin structures of cubic chiral magnets. Typical magnetic phase diagram (*center*) and schematic spin structures of the helical, the conical, the paramagnetic, and the field-polarized state. In a phase pocket (*red*) in finite fields just below the helimagnetic ordering temperature, T_c, a regular arrangement of topologically non-trivial spin whirls is observed, a so-called skyrmion lattice. Schematic depictions by Markus Garst and from [31]

The long-wavelength helimagnetic order observed in these compounds originates in a well-understood set of hierarchical energy scales, as already pointed out in Landau-Lifshitz, Vol. VIII, Sect. 52, [32]. On the strongest scale exchange interactions favor parallel spin alignment. On intermediate scales isotropic Dzyaloshinskii-Moriya spin-orbit interactions arise due to the lack of inversion symmetry of the crystal structure favoring perpendicular spin alignment [33–35]. In competition with the stronger exchange a helical modulation is stabilized [36, 37]. The chirality of the Dzyaloshinskii-Moriya interaction and thus of the helical modulation is fixed by the enantiomer of the crystal structure [38, 39]. Finally, on the weakest energy scale higher-order spin-orbit coupling terms, also referred to as crystal electric field effects or cubic anisotropies, determine the propagation direction of the helical modulations [40].

The hierarchy of energy scales is directly reflected in a rather universal magnetic phase diagram, as schematically depicted in Fig. 1.1. As summarized below, the same phase diagram is observed regardless whether the materials are metals, semiconductors, and insulators (MnGe is perhaps the only exception as discussed in Sect. 1.6). In particular, the phase diagram appears to be insensitive to the quantitative values of the transition temperatures, transition fields, and helix wavelengths, which vary by roughly two orders of magnitude between different compounds.

At sufficiently high temperatures the magnetic properties are characteristic of exchange-enhanced paramagnetism with large fluctuating moments [41]. At low temperatures and zero magnetic field multi-domain helical order is observed with equal domain populations, where the helical propagation vector is determined by weak cubic magnetic anisotropies, fourth-order in spin-orbit coupling. Under small applied magnetic fields the domain population changes, until the helical state undergoes a spin-flop transition at a transition field H_{c1} [42]. The spin-flop phase is broadly referred to as conical state, with a single-domain state of spin spirals propagating along the magnetic field direction. The expression conical phase alludes to the notion, that the spins tilt towards the field direction while twisting helically along to the field direction. When increasing the magnetic field further this conical angle closes and a transition takes place to a field-polarized state above H_{c2} [43]. We will return to a more detailed discussion of the transitions at H_{c1} and H_{c2} below.

In recent years the perhaps largest scientific interest has been attracted by a small phase pocket at intermediate fields just below the helimagnetic transition temperature, T_c. Historically this phase pocket has been referred to as A-phase. The existence of the A-phase, first discovered in MnSi, had already been reported in the 1970s [44, 45]. However, the detailed microscopic spin structure was only identified in 2008 (publication in 2009), when small-angle neutron scattering established the first realization of a skyrmion lattice in a bulk solid state system [18].

The skyrmion lattice consists of a regular hexagonal arrangement of spin whirls, that may essentially be described by the phase-locked superposition of three helices under 120° in a plane perpendicular to the applied magnetic field in combination with a ferromagnetic component along the field. Of particular interest is the non-trivial topology of this spin texture, meaning, it cannot be continuously transformed into a topologically trivial state such as a paramagnet, ferromagnet, or helimagnet. The associated winding number of the structure, Φ, is an integer and the integrated value of the skyrmion density, ϕ_i, per magnetic unit cell, given by

$$\phi_i = \frac{1}{8\pi} \epsilon_{ijk} \hat{\psi} \cdot \partial_j \hat{\psi} \times \partial_k \hat{\psi} \tag{1.1}$$

where, ϵ_{ijk} is the antisymmetric unit tensor and $\hat{\psi} = \boldsymbol{M}(\boldsymbol{r})/M(\boldsymbol{r})$ is the orientation of the local magnetization. Along the field direction the quasi two-dimensional spin structure repeats itself, forming skyrmion lines as depicted in the right panel of Fig. 1.1. Perhaps most intriguing, the interaction of each skyrmion with an electron spin corresponds to one quantum of emergent flux and an emergent electrodynamics presented in Sect. 1.5.

Experimentally, the existence of skyrmions was first recognized in the form of the skyrmion lattice as observed in reciprocal space using small-angle neutron scattering (SANS) in bulk samples [18, 19, 48–50]. Further detailed SANS studies on MnSi revealed the presence of weak higher-order scattering, indicating a weak particle-like character of the skyrmions. The evolution of this higher-order scattering as a function of temperature and field proved the long-range crystalline nature of the skyrmion lattice and, in particular, the phase-locked multi-Q nature of the modulation at heart of the non-trivial topological winding [46]. These measurements were soon followed-up by real-space imaging studies using Lorentz force transmission electron microscopy (LF-TEM). This method is sensitive to in-plane components of the magnetic moments. However, it may only be used to study thinned bulk samples [26, 47, 51, 52], whereas magnetic force microscopy (MFM) allowed the detection of the stray magnetic field above the surface of bulk samples [31]. As the most recent achievement of real-space imaging, the spin arrangement in the skyrmion lattice could even be reconstructed in three dimensions by means of electron holography [53].

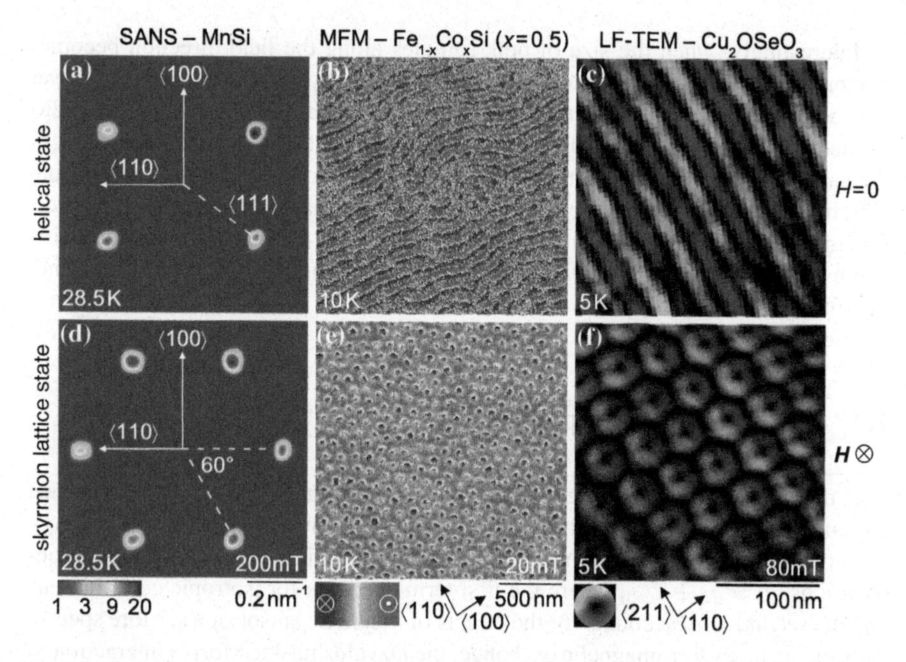

Fig. 1.2 Helical and skyrmion lattice state as observed in reciprocal and real space. **a–c** Helical state in zero magnetic field. **d–f** Skyrmion lattice state in finite field. Data from small angle neutron scattering (SANS) [18, 46], magnetic force microscopy (MFM) [31], and Lorentz force transmission electron microscopy (LF-TEM) [26, 47] are shown. The color-coded in-plane orientation in the LF-TEM data was obtained by a transport-of-intensity (TIE) analysis

Typical data from SANS, MFM, and LF-TEM recorded on different chiral magnets are shown for the helical and the skyrmion lattice state in Fig. 1.2. In the helical state at zero magnetic field SANS experiments show intensity maxima along the easy axes of the helical propagation vector $\boldsymbol{q}$, typically either ⟨100⟩ or ⟨111⟩ [36, 40]. Real-space images reveal stripy patterns with $\boldsymbol{q}$ perpendicular to the stripes [54]. The skyrmion lattice state in finite fields in SANS experiments, see Fig. 1.2d, is characterized by a sixfold scattering pattern in a plane perpendicular to the applied magnetic field that is only fully revealed if the magnetic field is applied parallel to the neutron beam. In earlier experiments the magnetic field and the neutron beam had been applied perpendicular to each other leading to erroneous interpretations [42, 55–57]. Note that the wave vector in the skyrmion lattice has the same absolute value as in the helical state, $q = 2\pi/\lambda_h$. Thus, due to the hexagonal packing of the skyrmions in real space, the distance between neighboring skyrmion cores is a factor of $2/\sqrt{3} \approx 1.15$ larger than the helix wavelength. In real-space images, see Fig. 1.2e, f, a hexagonal lattice of objects is observed. The magnetic moments in their cores are aligned antiparallel to the applied field, cf. blue color in Fig. 1.2e, i.e., the spin structure in the cubic chiral magnets in fact consists of anti-skyrmions.

Interestingly, when the size of bulk samples along the field direction becomes comparable to the helical modulation length, the skyrmion lattice extents over increasingly larger parts of the magnetic phase diagram as demonstrated in LF-TEM studies [51]. In contrast, the magnetic properties of epitaxially grown thin films of the same chiral magnets, forming equal crystalline domain populations with both chiralities in the same film, are still debated controversially [58–61]. Here, in addition to the effects resulting from the heterochirality and the reduced dimensionality, strain arising from the lattice mismatch with the substrate needs to be taken into account.

1.3 Theoretical Description

The thermodynamic properties of the cubic chiral magnets may be described extremely well in the framework of a Ginzburg-Landau model of the free energy density, see also chapter by Markus Garst. It is convenient to distinguish two contributions, $f = f_0 + f_{\text{cub}}$, where the first term accounts for isotropic contributions and the second term accounts for the effects of magnetic anisotropies. More specifically, f_0 includes ferromagnetic exchange, the Dzyaloshinskii-Moriya interaction as the highest-order (isotropic) spin-orbit coupling term, and the Zeeman term as the response on an external magnetic field. It may be written as:

$$f_0 = \frac{1}{2}\boldsymbol{\psi}(r - J\nabla^2)\boldsymbol{\psi} + D\boldsymbol{\psi}(\nabla \times \boldsymbol{\psi}) + \frac{u}{4!}(\boldsymbol{\psi}^2)^2 - \mu_0\mu\boldsymbol{\psi}\boldsymbol{H} \tag{1.2}$$

We choose the three component order parameter field, $\boldsymbol{\psi}$, with dimensionless units yielding a magnetization density $\boldsymbol{M} = \mu\boldsymbol{\psi}$ with $\mu = \mu_{\text{B}}/\text{f.u.}$, i.e., a single Bohr magneton per formula unit ($\mu_{\text{B}} > 0$). The parameter r tunes the distance to the phase transition, J is the exchange stiffness and u the lowest order mode-coupling parameter. The second term, $D\boldsymbol{\psi}(\nabla \times \boldsymbol{\psi})$, corresponds to the Dzyaloshinskii-Moriya interaction with the coupling constant D. This term is justified by the lack of inversion symmetry of the crystal structure. The last term describes the Zeeman coupling to an applied magnetic field $\boldsymbol{H}$. An ansatz for a single conical helix is:

$$\boldsymbol{\psi}(\boldsymbol{r}) = \psi_0\hat{\boldsymbol{\psi}}_0 + \Psi_{\text{hel}}\hat{\boldsymbol{e}}^-\text{e}^{\text{i}\boldsymbol{Q}\boldsymbol{r}} + \Psi^*_{\text{hel}}\hat{\boldsymbol{e}}^+\text{e}^{-\text{i}\boldsymbol{Q}\boldsymbol{r}} \tag{1.3}$$

Here, ψ_0 is the amplitude of the homogeneous magnetization and Ψ_{hel} is the complex amplitude of the helical order characterized by the pitch vector $\boldsymbol{Q}$. The vectors $\hat{\boldsymbol{e}}_1 \times \hat{\boldsymbol{e}}_2 = \hat{\boldsymbol{e}}_3$ form a normalized dreibein where $\hat{\boldsymbol{e}}^{\pm} = (\hat{\boldsymbol{e}}_1 \pm \text{i}\hat{\boldsymbol{e}}_2)/\sqrt{2}$ and $\boldsymbol{Q} = Q\hat{\boldsymbol{e}}_3$.

This brings us to the second term of the free energy density, f_{cub}, which contains spin-orbit coupling of second or higher order breaking the rotation symmetry of f_0 already in zero field.

$$f_{\text{cub}} = \frac{J_{\text{cub}}}{2}\left[(\partial_x \psi_x)^2 + (\partial_y \psi_y)^2 + (\partial_z \psi_z)^2\right] + \cdots \qquad (1.4)$$

This leading-order term of the cubic anisotropies, where $J_{\text{cub}} \ll J$, implies that the easy axis of the helical propagation vector is either a $\langle 100 \rangle$ or a $\langle 111 \rangle$ direction as explored by Bak and Jensen [40]. As the field is increased the Zeeman term gains importance and finally overcomes the cubic anisotropies, stabilizing the conical state with the propagation vector parallel to the magnetic field, in analogy to the spin-flop transition of a conventional antiferromagnet. In order to account for more subtle effects, further cubic anisotropies need to be considered consistent with the non-centrosymmetric space group $P2_13$.

While the contributions in f_0 and f_{cub} are sufficient to describe the helical, the conical, the field-polarized, and the paramagnetic ground states, specific issues require consideration of the higher-order spin-orbit coupling terms mentioned above and other contributions. For instance, for an universal account of the collective spin excitations it is necessary to include dipolar interactions [29]. Moreover, just above the paramagnetic-to-helimagnetic phase transition at T_c non-analytic corrections to the free energy functional arise from strong interactions between isotropic chiral fluctuations. These interactions suppress the correlation length and the second-order mean-field transition resulting in a fluctuation-disordered regime just above T_c and a fluctuation-induced first-order transition. The scenario relevant for cubic chiral magnets was originally predicted by Brazovskii [62] and recently demonstrated in MnSi by a study combining neutron scattering, susceptibility, and specific heat measurements [63]. Depending on the strength of the interaction between the fluctuations, for other chiral magnets an extended Bak-Jensen or a Wilson-Fischer scenario may be relevant [64–66].

As a hidden agenda the fluctuation-induced first-order transition underscores that the skyrmion lattice state is stabilized by thermal fluctuations, as depicted in Fig. 1.3a. The leading-order correction arise from Gaussian fluctuations around the mean-field spin configurations of the conical and the skyrmion lattice state, respectively.

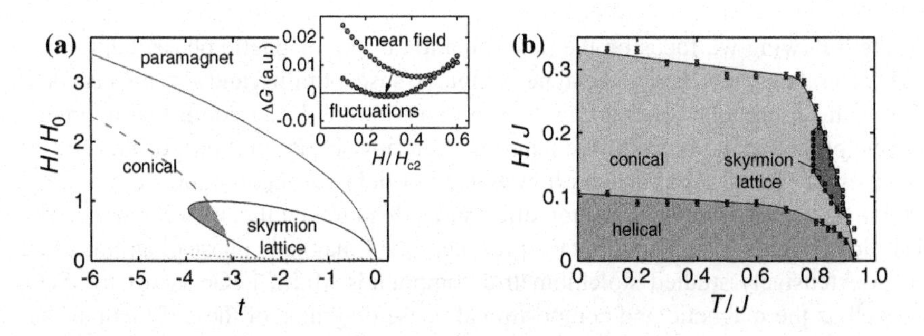

Fig. 1.3 Stabilization of the skyrmion lattice. **a** Theoretical magnetic phase diagram as obtained from a Ginzburg-Landau ansatz. The inset shows that thermal fluctuations already in Gaussian order stabilize the skyrmion lattice at intermediate fields [18]. **b** Magnetic phase diagram as obtained from Monte-Carlo simulations [67]

Interestingly, both short-range and long-range fluctuations favor the skyrmion lattice for intermediate magnetic fields [18]. Consistently, the skyrmion lattice forms rather independently from the orientation of the underlying crystalline lattice, where the cubic anisotropies only lead to a slightly anisotropic temperature and field range of the skyrmion lattice phase pocket [68, 69] and determine the precise orientation of the skyrmion lattice [18, 49].

Both the Brazovskii scenario and the stabilization of the skyrmion lattice by thermal fluctuations have recently been corroborated by classical Monte Carlo simulations [67]. Here, a fully non-perturbative study of a three-dimensional lattice spin model, i.e., going beyond Gaussian order, reproduced the thermodynamic signatures associated with a Brazovskii-type fluctuation-induced first-order phase transition and, as shown in Fig. 1.3b, the experimental magnetic phase diagram.

All of these recent advances compare and contrast with the seminal studies of Bogdanov and coworkers, who anticipated the existence of skyrmions in non-centrosymmetric materials with a uniaxial anisotropy and in the presence of a magnetic field [16, 17]. In particular, based on mean-field calculations ignoring the importance of thermal fluctuations, they concluded for cubic compounds that the skyrmion lattice would be metastable. Moreover, recently they predicted more complex magnetic phase diagrams comprising, besides the phases discussed so far, of meron textures and skyrmion liquids [70, 71]. Putative evidence for such complex phase diagrams has been reported in FeGe based on susceptibility [72, 73], specific heat [74], and SANS data [50]. However, as illustrated in Sect. 1.4, all data reported to date for all cubic chiral magnets are qualitatively extremely similar. Thus, when consistently inferring the transition fields and temperatures by virtue of the very same conditions, the magnetic phase diagrams of all compounds including FeGe are highly reminiscent of each other supporting strongly a rather universal scenario as described in the following without evidence of these complexities.

1.4 Magnetic Phase Diagrams

In the following we focus on the determination of the magnetic phase diagrams of cubic chiral magnets based on magnetization, ac susceptibility, and specific heat data, where the conditions for determining the transition fields are confirmed by microscopic probes, notably extensive neutron scattering. In the first part of this section we present typical data, explain how transition fields or temperatures are defined, and illustrate that demagnetization effects may lead to significant corrections. This is followed in the second part by the presentation of magnetic phase diagrams of the most-extensively studied stoichiometric compounds MnSi, FeGe, and Cu_2OSeO_3 as well as the magnetic and compositional phase diagrams of the most extensively studied doped compounds, namely $Mn_{1-x}Fe_xSi$ and $Fe_{1-x}Co_xSi$.

1.4.1 Phase Transitions in the Susceptibility and Specific Heat

The different magnetic states in the cubic chiral magnets and the phase transitions between them give rise to distinct signatures in various physical properties. Experimentally, the magnetic ac susceptibility and specific heat are easily accessible for most compounds and allow the determination of a very detailed magnetic phase diagram, based on feature-tracking. This provides the starting point for further studies and motivated us to concentrate on these quantities in the following. As an overview, we start with colormaps of the real and imaginary part of the ac susceptibility, Re χ_{ac} and Im χ_{ac}, in Fig. 1.4a, b, where blue shading corresponds to low and red shading to high values. As an example we show data for a cube-shaped single-crystal sample of MnSi measured at an excitation frequency of 120 Hz and an excitation amplitude of 0.5 mT. The field was applied after zero-field cooling along an $\langle 100 \rangle$ axis, i.e., along the hard direction for the helical propagation vector.

In Re χ_{ac} the conical state is characterized by a plateau of high and rather constant susceptibility (orange to red shading). The reduced value at low fields is associated with the helical state. Just below the helimagnetic ordering temperature, T_c, a plateau of reduced susceptibility in finite fields is characteristic for a single pocket of skyrmion lattice state (light blue shading). Just above T_c an area of relatively large sus-

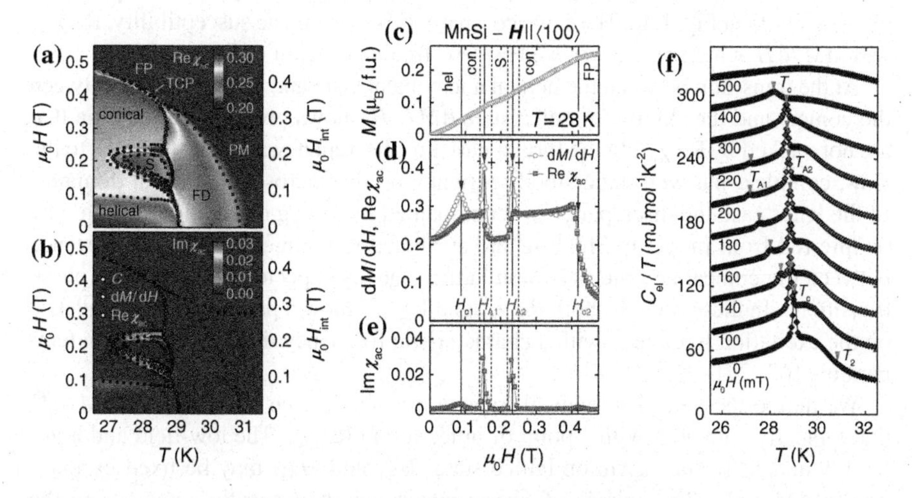

Fig. 1.4 Typical magnetization, ac susceptibility, and specific heat data of MnSi. **a** Color map of the real part of the ac susceptibility. We distinguish the following regimes; helical, conical, skyrmion lattice (S), fluctuation-disordered (FD), paramagnetic (PM), and field-polarized (FP). A field-induced tricritical point (TCP) is located at the high-field boundary of the FD regime. **b** Color map of the imaginary part revealing considerable dissipation only between the conical and the skyrmion lattice state. **c–e** Typical data of the magnetization, the susceptibility calculated from the magnetization, dM/dH, as well as the real and imaginary part of the ac susceptibility as a function of field. Note the definitions of the various transition fields. **f** Electronic contribution to the specific heat as a function of temperature for several applied magnetic fields. Data has been offset for clarity

ceptibility (green shading) is associated with the fluctuation-disordered (FD) regime that emerges as a consequence of the Brazovskii-type phase transition from paramagnetism to helimagnetism. At high temperatures or high fields, respectively, the system is in a paramagnetic (PM) or field-polarized (FP) state with low susceptibility (blue). A broad maximum observed in temperatures sweeps of Re χ_{ac} (not shown) marks the crossover between these two regimes [75]. Im χ_{ac} only shows contributions at the phase transitions and, in particular, between the skyrmion lattice and conical state. Here, the finite dissipation suggests a regime of phase coexistence where the nucleation process of topologically non-trivial skyrmions within the conical phase and vice versa eventually triggers a first-order transition [31, 68, 76]. In contrast, at the fluctuation-induced first-order transition between the skyrmion lattice and the fluctuation-disordered regime as a function of temperature no significant contribution to Im χ_{ac} is observed.

In order to define the different transition fields and temperatures, it is instructive to consider the typical field dependence of the magnetization, M, the susceptibility calculated from the measured magnetization, $\mathrm{d}M/\mathrm{d}H$, and the measured ac susceptibility for a temperature just below T_c as shown in Fig. 1.4c–e. Starting at $H = 0$, i.e., in the helical state, with increasing field the material undergoes transitions to the conical and the skyrmion lattice state before returning to the conical state and finally reaching the field-polarized state above H_{c2}. Below H_{c2} the magnetization increases almost linearly as shown in Fig. 1.4c, where the changes of slope at the different phase transitions are best resolved in the derivative $\mathrm{d}M/\mathrm{d}H$ depicted as open symbols in Fig. 1.4d. Here, we compare the measured ac susceptibility, Re χ_{ac}, with $\mathrm{d}M/\mathrm{d}H$ which may be viewed as zero-frequency limit of Re χ_{ac}.

At the transition between the helical and conical state and in the regimes between the conical and the skyrmion lattice state $\mathrm{d}M/\mathrm{d}H$ shows pronounced maxima that are not tracked by Re χ_{ac}. In the former case this discrepancy may be attributed to the slow, complex, but well-understood reorientation of macroscopic helical domains. In the latter case the discrepancy is accompanied by strong dissipation, which may be inferred from Im χ_{ac} in Fig. 1.4e and attributed to regimes of phase coexistence between the conical and the skyrmion lattice state as expected for first-order phase transitions. In these regimes both Re χ_{ac} and Im χ_{ac} show a pronounced dependence on the excitation frequency with a characteristic frequency that increases with temperature [68, 77].

We define the helical-to-conical transition at H_{c1} as the maximum of $\mathrm{d}M/\mathrm{d}H$ that typically coincides with a point of inflection in Re χ_{ac}. The low-field and high-field boundary of the skyrmion lattice state, H_{A1} and H_{A2}, may be fixed by maxima in $\mathrm{d}M/\mathrm{d}H$. The regimes of phase coexistence between the conical and the skyrmion lattice state are characterized by $\mathrm{d}M/\mathrm{d}H \neq$ Re χ_{ac} and Im $\chi_{ac} \gg 0$, where the corresponding boarders are labeled $H^{\pm}_{A1}$ and $H^{\pm}_{A2}$, respectively. For $H < H^{-}_{A1}$ and $H > H^{+}_{A2}$ the constant susceptibility of the conical phase is observed, while in the skyrmion lattice state for $H^{+}_{A1} < H < H^{-}_{A2}$ the system displays a plateau of lower susceptibility. The second-order transition from the conical to the field-polarized state belonging to the XY universality class is finally indicated by a point of inflection in both $\mathrm{d}M/\mathrm{d}H$ and $\mathrm{Re}\,\chi_{\mathrm{ac}}$. Similar criteria may be used to extract transition temperatures from data recorded as a function of temperature (not shown) [68].

Important related information on the nature of the phase transitions may be extracted from measurements of the specific heat. Using a quasi-adiabatic large heat pulse technique allows to determine transition temperatures with high precision [49, 76]. Figure 1.4f shows the electronic contribution to the specific heat, i.e., after subtraction of the phononic contribution, divided by temperature, C_{el}/T, as a function of temperature for different applied field values. In zero field a sharp symmetric peak marks the onset of helimagnetic order at the fluctuation-induced first-order transition at T_c. The peak resides on top of a broad shoulder that displays for small fields a so-called Vollhardt invariance [78] at T_2, i.e., an invariant crossing point of the specific heat, $\partial C/\partial H|_{T_2} = 0$, that coincides with a point of inflection in the magnetic susceptibility, $T\partial^2 M/\partial T^2|_{T_2} \approx TH\partial^2\chi/\partial T^2|_{T_2} = 0$ [75]. At intermediate fields two symmetric peaks, labeled T_{A1} and T_{A2}, track the phase boundaries of the skyrmion lattice state indicating two first-order transitions. In larger fields again one anomaly, labeled T_c, is observed. Increasing the field further causes a change of the shape of the anomaly from that of a slightly broadened symmetric delta peak to the asymmetric lambda anomaly of a second-order transition at a field-induced tricritical point (TCP). This field-induced change from first to second order is expected in the Brazovskii scenario, as the interactions between the chiral paramagnons become quenched under increasing magnetic fields.

In the magnetic phase diagram, see Fig. 1.4a, b, the crossovers between the fluctuation-disordered and the paramagnetic regime as well as between the paramagnetic and the field-polarized regime as observed in temperature sweeps of the susceptibility emanate from this TCP. An analysis of the entropy released at the phase transitions (not shown) also corroborates the position of the TCP. It suggests that the skyrmion lattice state possesses an entropy that is larger than the surrounding conical state, consistent with a stabilization by thermal fluctuations [76]. The latter is supported by the detailed shape of the phase boundary between the fluctuation-disordered and the long-range ordered states, where the skyrmion lattice extents to higher temperatures as compared to the conical state.

Following the detailed description of data recorded in MnSi with the magnetic field applied along $\langle 100\rangle$, we now turn to Fig. 1.5 illustrating typical susceptibility data as a function of field for different field directions and materials. Figure 1.5a shows data of MnSi for field applied along the major crystallographic axes after zero-field cooling measured on two cubes, i.e., with unchanged demagnetization effects. In general, the magnetic behavior is very isotropic. Changing the field direction only influences the weakest energy scale in the system, the cubic anisotropies, and has two well-understood consequences for the magnetic phase diagram. First, the helical-to-conical transition field is smallest for the easy axis of the helical propagation vector $\langle 111\rangle$ and largest for the hard axis $\langle 100\rangle$. In addition, the transition is only second-order if it is symmetry-breaking and otherwise represents a crossover. Second, the extent of the skyrmion lattice in both temperature and field decreases as the conical state is favored by the cubic anisotropies, i.e., in MnSi it is largest for field along $\langle 100\rangle$ and smallest for $\langle 111\rangle$. It is important to note, that even for field along the easy

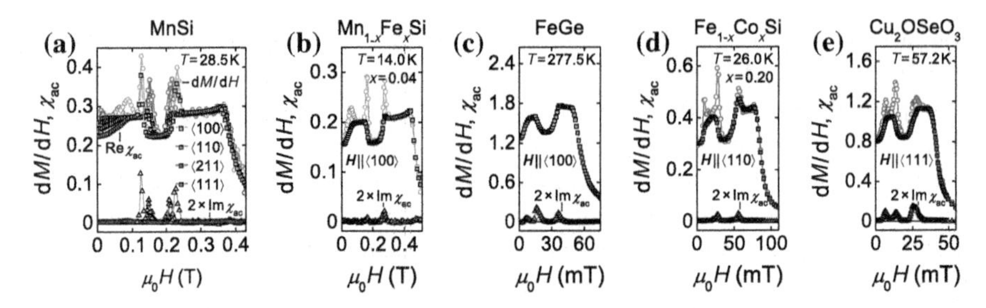

Fig. 1.5 Typical field dependence of the susceptibility for a temperature crossing the skyrmion lattice state. **a** Real and imaginary part of the ac susceptibility as well as susceptibility calculated from the magnetization, $\mathrm{d}M/\mathrm{d}H$, for MnSi and fields along major crystallographic directions. Besides well-understood anisotropies of the helical-to-conical transition and the extent of the skyrmion lattice phase pocket, the magnetic properties of MnSi are essentially isotropic. **b–e** Susceptibility for $Mn_{1-x}Fe_xSi$ ($x = 0.04$), FeGe, $Fe_{1-x}Co_xSi$ ($x = 0.20$), and Cu_2OSeO_3. Qualitatively very similar behavior is observed. Data in panel (**c**) taken from [73]

axis of the helix the skyrmion lattice is observed for all chiral magnets questioning a stabilization of the skyrmion lattice by cubic anisotropies only. In fact, for doped compounds such as $Fe_{1-x}Co_xSi$ or $Mn_{1-x}Fe_xSi$ the anisotropies are usually less pronounced or even completely suppressed, presumably due to the large amount of chemical disorder present in the system [19, 75], and yet the skyrmion lattice state represents nonetheless a well-defined stable phase.

Figure 1.5b–e show typical susceptibility data for $Mn_{1-x}Fe_xSi$ ($x = 0.04$), FeGe, $Fe_{1-x}Co_xSi$ ($x = 0.20$), and Cu_2OSeO_3 highlighting the universal aspects of different cubic chiral magnets. Despite the different temperature, field, length, and moment scales the susceptibilities of the different materials are qualitatively highly reminiscent. Omitting quantitative information on temperature, field, and susceptibility, even an expert would struggle to distinguish data between the different materials.

It is finally essential to account for demagnetization effects, for instance when data recorded on samples with different sample shapes are combined in a single magnetic phase diagram. In general, the internal magnetic field, $\boldsymbol{H}_{\mathrm{int}}$, is calculated as $\boldsymbol{H}_{\mathrm{int}} = \boldsymbol{H}_{\mathrm{ext}} - \mathbf{N}\boldsymbol{M}(\boldsymbol{H}_{\mathrm{ext}})$ with the externally applied magnetic field $\boldsymbol{H}_{\mathrm{ext}}$ and the 3×3 demagnetization matrix $\mathbf{N}$ that obeys $\mathrm{tr}\,\{\mathbf{N}\} = 1$ in SI units. While a proper treatment of the dipolar interactions in the cubic chiral magnets requires to take several matrix entries into account [29], in most cases consideration of the scalar equation $H_{\mathrm{int}} = H_{\mathrm{ext}} - NM(H_{\mathrm{ext}})$ is sufficient, in which for field along the z-direction the matrix entry N_{zz} is referred to as N. Note that for the measured ac susceptibility, $\chi_{\mathrm{ac}}^{\mathrm{ext}}$, not only the field scale but also the absolute value of the susceptibility depends on demagnetization effects via the applied excitation field $H_{\mathrm{ac}}^{\mathrm{ext}}$.

From a practical point of view many samples are essentially rectangular prisms for which effective demagnetization factors for fields applied along the edges may be calculated following [79]. In addition, in the cubic chiral magnets the susceptibility assumes essentially a constant value in the conical phase. Using the measured value, $\chi_{\mathrm{con}}^{\mathrm{ext}}$, as a first approximation for the entire helimagnetically ordered part of the magnetic phase diagram, i.e., for $T < T_c$ and $H < H_{c2}$, the magnetization may be expressed as $M(H_{\text{ext}}) = \chi_{\text{con}}^{\text{ext}} H_{\text{ext}} = \chi_{\text{con}}^{\text{int}} H_{\text{int}}$. Hence, the internal and the externally applied magnetic fields are related by:

$$H_{\text{int}} = H_{\text{ext}} \left(1 - N\chi_{\text{con}}^{\text{ext}}\right) = \frac{H_{\text{ext}}}{1 + N\chi_{\text{con}}^{\text{int}}} \tag{1.5}$$

We note that the internal value of the constant susceptibility of the conical state, $\chi_{\text{con}}^{\text{int}}$, is an important dimensionless measure for the effective strength of dipolar interactions in the chiral magnets [29]. If the magnetic properties and a second quantity, e.g., electrical resistivity, are determined on samples with differing demagnetization factors, N_1 and N_2, the formula to calculate the internal field of the second sample may be written as:

$$H_{\text{int},2} = H_{\text{ext},2} \left(1 - N_2 \frac{\chi_{\text{con},1}^{\text{ext}}}{1 - \chi_{\text{con},1}^{\text{ext}}(N_1 - N_2)}\right) \tag{1.6}$$

In the field-polarized state above H_{c2} one may, again in first approximation, assume the magnetization as saturated and thus $M(H_{\text{ext}}) = \chi_{\text{con}}^{\text{ext}} H_{c2}^{\text{ext}} = \chi_{\text{con}}^{\text{int}} H_{c2}^{\text{int}}$ leading to a constant offset, $H_{\text{int}} = H_{\text{ext}} - N\chi_{\text{con}}^{\text{ext}} H_{c2}^{\text{ext}}$.

Despite the rather crude approximation given above, this treatment proves to be sufficient to account for the most prominent effects of demagnetizing fields in the chiral magnets such as the shift of transition fields. Additionally, a smearing of phase transitions and very broad regimes of phase coexistence between the conical and the skyrmion lattice state may be observed in samples with large and, in particular, inhomogeneous demagnetization effects [68]. Such unfavorable sample shapes are, for instance, thin platelets with their short edge along the field or irregular shapes in general. Materials with a large absolute value of the susceptibility intensify the issue.

1.4.2 Magnetic Phase Diagrams for Different Materials

Using the definitions for the transition fields and temperature given in the previous subsection on susceptibility and specific heat data we have compiled magnetic and compositional phase diagrams of various cubic chiral magnets as shown in Fig. 1.6. Data extracted from measurements of the derivative of the magnetization, the ac susceptibility, and the specific heat are shown as circles, squares, and diamonds, respectively. Light and dark colors represent data from temperature and field sweeps, respectively. Magnetic fields were applied after zero-field cooling. All field values are given on internal field scales, i.e., after correcting for demagnetization effects. In general the magnetic phase diagrams of the cubic chiral magnets are qualitatively extremely similar. We distinguish the following six regimes; helical, conical, skyrmion lattice (S), fluctuation-disordered (FD), paramagnetic (PM), and field-polarized (FP). In addition, we mark the regime of phase coexistence between the conical and the skyrmion lattice state by a faint red shading. Solid and dashed lines

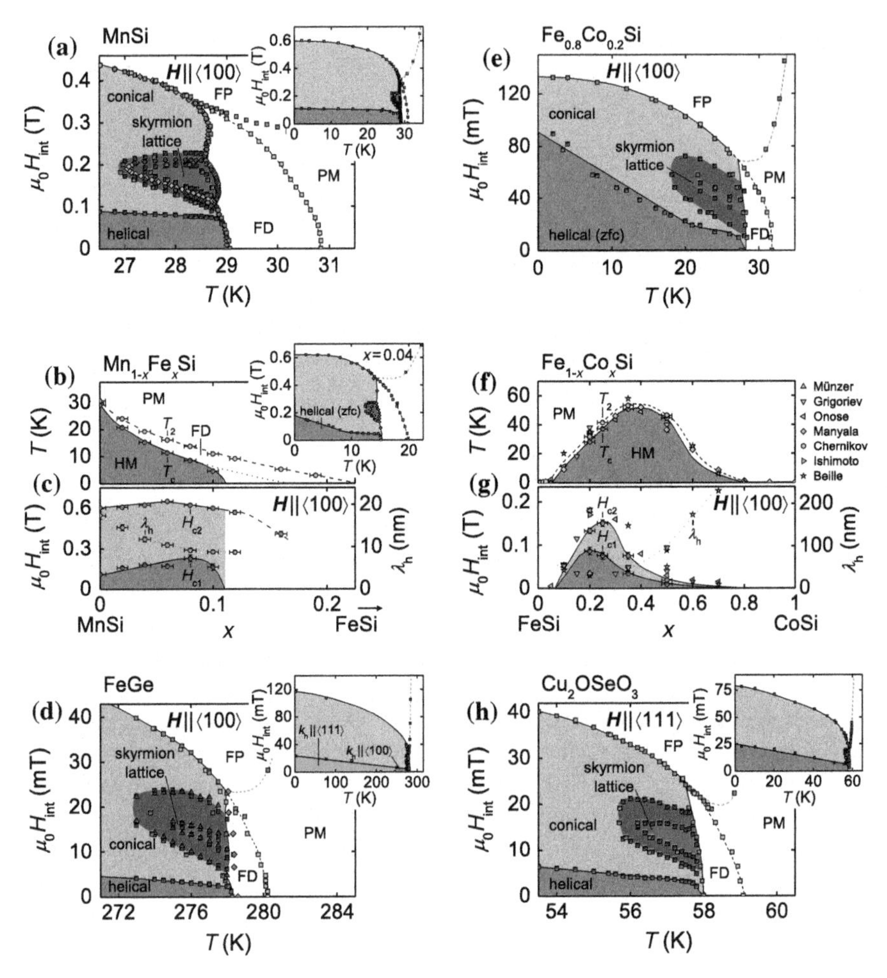

Fig. 1.6 Magnetic phase diagrams of selected cubic chiral magnets. **a** MnSi. **b**, **c** $Mn_{1-x}Fe_xSi$. Substitutional doping of MnSi with Fe leads to a suppression of the ordering temperature and a decrease of the helix wavelength, λ_h. The magnetic phase diagram, as shown in the inset for $x = 0.04$, stays qualitatively similar for $x \leq 0.10$. **d** FeGe. Susceptibility data from [72, 73], specific heat data from [74], and further information from [42, 80, 81] were analyzed in the same manner as for all other compounds. **e** $Fe_{0.8}Co_{0.2}Si$. **f**, **g** $Fe_{1-x}Co_xSi$. As a function of cobalt content x the characteristic temperature, field, and length scales may be varied over a large range. Values are taken from [19, 82–87]. **h** Cu_2OSeO_3. In contrast to the other materials, this local-moment insulator displays substantial magnetoelectric coupling. Still, the magnetic phase diagram is unchanged

indicate phase transitions and crossovers, respectively, while dotted lines represent guides to the eye.

Figure 1.6a reproduces the magnetic phase diagram of MnSi for field along ⟨100⟩, i.e., the hard axis for the helical propagation vector, as discussed in the previous subsection. The inset shows the phase diagram across the entire parameter range of long-range helimagnetic order. We note that the helix wavelength, λ_h, in MnSi increases from $\sim$165 Å at T_c to $\sim$180 Å at lowest temperatures [36, 57, 63].

Substitutional doping of iron at the manganese sites of MnSi results in a reduction of the helimagnetic ordering temperature while the critical field values in the zero-temperature limit change only weakly, cf. Fig. 1.6b, c. The magnetic phase diagram is qualitatively very similar to pure MnSi for $x \leq 0.10$ as shown in the inset of Fig. 1.6b for $Mn_{1-x}Fe_xSi$ with $x = 0.04$. The most notable difference concerns the helical state, which forms in $Mn_{1-x}Fe_xSi$ only properly after zero-field cooling. In addition, H_{c1} becomes essentially isotropic and increases with decreasing temperature. These effects, however, may be related to the increased amount of disorder present in the system. The helix wavelength and hence also the skyrmion lattice constant decreases by up to a factor of roughly 2 resulting in an increase of the skyrmion density by a factor of 4 [48, 88].

The complex quantum critical behavior that emerges at high iron concentrations, where static magnetic order is fully suppressed, is the topic of ongoing research [65, 75]. Doping with iron, cobalt, and nickel leads to an essentially identical modification of the magnetic behavior if scaled by the number valance electrons per formula unit [75, 89]. Doping with chromium, i.e., reducing the number of valance electrons, leads to a suppression of T_c comparable to iron doping [90]. This behavior is consistent with the notion that the main effects of chemical doping are due to a rigid shift the Fermi level, as recently inferred from a combined study of ab initio calculations and the electric transport properties in $Mn_{1-x}Fe_xSi$ [88].

We now turn to FeGe which is rather similar to MnSi, however, with a transition temperature near room temperature and $\lambda_h = 700$ Å. Around 230 K the easy direction of the helical pitch changes from $\langle 100 \rangle$ at high temperatures to $\langle 111 \rangle$ at low temperatures, where a large thermal hysteresis of $\sim$35 K is observed [81]. Recent publications [50, 72–74] claimed putative experimental evidence for the formation of a very complex magnetic phase diagram with multiple pockets and precursor phenomena around the skyrmion lattice state. The authors concluded that these findings prove that the skyrmion lattice is in fact not stabilized by thermal fluctuations but by a combination of uniaxial anisotropies and a softened modulus of the magnetization.

In stark contrast, applying accurately the same definitions given in the previous subsection to the data published in [72–74] provides the phase diagram shown in Fig. 1.6d. This phase diagram strongly resembles that of the other cubic chiral magnets. The broad regimes of phase coexistence may be attributed to large demagnetization effects as a consequence of the relatively large absolute value of the susceptibility in FeGe and the shape of the samples used in these studies; we extract $\chi_{\mathrm{con}}^{\mathrm{ext}} = 1.6$ and $N \approx 0.33$ from [73] yielding $\chi_{\mathrm{con}}^{\mathrm{int}} = 3.4$. Most importantly, however, we observe no signatures of additional phase pockets or mesophases. We finally note that a temperature discrepancy of the maximum in the specific heat in [73, 74] indicates that care has to be taken when combining data from different samples or measurement setups.

Figure 1.6e–g are dedicated to $Fe_{1-x}Co_xSi$, a pseudo-binary $B20$ system that displays helimagnetism in a large composition range, $0.05 \lesssim x \lesssim 0.8$ [82, 85, 91], albeit the parent compounds FeSi and CoSi are a paramagnetic insulator [92] and a diamagnetic metal [93], respectively. Starting from the strongly correlated insulator

FeSi [94], an insulator-to-metal transition takes place around $x \approx 0.02$ [84]. However, due to the comparatively high absolute value of the electrical resistivity and an upturn at low temperatures helimagnetic $Fe_{1-x}Co_xSi$ is typically referred to as a strongly doped semiconductor [82, 86, 95].

Compared to the stoichiometric helimagnets, $Fe_{1-x}Co_xSi$ offers the opportunity to vary the characteristic parameters of the helimagnetism over a wide range by compositional tuning while the magnetic phase diagrams stays that of a typical cubic chiral magnet, cf. Fig. 1.6e. As summarized in Fig. 1.6f, g, the helimagnetic transition temperature reaches up to $\sim$50 K, the critical fields assume values up to $\sim$150 mT, and the helix wavelength ranges from about 300 Å to more than 2000 Å. As for doped MnSi, a proper helical state is observed only after zero-field cooling. $Fe_{1-x}Co_xSi$ displays easy $\langle 100 \rangle$ axes that, especially for larger cobalt contents, are less pronounced than for other cubic chiral helimagnets [87]. For $x = 0.20$ a helical pitch along $\langle 110 \rangle$ was identified in [19]. The latter study also revealed the existence of a skyrmion lattice in $Fe_{1-x}Co_xSi$ that is sensitive to the field and temperature history. While the reversible pocket of skyrmion lattice state is comparable to other systems, field cooling may result in a metastable extension down to lowest temperatures allowing for conceptionally new types of experiments [31]. A similar behavior was later also discovered in low-quality MnSi samples under applied pressure [96]. Moreover, depending on the field direction, two Skyrmion lattice domains with different in-plane orientations were observed leading to a twelvefold small-angle scattering pattern [97].

Figure 1.6h finally shows the magnetic phase diagram of copper-oxo-selenite, Cu_2OSeO_3. The crystalline structure of this compound is more complex than that of the $B20$ transition metal systems, but also belongs to space group $P2_13$ [98]. Magnetically, on the strongest scale Cu_2OSeO_3 shows local-moment ferrimagnetic order of the spin-$\frac{1}{2}$ Cu^{2+} ions. Here, the ferromagnetically aligned moments on the Cu^{I} sites couple antiferromagnetically to the ions on the Cu^{II} sites leading to a 3:1 ratio [99] with exchange constants $J_{FM} = -50$ K and $J_{AFM} = 65$ K [100]. No breaking of the ferrimagnetic coupling is observed up to 55 T [101]. The ferrimagnetism is superimposed by a long-wavelength helical modulation based on the Dzyaloshinskii-Moriya interaction with $\lambda_h = 620$ Å [26]. The resulting magnetic phase diagram is highly reminiscent of the helimagnetic $B20$ compounds with an easy $\langle 100 \rangle$ for the helical propagation vector and a delicate pinning within the skyrmion lattice state [49, 102]. A study using resonant soft x-ray diffraction further suggested that the Cu^{I} and Cu^{II} sites may form individual but coupled skyrmion lattices that are rotated by a few degree with respect to each other giving rise to a moiré pattern [103]. More recent work reveals, however, that this conjecture may be wrong.

Cu_2OSeO_3, albeit being a non-polar insulator, possesses a magnetically induced electrical polarization in finite fields and, in particular, within the skyrmion lattice state [69]. The polarization resulting from this magnetoelectric coupling may be described in a d-p hybridization model [104], where the covalency between copper d and oxygen p orbitals is modulated according to the local magnetization direction via the spin-orbit interaction leading to a local electric dipole along the bond direction [69]. Hence, though Cu_2OSeO_3 is actually a (heli-)ferrimagnetic magnetoelectric, it is often erroneously referred to as a multiferroic. The origin of this notion may be seen in the hitherto unique opportunity to manipulate a topologically non-trivial entity of magnetoelectric nature using various external control parameters, see for example [28, 105–108].

1.5 Emergent Electrodynamics

A particularly exciting consequence of the non-trivial topology of the skyrmions concerns their coupling to spin currents. In the following we focus on the consequences in metallic compounds and we refer to by Markus Garst for a more detailed account. The spin structure of the skyrmion, as seen from the point of view of an electron traversing it, gives rise to real-space Berry phases which may be expressed as emergent magnetic and electric fields, $B_i^e = \frac{\hbar}{2}\epsilon_{ijk}\hat{\psi} \cdot \partial_j \hat{\psi} \times \partial_k \hat{\psi}$ and $E_i^e = \hbar \hat{\psi} \cdot \partial_i \hat{\psi} \times \partial_t \hat{\psi}$, respectively, with $\partial_i = \partial/\partial r_i$ and $\partial_t = \partial/\partial t$ [111]. As a consequence an additional topological contribution to the Hall effect may be observed in the skyrmion lattice state as illustrated in Fig. 1.7a [109].

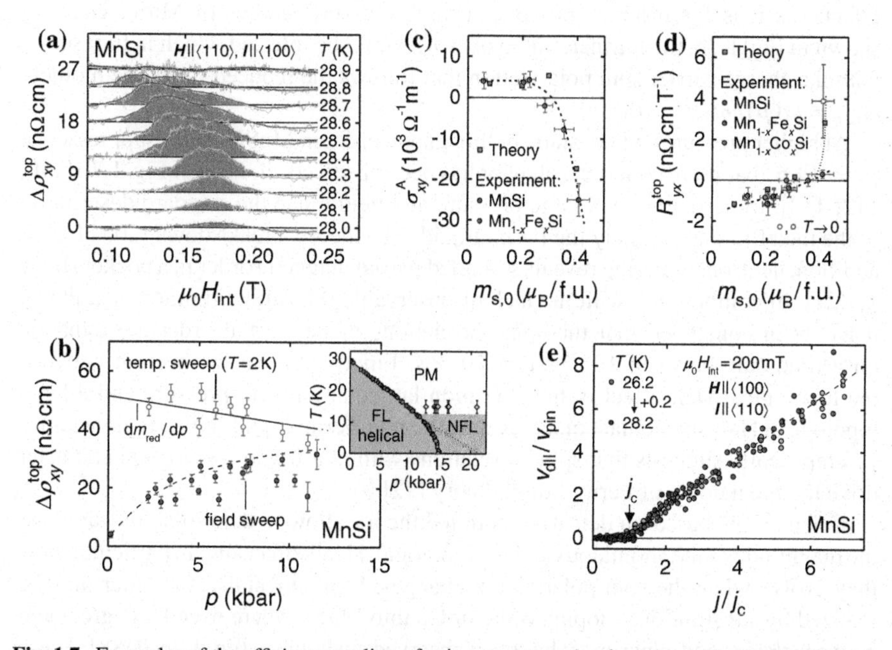

Fig. 1.7 Examples of the efficient coupling of spin currents to the skyrmion lattice. **a** Topological Hall contribution, $\Delta\rho_{xy}^{top}$, in MnSi as a function of field [109]. **b** Topological Hall contribution, $\Delta\rho_{xy}^{top}$, as a function of hydrostatic pressure in MnSi [96]. The intrinsic size (*open symbols*) may only be observed after field-cooling down to the lowest temperatures. The inset shows the pressure-temperature phase diagram of MnSi highlighting the extended regime of non-Fermi liquid (NFL) behavior [22, 110]. **c, d** Anomalous Hall conductivity, σ_{xy}^{A}, and topological Hall constant, R_{yx}^{top}, as a function of the magnetic moment as varied, e.g., by iron or cobalt doping. First-principle calculations and experimental data are in excellent agreement [88]. **e** Drift velocity of the skyrmion lattice, $v_{d\parallel}$, as a function of current density, j. Ultra-low current densities in the order of $j_c \sim 10^6$ A/m^2 unpin the skyrmion lattice [24, 111]

Using the charge carrier spin polarization P and assuming the absence of spin-flip scattering, while non-spin-flip scattering is captured by the normal Hall constant R_0, the topological Hall contribution may be estimated as $\Delta\rho_{xy}^{\mathrm{top}} = P R_0 B_{\mathrm{eff}}$. The effective emergent field, B_{eff}, is topologically quantized in the sense that it is given by the product of the emergent flux quantum that each skyrmion supports, $\phi_0 = h/e$, and the skyrmion density ϕ. Thus, the sign of the topological Hall contribution allows to distinguish, in principle, between skyrmions ($\Phi = +1$) and anti-skyrmions ($\Phi = -1$), such as in MnSi, provided the normal Hall constant R_0 is sufficient to express the details of the band structure [109].

In real materials the electronic structure at the Fermi surface may contribute in different ways and the spin polarization as well as the skyrmion lattice constant may change as a function of temperature or field. In addition, processes such as spin-flip scattering may cause a reduction compared to the intrinsic value of $\Delta\rho_{xy}^{\mathrm{top}}$. For instance, in MnSi the topological Hall contribution in the skyrmion lattice is of the order of 4 nΩ cm whereas an intrinsic topological Hall signal of the order of 50 nΩ cm is expected for its emergent field of $B_{\mathrm{eff}} = -13$ T [96]. Field-cooling the skyrmion lattice down to low temperatures allows to reduce the finite temperature effects, as it is for instance possible in high-pressure studies of MnSi. Here, as shown in Fig. 1.7b, the intrinsic value of $\Delta\rho_{xy}^{\mathrm{top}}$ could be inferred which in turn scales with the charge carrier spin polarization that follows the reduced magnetic moment $m_{\mathrm{red}} = m(p)/m(p = 0)$.

At higher pressures where static helimagnetic order in MnSi is fully suppressed at $p_c = 14.6$ kbar more complex behavior has been observed, cf. inset of Fig. 1.7b [110, 112, 113]. In particular, in a large pressure and field range the standard description of the metallic state, namely the Fermi liquid (FL) theory, breaks down [20, 114]. In addition, neutron scattering reveals so-called partial magnetic order in a pocket above p_c [21]. In combination with the lack of observable relaxation in muon data [115], it has been concluded that the spin correlations of the partial order are dynamic on a timescale between 10^{-10} s and 10^{-11} s. Finally, a clear connection between the topological Hall effect in the skyrmion lattice at ambient pressure and a large topological Hall signal that coincides with the non-Fermi liquid (NFL) regime above p_c empirically suggests that spin correlations with non-trivial topological character drive the breakdown of Fermi liquid theory [22].

Calculations based on density functional theory allow to determine the sign and the magnitude of the anomalous and the topological Hall effect and, in particular, how they evolve when the spin polarization changes. Experimentally, the latter may be realized by substitutional doping of Fe or Co into MnSi, where excellent agreement between theory and experiment has been observed as shown in Fig. 1.7c, d [88]. These results provide the quantitative microscopic underpinning that, while the anomalous Hall effect is due to the reciprocal-space Berry curvature [116], the topological Hall effect originates in real-space Berry phases. As a theoretical prediction that awaits further confirmation even contributions arising from mixed phase-space Berry phases have been proposed [96, 117].

The efficient coupling of spin currents to the magnetic structure, together with the exceptional long-range order of the skyrmion lattice [46] and the resulting very weak collective pinning to defects, causes a sizeable response of the magnetic textures at ultra-low current densities. Above an exceptionally low threshold current density of the order of $j_c \sim 10^6\,\mathrm{A/m^2}$ the skyrmion lattice unpins and begins to drift [24, 118]. Numerical simulations revealed that the skyrmion motion exhibits a universal current-velocity relation that is (on the scale of the study) unaffected by impurities and non-adiabatic effects [119]. Flexible shape-deformations of individual skyrmions and the skyrmion lattice permit to avoid pinning centers.

Theoretically, the spin transfer torques in the cubic chiral magnets may be accounted for in the framework of a Landau-Lifshitz-Gilbert equation using the Thiele approach [120, 121]. Here, a Magnus force perpendicular to the current direction and a dissipative drag force along it are balanced by pinning forces, e.g., due to defects. The Magnus force represents the effective Lorentz force arising from the emergent magnetic field $\boldsymbol{B}^e$ and leads to a certain angle between the current direction and the drift direction of the skyrmion lattice. According to Faraday's law of induction, a moving skyrmion, which supports exactly one quantum of emergent magnetic flux, may then induce an emergent electric field $\boldsymbol{E}^e$ that inherits the topological quantization [122]. These electric fields have been observed directly [111]. A scaling plot as depicted in Fig. 1.7e reveals a universal relation between the current density, j, and the drift velocity of the skyrmion lattice, $v_{\mathrm{d}\parallel}$, where typical pinning velocities are of the order of 0.1 mm/s, i.e., the drift velocity of conduction electrons.

1.6 Conclusions and Outlook

Taken together, cubic chiral magnets with non-centrosymmetric space group $P2_13$ represent a class of materials that share a universal magnetic phase diagram. The skyrmion lattice state occupies a single phase pocket and the entire magnetic phase diagram is well accounted for by a Ginzburg-Landau approach including the effects of thermal fluctuations. Depending on the specific material, key parameters such as the transition temperatures, critical fields, or the helix wavelength may be varied by two orders of magnitude. With compounds ranging from pure metals to magneto-electric insulators, this material class provides well-understood model systems for experiments, theory, and simulations. In recent studies, for instance, aspects were addressed such as the topological unwinding at the transition to conventional helimagnetic order [31] or the collective excitations of the different spin structures [27, 129–131].

Current research activities on topologically non-trivial spin states, however, are not restricted to cubic chiral magnets. In thin films or monolayers, where the inversion symmetry is broken by the surface, skyrmions may be stabilized by the Dzyaloshinskii-Moriya interaction as combined with four-spin exchange interactions [132, 133]. Another route towards skyrmionic textures may be long-range magnetodipolar interactions [134]. In such systems, it was already demonstrated that skyrmions may be created and annihilated individually using spin-polarized currents of a scanning tunneling microscope [133] or laser pulses [136]. The creation, manipulation, and the dynamics of skyrmions in thin films, nanowires, and patterned nanostructures offer great potential for future applications, see for instance [23, 136–143]. The efficient gyromagnetic coupling, the topological stability, and the small size of the skyrmions promise devices for ultra-dense information storage and spintronics [25], while their unique collective excitations may be exploited for the design of conceptually new microwave devices [28, 29, 144].

In parallel, topologically non-trivial spin states have been identified in a rapidly growing number of bulk compounds suggesting that these complex magnetic structures may be in fact rather common. In Fig. 1.8 we summarize three recent examples. The first material, CoZn, crystallizes in the cubic space group $P4_132$ or $P4_332$, depending on the handedness, and orders magnetically well above room temperature [145]. Doping manganese into the system, see Fig. 1.8a, reduces the transition temperature. Figure 1.8b shows the magnetic phase diagram of $Co_8Zn_9Mn_3$ extracted from the magnetic susceptibility. It is highly reminiscent to that of the cubic chiral magnets including a pocket of skyrmion lattice state as identified by LF-TEM and

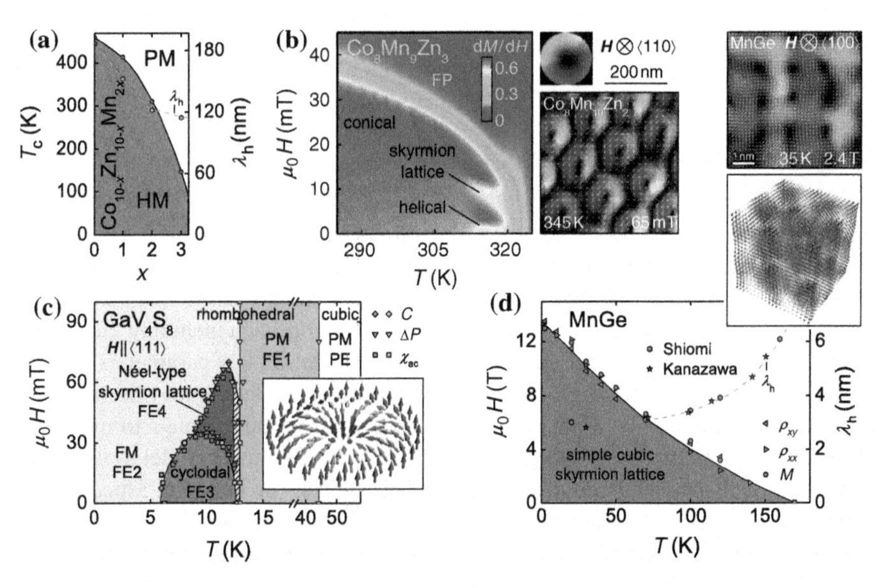

Fig. 1.8 Topologically non-trivial spin structures in further bulk materials. **a** Part of the compositional phase diagram of the system $Co_{10-x}Zn_{10-y}Mn_{x+y}$. Long-wavelength helimagnetic order with transition temperatures exceeding room temperature has been reported [123]. **b** Colormap of the susceptibility of $Co_8Zn_9Mn_3$ revealing a skyrmion lattice state and corresponding real-space spin structure in $Co_8Zn_{10}Mn_2$ as obtained by LF-TEM [123]. The behavior is highly reminiscent of the cubic chiral magnets. **c** Magnetic phase diagram of GaV_4S_8 exhibiting a Néel-type skyrmion lattice and various types of ferroelectric order [30, 124]. **d** Magnetic phase diagram of MnGe [125–127] giving rise to a simple cubic lattice of spin whirls as recently observed by LF-TEM [128]

SANS measurements [123]. Hence, the material system $Co_{10-x}Zn_{10-y}Mn_{x+y}$ is not only the first bulk compound with a space group other than $P2_13$ that exhibits a skyrmion lattice state, but also the first compound stabilizing skyrmions above room temperature.

Another important example is shown in Fig. 1.8c, which depicts the magnetic phase diagram of the lacunar spinel GaV_4S_8. This system crystallizes in the cubic space group $F\bar{4}3m$ at room temperature. At $T_{JT} = 44\,K$ GaV_4S_8 shows a structural phase transition [146] into the rhombohedral space group $R3m$ driven by Jahn-Teller orbital order and accompanied by an onset of ferroelectricity (FE). The structural transition creates a multi-domain state with submicron-thick sheets of the four different rhombohedral domains [30]. Below $T_C = 13\,K$ magnetic order sets in [147] and as a function of temperature and field a rich magnetic phase diagram unfolds with various magnetically ordered states of multiferroic nature [124]. This phase diagram hosts a pocket of ferroelectric spin vortices forming a hexagonal skyrmion lattice as identified by means of force microscopy and SANS [30]. However, in contrast to the cubic chiral magnets or $Co_{10-x}Zn_{10-y}Mn_{x+y}$ where Bloch-type chiral skyrmions are described in terms of spin helices, in GaV_4S_8 Néel-type non-chiral skyrmions are addressed in form of a superposition of spin cycloids. Moreover, while in the cubic chiral magnets the skyrmion lines are always essentially parallel to the applied magnetic field, in GaV_4S_8 the vortex cores are confined along an $\langle 111 \rangle$ axis. In combination with the multiferroic nature of this polar magnetic semiconductor new ways of controlling and manipulating skyrmions may be possible.

Last but not least, we return to MnGe which is isostructural to the cubic chiral magnets with a magnetic phase diagram that differs from the ones described so far. In this compound, measurements of the topological Hall effect [125] and the topological Nernst effect [127] as well as data from SANS [126] and LF-TEM [128] consistently suggest the formation of a simple cubic lattice of spin whirls in zero and finite field. The magnetic lattice vectors are oriented along the $\langle 100 \rangle$ axes of the crystal lattice. The resulting magnetic phase diagram is depicted in Fig. 1.8c, where the inset schematically shows the spin structure and the upper panel shows the in-plane distribution of magnetic moments as obtained from LF-TEM. Compared to the cubic chiral magnets the corresponding lattice period is relatively small and exhibits a strong increase from 3 nm at low temperatures to 6 nm close to $T_c = 170\,K$. To what extent this marks the starting point of a new generic understanding of complex spin textures remains to be seen.

We are deeply indebted to our co-workers, in particular: T. Adams, R. Bamler, G. Benka, H. Berger, S. Blügel, P. Böni, G. Brandl, S. Buhrandt, A. Chacon, C. Duvinage, H.-M. Eiter, L. M. Eng, K. Everschor, C. Franz, F. Freimuth, M. Gangl, M. Garst, R. Georgii, D. Grundler, R. Hackl, M. Halder, F. Haslbeck, W. Häußler, T. Hesjedal, J. P. Hinton, C. Hugenschmidt, M. Janoschek, F. Jarzembeck, P. Jaschke, F. Jonietz, J. Kindervater, D. Köhler, J. D. Koralek, P. Krautscheid, M. Kugler, A. Kusmartseva, P. Lemmens, N. Martin, S. Mayr, D. Meier, M. Meven, P. Milde, Y. Mokrousov, S. Mühlbauer, J. Orenstein, S. A. Parameswaran, B. Pedersen, R. Ramesh, T. Reimann, M. Reiner, R. Ritz, A. Rosch, F. Rucker, S. Säubert, C. Schnarr, R. W. Schoenlein, T. Schröder, S. Schulte, T. Schulz, C. Schütte,

T. Schwarze, K. Seemann, J. Seidel, A. Senyshyn, I. Stasinopoulos, A. Vishwanath, M. Wagner, J. Waizner, T. Weber, S. Weichselbaumer, B. Wiedemann, S. Zhang, and the team of FRM II. Financial support through DFG FOR960, DFG TRR80, and ERC advanced grant 291079 (TOPFIT) is gratefully acknowledged.

References

1. T.H.R. Skyrme, A non-linear field theory. Proc. R. Soc. Lond. A **260**, 127 (1961)
2. T.H.R. Skyrme, Particle states of a quantized meson field. Proc. R. Soc. Lond. A **262**, 237 (1961)
3. T.H.R. Skyrme, A unified field theory of mesons and baryons. Nucl. Phys. **31**, 556 (1962)
4. G.S. Adkins, C.R. Nappi, E. Witten, Static properties of nucleons in the Skyrme model. Nucl. Phys. B **228**, 552 (1983)
5. I. Zahed, G.E. Brown, The Skyrme model. Phys. Rep. **142**, 1 (1986)
6. D. Diakonov, V. Petrov, M. Polyakov, Exotic anti-decuplet of baryons: prediction from chiral solitons. Z. Phys. A **359**, 305 (1997)
7. E. Chabanat, P. Bonche, P. Haensel, J. Meyer, R. Schaeffer, A Skyrme parametrization from subnuclear to neutron star densitie. Nucl. Phys. A **627**, 710 (1997)
8. E. Chabanat, P. Bonche, P. Haensel, J. Meyer, R. Schaeffer, A Skyrme parametrization from subnuclear to neutron star densities Part II. Nuclei far from stabilities. Nucl. Phys. A **635**, 231 (1998)
9. S.L. Sondhi, A. Karlhede, S.A. Kivelson, E.H. Rezayi, Skyrmions and the crossover from the integer to fractional quantum Hall effect at small Zeeman energies. Phys. Rev. B **47**, 16419 (1993)
10. A. Schmeller, J.P. Eisenstein, L.N. Pfeiffer, K.W. West, Evidence for skyrmions and single spin flips in the integer quantized Hall effect. Phys. Rev. Lett. **75**, 4290 (1995)
11. K. Yang, S. Das Sarma, A.H. MacDonald, Collective modes and skyrmion excitations in graphene $SU(4)$ quantum Hall ferromagnets. Phys. Rev. B **74**, 075423 (2006)
12. T.-L. Ho, Spinor Bose condensates in optical traps. Phys. Rev. Lett. **181**, 742 (1998)
13. U.A. Khawaja, H. Stoof, Skyrmions in a ferromagnetic Bose-Einstein condensate. Nature (London) **411**, 918 (2001)
14. L.S. Leslie, A. Hansen, K.C. Wright, B.M. Deutsch, N.P. Bigelow, Creation and detection of Skyrmions in a Bose-Einstein condensate. Phys. Rev. Lett. **103**, 250401 (2009)
15. J. Fukuda, S. Žumer, Quasi-two-dimensional Skyrmion lattices in a chiral nematic liquid crystal. Nat. Commun. **2**, 246 (2011)
16. A.N. Bogdanov, D.A. Yablonskii, Thermodynamically stable "vortices" in magnetically ordered crystals. The mixed state of magnets. Sov. Phys. JETP **95**, 178 (1989)
17. A. Bogdanov, A. Hubert, Thermodynamically stable magnetic vortex states in magnetic crystals. J. Magn. Magn. Mater. **138**, 255 (1994)
18. S. Mühlbauer, B. Binz, F. Jonietz, C. Pfleiderer, A. Rosch, A. Neubauer, R. Georgii, P. Böni, Skyrmion lattice in a chiral magnet. Science **323**, 915 (2009)
19. W. Münzer, A. Neubauer, T. Adams, S. Mühlbauer, C. Franz, F. Jonietz, R. Georgii, P. Böni, B. Pedersen, M. Schmidt, A. Rosch, C. Pfleiderer, Skyrmion lattice in the doped semiconductor $Fe_{1-x}Si$. Phys. Rev. B **81**, 041203 (R) (2010)
20. C. Pfleiderer, S.R. Julian, G.G. Lonzarich, Non-Fermi-liquid nature of the normal state of itinerant-electron ferromagnets. Nature (London) **414**, 427 (2001)
21. C. Pfleiderer, D. Reznik, L. Pintschovius, H.V. Löhneysen, M. Garst, A. Rosch, Partial order in the non-Fermi-liquid phase of MnSi. Nature (London) **427**, 227 (2004)
22. R. Ritz, M. Halder, M. Wagner, C. Franz, A. Bauer, C. Pfleiderer, Formation of a topological non-Fermi liquid in MnSi. Nature (London) **497**, 231 (2013)

23. N. Nagaosa, Y. Tokura, Topological properties and dynamics of magnetic skyrmions. Nat. Nanotechnol. **8**, 899 (2013)
24. F. Jonietz, S. Mühlbauer, C. Pfleiderer, A. Neubauer, W. Münzer, A. Bauer, T. Adams, R. Georgii, P. Böni, R.A. Duine, K. Everschor, M. Garst, A. Rosch, Spin transfer torques in MnSi at ultralow current densities. Science **330**, 1648 (2010)
25. A. Fert, V. Cros, J. Sampaio, Skyrmions on the track. Nat. Nanotechnol. **8**, 152 (2013)
26. S. Seki, X.Z. Yu, S. Ishiwata, Y. Tokura, Observation of skyrmions in a multiferroic material. Science **336**, 198 (2012)
27. M. Mochizuki, Spin-wave modes and their intense excitation effects in skyrmion crystals. Phys. Rev. Lett. **108**, 017601 (2012)
28. Y. Okamura, F. Kagawa, M. Mochizuki, M. Kubota, S. Seki, S. Ishiwata, M. Kawasaki, Y. Onose, Y. Tokura, Microwave magnetoelectric effect via Skyrmion resonance modes in a helimagnetic multiferroic. Nat. Commun. **4**, 2391 (2013)
29. T. Schwarze, J. Waizner, M. Garst, A. Bauer, I. Stasinopoulos, H. Berger, A. Rosch, C. Pfleiderer, D. Grundler, Universal helimagnon and skyrmion excitations in metallic, semiconducting and insulating chiral magnets. Nat. Mater. **14**, 478 (2015)
30. I.Kézsmárki, S. Bordács, P. Milde, E. Neuber, L.M. Eng, J.S. White, H.M. Rønnow, C.D. Dewhurst, M. Mochizuki, K. Yanai, H. Nakamura, D. Ehlers, V. Tsurkan, A. Loidl, Néel-type Skyrmion Lattice with Confined Orientation in the Polar Magnetic Semiconductor GaV_4, arXiv:1502.08049 (2015)
31. P. Milde, D. Köhler, J. Seidel, L.M. Eng, A. Bauer, A. Chacon, J. Kindervater, S. Mühlbauer, C. Pfleiderer, S. Buhrandt, C. Schütte, A. Rosch, Unwinding of a skyrmion lattice by magnetic monopoles. Science **340**, 1076 (2013)
32. L.D. Landau, E.M. Lifshitz, *Course of Theoretical Physics* (Pergamon Press, New York, 1980)
33. I.E. Dzyaloshinsky, Thermodynamic theory of weak ferromagnetism in antiferromagnetic substances. Sov. Phys. JETP **5**, 1259 (1957)
34. T. Moriya, Anisotropic superexchange interaction and weak ferromagnetism. Phys. Rev. **120**, 91 (1960)
35. I.E. Dzyaloshinsky, Theory of helicoidal structures in antiferromagnets. Sov. Phys. JETP **19**, 960 (1964)
36. Y. Ishikawa, K. Tajima, D. Bloch, M. Roth, Helical spin structure in manganese silicide MnSi. Solid State Commun. **19**, 525 (1976)
37. K. Motoya, H. Yasuoka, Y. Nakamura, J. Wernick, Helical spin structure in MnSi-NMR studies. Solid State Commun. **19**, 529 (1976)
38. G. Shirane, R. Cowley, C. Majkrzak, J.B. Sokoloff, B. Pagonis, C.H. Perry, Y. Ishikawa, Spiral magnetic correlation in cubic MnSi. Phys. Rev. B **28**, 6251 (1983)
39. M. Ishida, Y. Endoh, S. Mitsuda, Y. Ishikawa, M. Tanaka, Crystal chirality and helicity of the helical spin density wave in MnSi. II. Polarized neutron diffraction. J. Phys. Soc. Jpn. **54**, 2975 (1985)
40. P. Bak, M.H. Jensen, Theory of helical magnetic structures and phase transitions in MnSi and FeGe. J. Phys. C: Solid State **13**, L881 (1980)
41. Y. Ishikawa, Y. Noda, Y.J. Uemura, C.F. Majkrzak, G. Shirane, Paramagnetic spin fluctuations in the weak itinerant-electron ferromagnet MnSi. Phys. Rev. B **31**, 5884 (1985)
42. B. Lebech, *Recent Advances in Magnetism of Transition Metal Compounds* (World Scientific, Singapore, 1993), p. 167
43. D. Bloch, J. Voiron, V. Jaccarino, J.H. Wernick, The high field-high pressure magnetic properties of MnSi. Phys. Lett. A **51**, 259 (1975)
44. E. Fawcett, J.P. Maita, J.H. Wernick, Magnetoelastic and thermal properties of MnSi. Int. J. Magn. **1**, 29 (1970)
45. S. Kusaka, K. Yamamoto, T. Komatsubara, Y. Ishikawa, Ultrasonic study of magnetic phase diagram of MnSi. Solid State Commun. **20**, 925 (1976)
46. T. Adams, S. Mühlbauer, C. Pfleiderer, F. Jonietz, A. Bauer, A. Neubauer, R. Georgii, P. Böni, U. Keiderling, K. Everschor, M. Garst, A. Rosch, Long-range crystalline nature of the skyrmion lattice in MnSi. Phys. Rev. Lett. **107**, 217206 (2011)

47. X.Z. Yu, Y. Onose, N. Kanazawa, J.H. Park, J.H. Han, Y. Matsui, N. Nagaosa, Y. Tokura, Real-space observation of a two-dimensional skyrmion crystal. Nature (London) **465**, 901 (2010)
48. C. Pfleiderer, T. Adams, A. Bauer, W. Biberacher, B. Binz, F. Birkelbach, P. Böni, C. Franz, R. Georgii, M. Janoschek, F. Jonietz, T. Keller, R. Ritz, S. Mühlbauer, W. Münzer, A. Neubauer, B. Pedersen, A. Rosch, Skyrmion lattices in metallic and semiconducting B20 transition metal compounds. J. Phys.: Condens. Matter **22**, 164207 (2010)
49. T. Adams, A. Chacon, M. Wagner, A. Bauer, G. Brandl, B. Pedersen, H. Berger, P. Lemmens, C. Pfleiderer, Long-wavelength helimagnetic order and skyrmion lattice phase in Cu_2OSeO_3. Phys. Rev. Lett. **108**, 237204 (2012)
50. E. Moskvin, S. Grigoriev, V. Dyadkin, H. Eckerlebe, M. Baenitz, M. Schmidt, H. Wilhelm, Complex chiral modulations in FeGe close to magnetic ordering. Phys. Rev. Lett. **110**, 077207 (2013)
51. X.Z. Yu, N. Kanazawa, Y. Onose, K. Kimoto, W.Z. Zhang, S. Ishiwata, Y. Matsui, Y. Tokura, Near room-temperature formation of a skyrmion crystal in thin-films of the helimagnet FeGe. Nat. Mater. **10**, 106 (2011)
52. A. Tonomura, X. Yu, K. Yanagisawa, T. Matsuda, Y. Onose, N. Kanazawa, H.S. Park, Y. Tokura, Real-space observation of skyrmion lattice in helimagnet MnSi thin samples. Nano Lett. **12**, 1673 (2012)
53. H.S. Park, X. Yu, S. Aizawa, T. Tanigaki, T. Akashi, Y. Takahashi, T. Matsuda, N. Kanazawa, Y. Onose, D. Shindo, A. Tonomura, Y. Tokura, Observation of the magnetic flux and three-dimensional structure of skyrmion lattices by electron holography. Nat. Nano. **9**, 337 (2014)
54. M. Uchida, Y. Onose, Y. Matsui, Y. Tokura, Real-space observation of helical spin order. Science **311**, 359 (2006)
55. Y. Ishikawa, M. Arai, Magnetic phase diagram of MnSi near critical temperature studied by neutron small angle scattering. J. Phys. Soc. Jpn. **53**, 2726 (1984)
56. B. Lebech, P. Harris, J.S. Pedersen, K. Mortensen, C.I. Gregory, N.R. Bernhoeft, M. Jermy, S.A. Brown, Magnetic phase diagram of MnSi. J. Magn. Magn. Mater. **140–144**, 119 (1995)
57. S.V. Grigoriev, S.V. Maleyev, A.I. Okorokov, Y.O. Chetverikov, H. Eckerlebe, Field-induced reorientation of the spin helix in MnSi near T_c. Phys. Rev. B **73**, 224440 (2006)
58. S.X. Huang, C.L. Chien, Extended skyrmion phase in epitaxial FeGe(111) thin films. Phys. Rev. Lett. **108**, 267201 (2012)
59. Y. Li, N. Kanazawa, X.Z. Yu, A. Tsukazaki, M. Kawasaki, M. Ichikawa, X.F. Jin, F. Kagawa, Y. Tokura, Robust formation of skyrmions and topological Hall effect anomaly in epitaxial thin films of MnSi. Phys. Rev. Lett. **110**, 117202 (2013)
60. P. Sinha, N.A. Porter, C.H. Marrows, Strain-induced effects on the magnetic and electronic properties of epitaxial $Fe_{1-x}Co_xSi$ thin films. Phys. Rev. B **89**, 134426 (2014)
61. M.N. Wilson, A.B. Butenko, A.N. Bogdanov, T.L. Monchesky, Chiral skyrmions in cubic helimagnet films: the role of uniaxial anisotropy. Phys. Rev. B **89**, 094411 (2014)
62. S.A. Brazovskii, Phase transition of an isotropic system to a nonuniform state. Sov. Phys. JETP **41**, 85 (1975)
63. M. Janoschek, M. Garst, A. Bauer, P. Krautscheid, R. Georgii, P. Böni, C. Pfleiderer, Fluctuation-induced first-order phase transition in Dzyaloshinskii-Moriya helimagnets. Phys. Rev. B **87**, 134407 (2013)
64. S.V. Grigoriev, S.V. Maleyev, E.V. Moskvin, V.A. Dyadkin, P. Fouquet, H. Eckerlebe, Crossover behavior of critical helix fluctuations in MnSi. Phys. Rev. B **81**, 144413 (2010)
65. S.V. Grigoriev, E.V. Moskvin, V.A. Dyadkin, D. Lamago, T. Wolf, H. Eckerlebe, S.V. Maleyev, Chiral criticality in the doped helimagnets $Mn_{1-y}Fe_ySi$. Phys. Rev. B **83**, 224411 (2011)
66. I. Živković, J.S. White, H.M. Rønnow, K. Prša, H. Berger, Critical scaling in the cubic helimagnet Cu_2OSeO_3. Phys. Rev. B **89**, 060401 (2014)
67. S. Buhrandt, L. Fritz, Skyrmion lattice phase in three-dimensional chiral magnets from Monte Carlo simulations. Phys. Rev. B **88**, 195137 (2013)
68. A. Bauer, C. Pfleiderer, Magnetic phase diagram of MnSi inferred from magnetization and ac susceptibility. Phys. Rev. B **85**, 214418 (2012)

69. S. Seki, S. Ishiwata, Y. Tokura, Magnetoelectric nature of skyrmions in a chiral magnetic insulator Cu_2OSeO_3. Phys. Rev. B **86**, 060403 (2012)
70. U.K. Rößler, A.N. Bogdanov, C. Pfleiderer, Spontaneous skyrmion ground states in magnetic metals. Nature (London) **442**, 797 (2006)
71. A.B. Butenko, A.A. Leonov, U.K. Rößler, A.N. Bogdanov, Stabilization of skyrmion textures by uniaxial distortions in noncentrosymmetric cubic helimagnets. Phys. Rev. B **82**, 052403 (2010)
72. H. Wilhelm, M. Baenitz, M. Schmidt, U.K. Rößler, A.A. Leonov, A.N. Bogdanov, Precursor phenomena at the magnetic ordering of the cubic helimagnet FeGe. Phys. Rev. Lett. **107**, 127203 (2011)
73. H. Wilhelm, M. Baenitz, M. Schmidt, C. Naylor, R. Lortz, U.K. Rößler, A.A. Leonov, A.N. Bogdanov, Confinement of chiral magnetic modulations in the precursor region of FeGe. J. Phys.: Condens. Matter **24**, 294204 (2012)
74. L. Cevey, H. Wilhelm, M. Schmidt, R. Lortz, Thermodynamic investigations in the precursor region of FeGe. Phys. Status Solidi B **250**, 650 (2013)
75. A. Bauer, A. Neubauer, C. Franz, W. Münzer, M. Garst, C. Pfleiderer, Quantum phase transitions in single-crystal $Mn_{1-x}Fe_xSi$ and $Mn_{1-x}Co_xSi$: crystal growth, magnetization, ac susceptibility, and specific heat. Phys. Rev. B **82**, 064404 (2010)
76. A. Bauer, M. Garst, C. Pfleiderer, Specific heat of the skyrmion lattice phase and field-induced tricritical point in MnSi. Phys. Rev. Lett. **110**, 177207 (2013)
77. I. Levatić, V. Šurija, H. Berger, I. Živković, Dissipation processes in the insulating skyrmion compound Cu_2OSeO_3. Phys. Rev. B **90**, 224412 (2014)
78. D. Vollhardt, Characteristic crossing points in specific heat curves of correlated systems. Phys. Rev. Lett. **78**, 1307 (1997)
79. A. Aharoni, Demagnetizing factors for rectangular ferromagnetic prisms. J. Appl. Phys. **83**, 3432 (1998)
80. L. Ludgren, O. Beckman, V. Attia, S.P. Bhattacheriee, M. Richardson, Helical spin arrangement in cubic FeGe. Phys. Scr. **1**, 69 (1970)
81. B. Lebech, J. Bernhard, T. Freltoft, Magnetic structures of cubic FeGe studied by small-angle neutron scattering. J. Phys.: Condens. Matter **1**, 6105 (1989)
82. J. Beille, J. Voiron, M. Roth, Long period helimagnetism in the cubic B20 $Fe_xCo_{1-x}Si$ and $Co_xMn_{1-x}Si$ alloys. Solid State Commun. **47**, 399 (1983)
83. K. Ishimoto, Y. Yamaguchi, J. Suzuki, M. Arai, M. Furusaka, Y. Endoh, Small-angle neutron diffraction from the helical magnet $Fe_{0.8}Co_{0.2}Si$. Phys. B **213–214**, 381 (1995)
84. M.A. Chernikov, L. Degiorgi, E. Felder, S. Paschen, A.D. Bianchi, H.R. Ott, J.L. Sarrao, Z. Fisk, D. Mandrus, Low-temperature transport, optical, magnetic and thermodynamic properties of $Fe_{1-x}Co_xSi$. Phys. Rev. B **56**, 1366 (1997)
85. N. Manyala, Y. Sidis, J.F. DiTusa, G. Aeppli, D.P. Young, Z. Fisk, Magnetoresistance from quantum interference effects in ferromagnets. Nature (London) **404**, 581 (2000)
86. Y. Onose, N. Takeshita, C. Terakura, H. Takagi, Y. Tokura, Doping dependence of transport properties in $Fe_{1-x}Co_xSi$. Phys. Rev. B **72**, 224431 (2005)
87. S.V. Grigoriev, V.A. Dyadkin, D. Menzel, J. Schoenes, Y.O. Chetverikov, A.I. Okorokov, H. Eckerlebe, S.V. Maleyev, Magnetic structure of $Fe_{1-x}Co_xSi$ in a magnetic field studied via small-angle polarized neutron diffraction. Phys. Rev. B **76**, 224424 (2007)
88. C. Franz, F. Freimuth, A. Bauer, R. Ritz, C. Schnarr, C. Duvinage, T. Adams, S. Blügel, A. Rosch, Y. Mokrousov, C. Pfleiderer, Real-space and reciprocal-space Berry phases in the Hall effect of $Mn_{1-x}Fe_xSi$. Phys. Rev. Lett. **112**, 186601 (2014)
89. J. Teyssier, E. Giannini, V. Guritanu, R. Viennois, D. van der Marel, A. Amato, S.N. Gvasaliya, Spin-glass ground state in $Mn_{1-x}Co_xSi$. Phys. Rev. B **82**, 064417 (2010)
90. E.W. Achu, H.J. Al-Kanani, J.G. Booth, M.M.R. Costa, B. Lebech, Studies of the incommensurate structures of B20 alloys. J. Magn. Magn. Mater. **177–181**, 779 (1998)
91. M. Motokawa, S. Kawarazaki, H. Nojiri, T. Inoue, Magnetization measurements of $Fe_{1-x}Co_xSi$. J. Magn. Magn. Mater. **70**, 245 (1987)

92. V. Jaccarino, G.K. Wertheim, J.H. Wernick, L.R. Walker, S. Arajs, Paramagnetic excited state of FeSi. Phys. Rev. **160**, 476 (1967)
93. D. Shinoda, S. Asanabe, Magnetic properties of silicides of iron group transition elements. J. Phys. Soc. Jpn. **21**, 555 (1966)
94. M. Arita, K. Shimada, Y. Takeda, M. Nakatake, H. Namatame, M. Taniguchi, H. Negishi, T. Oguchi, T. Saitoh, A. Fujimori, T. Kanomata, Angle-resolved photoemission study of the strongly correlated semiconductor FeSi. Phys. Rev. B **77**, 205117 (2008)
95. N. Manyala, Y. Sidis, J.F. DiTusa, G. Aeppli, D.P. Young, Z. Fisk, Large anomalous Hall effect in a silicon-based magnetic semiconductor. Nat. Mater. **3**, 255 (2004)
96. R. Ritz, M. Halder, C. Franz, A. Bauer, M. Wagner, R. Bamler, A. Rosch, C. Pfleiderer, Giant generic topological Hall resistivity of MnSi under pressure. Phys. Rev. B **87**, 134424 (2013)
97. T. Adams, S. Mühlbauer, A. Neubauer, W. Münzer, F. Jonietz, R. Georgii, B. Pedersen, P. Böni, A. Rosch, C. Pfleiderer, Skyrmion lattice domains in $Fe_{1-x}Co_xSi$. J. Phys.: Conf. Ser. **200**, 032001 (2010)
98. G. Meunier, M. Bertaud, Constantes cristallographiques de $CuSe_2O_5$, $CuSeO_3$ et Cu_2SeO_4. J. Appl. Cryst. **9**, 364 (1976)
99. K. Kohn, A new ferrimagnet Cu_2SeO_4. J. Phys. Soc. Jpn **42**, 2065 (1977)
100. M. Belesi, I. Rousochatzakis, H.C. Wu, H. Berger, I.V. Shvets, F. Mila, J.P. Ansermet, Ferrimagnetism of the magnetoelectric compound Cu_2OSeO_3 probed by ^{77}Se NMR. Phys. Rev. B **82**, 094422 (2010)
101. C.L. Huang, K.F. Tseng, C.C. Chou, S. Mukherjee, J.L. Her, Y.H. Matsuda, K. Kindo, H. Berger, H.D. Yang, Observation of a second metastable spin-ordered state in ferrimagnet Cu_2OSeO_3. Phys. Rev. B **83**, 052402 (2011)
102. S. Seki, J.-H. Kim, D. S. Inosov, R. Georgii, B. Keimer, S. Ishiwata, Y. Tokura, Formation and rotation of skyrmion crystal in the chiral-lattice insulator Cu_2. Phys. Rev. B **85**, 220406 (R) (2012)
103. M.C. Langner, S. Roy, S.K. Mishra, J.C.T. Lee, X.W. Shi, M.A. Hossain, Y.-D. Chuang, S. Seki, Y. Tokura, S.D. Kevan, R.W. Schoenlein, Coupled skyrmion sublattices in Cu_2OSeO_3. Phys. Rev. Lett. **112**, 167202 (2014)
104. C. Jia, S. Onoda, N. Nagaosa, J.H. Han, Bond electronic polarization induced by spin. Phys. Rev. B **74**, 224444 (2006)
105. J.S. White, I. Levatić, A.A. Omrani, N. Egetenmeyer, K. Prša, I. Živković, J.L. Gavilano, J. Kohlbrecher, M. Bartkowiak, H. Berger, H.M. Rønnow, Electric field control of the skyrmion lattice in Cu_2OSeO_3. J. Phys.: Condens. Matter **24**, 432201 (2012)
106. M. Mochizuki, S. Seki, Magnetoelectric resonances and predicted microwave diode effect of the skyrmion crystal in a multiferroic chiral-lattice magnet. Phys. Rev. B **87**, 134403 (2013)
107. M. Mochizuki, X.Z. Yu, S. Seki, N. Kanazawa, W. Koshibae, J. Zang, M. Mostovoy, Y. Tokura, N. Nagaosa, Thermally driven ratchet motion of a skyrmion microcrystal and topological magnon Hall effect. Nat. Mater. **13**, 241 (2014)
108. Y. Okamura, F. Kagawa, S. Seki, M. Kubota, M. Kawasaki, Y. Tokura, Microwave magnetochiral dichroism in the chiral-lattice magnet Cu_2OSeO_3. Phys. Rev. Lett **114**, 197202 (2015)
109. A. Neubauer, C. Pfleiderer, B. Binz, A. Rosch, R. Ritz, P.G. Niklowitz, P. Böni, Topological Hall effect in the *A* phase of MnSi. Phys. Rev. Lett. **102**, 186602 (2009)
110. C. Pfleiderer, P. Böni, T. Keller, U.K. Rößler, A. Rosch, Non-Fermi liquid metal without quantum criticality. Science **316**, 1871 (2007)
111. T. Schulz, R. Ritz, A. Bauer, M. Halder, M. Wagner, C. Franz, C. Pfleiderer, K. Everschor, M. Garst, A. Rosch, Emergent electrodynamics of skyrmions in a chiral magnet. Nat. Phys. **8**, 301 (2012)
112. C. Pfleiderer, G.J. McMullan, S.R. Julian, G.G. Lonzarich, Magnetic quantum phase transition in MnSi under hydrostatic pressure. Phys. Rev. B **55**, 8330 (1997)
113. C. Thessieu, C. Pfleiderer, A.N. Stepanov, J. Flouquet, Field dependence of the magnetic quantum phase transition in MnSi. J. Phys.: Condens. Matter **9**, 6677 (1997)

114. N. Doiron-Leyraud, I.R. Walker, L. Taillefer, M.J. Steiner, S.R. Julian, G.G. Lonzarich, Fermi-liquid breakdown in the paramagnetic phase of a pure metal. Nature (London) **425**, 595 (2003)
115. Y.J. Uemura, T. Goko, I.M. Gat-Malureanu, J.P. Carlo, P.L. Russo, A.T. Savici, A. Aczel, G.J. MacDougall, J.A. Rodriguez, G.M. Luke, S.R. Dunsiger, A. McCollam, J. Arai, C. Pfleiderer, P. Böni, K. Yoshimura, E. Baggio-Saitovitch, M.B. Fontes, J. Larrea, Y.V. Sushko, J. Sereni, Phase separation and suppression of critical dynamics at quantum phase transitions of MnSi and $(Sr_{1-x}Ca_x)RuO_3$. Nat. Phys. **3**, 29 (2007)
116. N. Nagaosa, J. Sinova, S. Onoda, A.H. MacDonald, N.P. Ong, Anomalous Hall effect. Rev. Mod. Phys. **82**, 1539 (2010)
117. F. Freimuth, R. Bamler, Y. Mokrousov, A. Rosch, Phase-space Berry phases in chiral magnets: Dzyaloshinskii-Moriya interaction and the charge of skyrmions. Phys. Rev. B **88**, 214409 (2013)
118. X.Z. Yu, N. Kanazawa, W.Z. Zhang, T. Nagai, T. Hara, K. Kimoto, Y. Matsui, Y.O.Y. Tokura, Skyrmion flow near room temperature in an ultralow current density. Nat. Commun. **3**, 988 (2012)
119. J. Iwasaki, M. Mochizuki, N. Nagaosa, Universal current-velocity relation of skyrmion motion in chiral magnets. Nat. Commun. **4**, 1463 (2013)
120. A.A. Thiele, Steady-state motion of magnetic domains. Phys. Rev. Lett. **30**, 230 (1973)
121. K. Everschor, M. Garst, R.A. Duine, A. Rosch, Current-induced rotational torques in the skyrmion lattice phase of chiral magnets. Phys. Rev. B **84**, 064401 (2011)
122. J. Zang, M. Mostovoy, J.H. Han, N. Nagaosa, Dynamics of Skyrmion crystals in metallic thin films. Phys. Rev. Lett. **107**, 136804 (2011)
123. Y. Tokunaga, X. Z. Yu, J. S. White, H. M. Rønnow, D. Morikawa, Y. Taguchi, Y. Tokura, A new class of chiral materials hosting magnetic skyrmions beyond room temperature, arXiv:1503.05651 (2015)
124. E. Ruff, S. Widmann, P. Lunkenheimer, V. Tsurkan, S. Bordács, I. Kézsmárki, A. Loidl, Ferroelectric skyrmions and a zoo of multiferroic phases in GaV_4, arXiv:1504.00309 (2015)
125. N. Kanazawa, Y. Onose, T. Arima, D. Okuyama, K. Ohoyama, S. Wakimoto, K. Kakurai, S. Ishiwata, Y. Tokura, Large topological Hall effect in a short-period helimagnet MnGe. Phys. Rev. Lett. **106**, 156603 (2011)
126. N. Kanazawa, J.-H. Kim, D.S. Inosov, J.S. White, N. Egetenmeyer, J.L. Gavilano, S. Ishiwata, Y. Onose, T. Arima, B. Keimer, Y. Tokura, Possible skyrmion-lattice ground state in the $B20$ chiral-lattice magnet MnGe as seen via small-angle neutron scattering. Phys. Rev. B **86**, 134425 (2012)
127. Y. Shiomi, N. Kanazawa, K. Shibata, Y. Onose, Y. Tokura, Topological Nernst effect in a three-dimensional skyrmion-lattice phase. Phys. Rev. B **88**, 064409 (2013)
128. T. Tanigaki, K. Shibata, N. Kanazawa, X. Z. Yu, S. Aizawa, Y. Onose, H.S. Park, D. Shindo, Y. Tokura, Real-space observation of short-period cubic lattice of skyrmions in MnGe, arXiv:1503.03945 (2015)
129. M. Janoschek, F. Bernlochner, S. Dunsiger, C. Pfleiderer, P. Böni, B. Roessli, P. Link, A. Rosch, Helimagnon bands as universal excitations of chiral magnets. Phys. Rev. B **81**, 214436 (2010)
130. J.D. Koralek, D. Meier, J.P. Hinton, A. Bauer, S.A. Parameswaran, A. Vishwanath, R. Ramesh, R.W. Schoenlein, C. Pfleiderer, J. Orenstein, Observation of coherent helimagnons and Gilbert damping in an itinerant magnet. Phys. Rev. Lett. **109**, 247204 (2012)
131. Y. Onose, Y. Okamura, S. Seki, S. Ishiwata, Y. Tokura, Observation of magnetic excitations of skyrmion crystal in a helimagnetic insulator Cu_2OSeO_3. Phys. Rev. Lett. **109**, 037603 (2012)
132. S. Heinze, K. v. Bergmann, M. Menzel, J. Brede, A. Kubetzka, R. Wiesendanger, G. Bihlmayer, S. Blügel, Spontaneous atomic-scale magnetic skyrmion lattice in two dimensions. Nat. Phys. **7**, 713 (2011)
133. N. Romming, C. Hanneken, M. Menzel, J.E. Bickel, B. Wolter, K. von Bergmann, A. Kubetzka, R. Wiesendanger, Writing and deleting single magnetic skyrmions. Science **341**, 636 (2013)
134. X. Yu, M. Mostovoy, Y. Tokunaga, W. Zhang, K. Kimoto, Y. Matsui, Y. Kaneko, N. Nagaosa, Y. Tokura, Magnetic stripes and skyrmions with helicity reversals. Proc. Natl. Acad. Sci. USA **109**, 8856 (2012b)

135. S.-Z. Lin, C. Reichhardt, C.D. Batista, A. Saxena, Driven skyrmions and dynamical transitions in chiral magnets. Phys. Rev. Lett. **110**, 207202 (2013)
136. M. Finazzi, M. Savoini, A.R. Khorsand, A. Tsukamoto, A. Itoh, L. Duò, A. Kirilyuk, T. Rasing, M. Ezawa, Laser-induced magnetic nanostructures with tunable topological properties. Phys. Rev. Lett. **110**, 177205 (2013)
137. J. Sampaio, V. Cros, S. Rohart, A. Thiaville, A. Fert, Nucleation, stability and current-induced motion of isolated magnetic skyrmions in nanostructures. Nat. Nano. **8**, 839 (2013)
138. X. Yu, J.P. DeGrave, Y. Hara, T. Hara, S. Jin, Y. Tokura, Observation of the magnetic skyrmion lattice in a MnSi nanowire by Lorentz TEM. Nano Lett. **13**, 3755 (2013)
139. J. Iwasaki, M. Mochizuki, N. Nagaosa, Current-induced skyrmion dynamics in constricted geometries. Nat. Nano. **8**, 742 (2013)
140. S.-Z. Lin, C. Reichhardt, C.D. Batista, A. Saxena, Driven skyrmions and dynamical transitions in chiral magnets. Phys. Rev. Lett. **110**, 207202 (2013)
141. J. Iwasaki, W. Koshibae, N. Nagaosa, Colossal spin transfer torque effect on skyrmion along the edge. Nano Lett. **14**, 4432 (2014)
142. J. Müller, A. Rosch, Capturing of a magnetic skyrmion with a hole. Phys. Rev. B **91**, 054410 (2015)
143. X. Zhang, G.P. Zhao, H. Fangohr, J.P. Liu, W.X. Xia, J. Xia, F.J. Morvan, Skyrmion-skyrmion and skyrmion-edge repulsions in skyrmion-based racetrack memory. Sci. Rep. **5**, 7643 (2015)
144. N. Ogawa, S. Seki, Y. Tokura, Ultrafast optical excitation of magnetic skyrmions. Sci. Rep. **5**, 9552 (2015)
145. W. Xie, S. Thimmaiah, J. Lamsal, J. Liu, T.W. Heitmann, D. Quirinale, A.I. Goldman, V. Pecharsky, G.J. Miller, β-Mn-Type $Co_{8+x}Zn_{12x}$ as a defect cubic laves phase: site preferences, magnetism, and electronic structure. Inorg. Chem. **52**, 9399 (2013)
146. R. Pocha, D. Johrendt, R. Pöttgen, Electronic and structural instabilities in GaV_4S_8 and $GaMo_4S_8$. Chem. Mater. **12**, 2882 (2010)
147. C.S. Yadav, A.K. Nigam, A.K. Rastogi, Thermodynamic properties of ferromagnetic Mott-insulator GaV_4S_8. Phys. B **403**, 1474 (2008)

Chapter 2
Topological Skyrmion Dynamics in Chiral Magnets

Markus Garst

Abstract The cubic chiral magnets form topological skyrmion textures due to the presence of the Dzyaloshinskii-Moriya interaction. We briefly review in this chapter the dynamics of these magnetic textures and their interaction with magnon and electron currents, which is fundamentally influenced by their topological origin. In particular, the effective Thiele equation of motion of the skyrmion is governed by a gyrotropic force that is proportional to its topological charge. Moreover, the continuity equation associated with the conservation of topological charge can be interpreted as the Maxwell-Faraday equation of an emergent electrodynamics. As a consequence, magnons as well as electron excitations experience, for example, a topological Hall effect as well as a skyrmion-flow Hall effect.

2.1 Introduction

Ferromagnets in a cubic crystal host that possesses a chiral, non-centrosymmetric point group become unstable in the presence of a weak spin-orbit coupling, λ_{SOC}. Instead of forming homogeneously polarized magnetic domains, the magnetization instead tends to twist on long length scales inversely proportional to λ_{SOC}. This twist of the magnetization is favored by the Dzyaloshinskii-Moriya interaction that competes with the usual exchange interaction and gives rise to modulated chiral magnetic textures like magnetic helices and chiral magnetic skyrmions, see Fig. 2.1.

Interestingly, in the limit of small spin-orbit coupling, λ_{SOC}, this competition is captured within an effective low-energy theory that is specified by only a few parameters resulting in a universal description of magnetic phenomena. The combination of weak λ_{SOC} and chiral but cubic crystal symmetry thus gives rise to a magnetic universality. This probably applies to a whole class of chiral magnets with space group $P2_13$ (B20) that encompasses the metals MnSi and FeGe, the semiconductor

M. Garst (✉)
Institut Für Theoretische Physik, Universität Zu Köln,
Zülpicher Str. 77a, 50937 Cologne, Germany
e-mail: mgarst@uni-koeln.de

J. Seidel (ed.), *Topological Structures in Ferroic Materials*, Springer Series in Materials Science 228, DOI 10.1007/978-3-319-25301-5_2

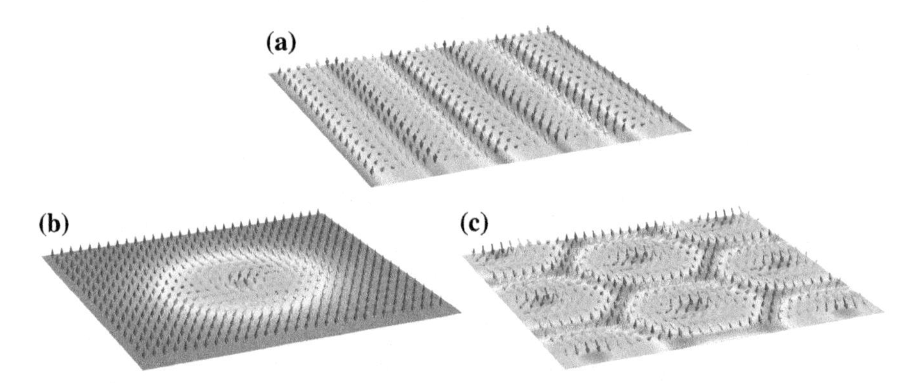

Fig. 2.1 Illustration of **a** a magnetic helix, **b** a single chiral skyrmion in a ferromagnetic background and **c** a magnetic skyrmion crystal

$Fe_{1-x}Co_xSi$, and the insulator Cu_2OSeO_3, which indeed share a common magnetic phase diagram irrespective of their electronic properties.

The material MnSi is especially well-studied and known to exhibit a hierarchy of energy scales that reflect this universality [1] as discussed in Chap. 1. The strongest energy scale is the magnetic exchange that aligns the magnetic moment on short length scales while the weak spin-orbit coupling results at zero magnetic field, $\mathbf{H} = 0$, in the formation of a magnetic helix. The weakest energy scale is associated with the cubic anisotropies that align the helix with a crystallographic $\langle 111 \rangle$ direction. A finite field $\mathbf{H} \neq 0$ competes with the cubic anisotropies and at a critical field H_{c1} the helix is oriented in the field direction so that Zeeman energy can be gained by canting the magnetic moments towards the field resulting in a conical helix configuration. Finally, at a second critical field H_{c2} a transition into the fully field-polarized state occurs, see the phase diagram in Fig. 1.1.

The magnetic phase transitions in the cubic chiral magnets that obtain as a function of temperature T and field H attract attention since the early work of Bak and Jensen [2]. At zero field, the transition as a function of T is weakly first-order [3, 4] with a small latent heat [5]. The critical chiral paramagnons are strongly interacting and, in fact, drive the transition first-order [6–8]. The formation of the long-ranged ordered magnetic helix with pitch $\mathbf{Q}$—a one-dimensional magnetic crystal with periodicity $2\pi/Q$—can be viewed as a weak crystallization process [9] that is accompanied with strongly interacting critical fluctuations. The resulting strong correlations of critical paramagnons is reflected in a substantial renormalization of their correlation length that is quantitatively described by Brazovskii theory [6, 10]. At finite field, correlations of the magnetization involving a set of momentum vectors that form equilateral triangles become important and eventually result in the formation of a two-dimensional magnetic crystal in a small magnetic field range close to the critical temperature. Here, the magnetization does not vary along the field direction $\hat{H}$ but only within the plane perpendicular to $\hat{H}$. It turns out that this magnetic crystal can be identified as a lattice of chiral magnetic skyrmions [11–13].

The importance of magnetic skyrmion configurations in the cubic chiral magnets has been theoretically anticipated by Bogdanov and collaborators in early seminal work [14–16]. The direction of the magnetization, that we denote by $\hat{n}$ in the following, covers the 2-sphere, S^2, which is the order parameter space of the magnet. As the second homotopy group $\pi_2(S^2) = \mathbb{Z}$ is non-trivial, there exist topological textures in two spatial dimensions $d = 2$. The associated topological charge density within the (x, y) plane reads

$$\rho_{\text{top}} = \frac{1}{4\pi}\hat{n}(\partial_x \hat{n} \times \partial_y \hat{n}). \tag{2.1}$$

The two-dimensional spatial integral over the charge density is quantized, for example, if the unit vector $\hat{n}$ on the boundary of the integration area points in a common direction,

$$\int dxdy\, \rho_{\text{top}} = W \in \mathbb{Z}, \tag{2.2}$$

where W is the winding number of the texture. As the total topological charge in the system is quantized, topologically non-trivial configurations can be counted. We will identify a magnetic configuration with a finite winding number, in general, as a skyrmion. In the absence of singularities of the $\hat{n}$ field, i.e., hedgehog defects in $2 + 1$ space-time, there exists a continuity equation for the topological charge,

$$\partial_t \rho_{\text{top}} + \partial_\alpha j_\alpha^{\text{top}} = 0, \tag{2.3}$$

with $\alpha = x, y$. The topological charge current is defined by

$$j_\alpha^{\text{top}} = \frac{1}{4\pi}\epsilon_{0\alpha\beta}\hat{n}(\partial_\beta \hat{n} \times \partial_t \hat{n}), \tag{2.4}$$

with $\alpha, \beta = x, y$ and $\epsilon_{0\alpha\beta}$ is the totally antisymmetric tensor with $\epsilon_{0xy} = 1$. The conservation law (2.3) directly follows from the property $\hat{n}\partial_\mu \hat{n} = 0$ valid for the unit vector $\hat{n}$ for all $\mu = x, y, t$.

Within the skyrmion lattice phase of the cubic chiral magnets, the magnetization varies only within the plane perpendicular to the field, which identifies an effective two-dimensional system where skyrmion configurations can be defined according to (2.2). Integrating the topological charge ρ_{top} over a single primitive unit cell of the two-dimensional magnetic crystal yields just unity (up to a sign) so that each unit cell houses a single magnetic skyrmion, as will be discussed in more detail below. As the magnetic configuration is translationally invariant along the field, each skyrmion extends along the field direction giving rise, in fact, to skyrmion strings, see Fig. 1.1 for an illustration.

It is their finite topological charge (2.1) that makes the topic of magnetic skyrmions so appealing, see also [17] for a recent review article. As will be explained in the following subsections, the finite topological winding of skyrmions leads to various fascinating topological transport phenomena like a skew scattering of magnons resulting in a magnon Hall effect, a spin-Magnus force that governs the dynamics of skyrmion configurations, and an emergent electrodynamics for electrons traversing the topological skyrmion texture that gives rise to a topological Hall effect and a skyrmion-flow Hall effect.

2.2 Effective Theory of Cubic Chiral Magnets

The magnetism in the cubic chiral magnets, in general, has different origins. Whereas in the metallic systems the magnetization might be attributed to the magnetic moments of the itinerant electrons, in the insulating compounds the magnetization derives from localized spins. Nevertheless, the effective theory describing the magnetic properties of all these materials turn out to be the same in the limit of weak spin-orbit coupling.

For the metallic systems, it is naturally to start from an effective Landau-Ginzburg functional for the magnetization $\boldsymbol{M} = \mu \boldsymbol{\phi}$ with the magnetic moment density μ and a dimensionless field $\boldsymbol{\phi}$. The free energy density functional $f = f_0 + f_{\text{cubic}} + f_{\text{dipolar}} + \cdots$ contains various contributions; the first term reads [1, 2]

$$f_0 = \frac{J}{2}(\nabla_i \phi_j)^2 + D\boldsymbol{\phi}(\nabla \times \boldsymbol{\phi}) + \frac{r}{2}\boldsymbol{\phi}^2 + \frac{u}{4!}(\boldsymbol{\phi}^2)^2 - \mu_0 \mu \boldsymbol{\phi} \boldsymbol{H}. \tag{2.5}$$

It is the standard low-energy continuum ϕ^4-model for ferromagnetism but supplemented by the Dzyaloshinskii-Moriya interaction D. This interaction D requires lack of inversion symmetry as it is linear in the gradient. Furthermore, it is attributed to spin-orbit coupling, $D \sim \lambda_{\text{SOC}}$, as it mixes the spatial indices of the gradient with indices of the magnetization field. Note, however, that the Dzyaloshinskii-Moriya interaction is still symmetric with respect to a combined rotation of real and spin space due to the high symmetry of the cubic crystal class [14]. The sign of D defines the chirality of the system, and it is directly related to the underlying chirality of the atomic crystal. In the following, we assume without loss of generality that $D > 0$ favouring right-handed magnetic helices.

The term f_{cubic} accounts for contributions that break the rotation symmetry due to cubic anisotropies. A representative contribution reads [2, 18]

$$f_{\text{cubic}} = \frac{J_{\text{cubic}}}{2}\Big[(\partial_x \phi_x)^2 + (\partial_y \phi_y)^2 + (\partial_z \phi_z)^2\Big] + \cdots . \tag{2.6}$$

The cubic anisotropies f_{cubic} are important for the description of the transition at the first critical field H_{c1}. In the following, however, the term f_{cubic} will be of only minor importance and not be further discussed. Finally, the term f_{dipolar} contains the dipolar interaction between the magnetic moments. These dipolar interactions are important, for example, for the quantitative description of magnetic resonances [19].

In thin films, they also give rise to magnetic anisotropies that influences, in particular, the stability of the skyrmion crystal phase [15]. In order to simplify the discussion, we will however neglect them in the following.

As the Dzyaloshinskii-Moriya interaction is linear in the gradient the magnetization can gain energy by allowing a spatial modulation. The competition with the exchange energy J eventually result in modulated magnetic textures with typical wavevectors $Q \sim D/J$ that are small in spin-orbit coupling $Q \sim \lambda_{\text{SOC}}$. Importantly, in the limit of small λ_{SOC} this competition is thus fully captured by the low-energy continuum description (2.5) allowing to neglect higher order gradient terms like $(\nabla^2 \boldsymbol{\phi})^2$.

Approximating $\boldsymbol{\phi}(\boldsymbol{r}) = \phi\, \hat{n}(\boldsymbol{r})$ with a constant amplitude ϕ, the free energy density functional f_0 simplifies to $f_0 = \mathcal{L} + \frac{r\phi^2}{2} + \frac{u\phi^4}{4!}$ with the non-linear sigma model

$$\mathcal{L} = \frac{\rho_s}{2}\Big[(\nabla_i \hat{n}_j)^2 + 2Q\hat{n}(\nabla \times \hat{n}) - 2\kappa^2 \hat{n}\hat{H}\Big] \tag{2.7}$$

where we introduced the stiffness $\rho_s = J\phi^2$, the pitch length $Q = D/J$ and $\kappa^2 = \frac{\mu_0 \mu H}{J\phi}$ parametrizes the strenght of the applied magnetic field. The same continuum description in terms of a unit vector $\hat{n}$ also arises in the classical limit of localized spins in the insulating cubic chiral magnets. Many of the interesting aspects of chiral magnets are captured by the theory (2.7).

With the help of (2.7), the favouring of modulated textures can be easily illustrated. For a magnetic field along the z-axis, $\hat{H} = \hat{z}$, we use the parametrization for the unit vector

$$\hat{n}^T = (\sin\theta\cos\varphi, \sin\theta\sin\varphi, \cos\theta) \tag{2.8}$$

so that the Lagrangian can be written in the form

$$\mathcal{L} = \frac{\rho_s}{2}\Big[(\partial_i\theta - Q\hat{e}_{\varphi i})^2 + (\sin\theta\partial_i\varphi - Q\hat{e}_{\theta i})^2 - 2\kappa^2\cos\theta - 2Q^2\Big] \tag{2.9}$$

where $\hat{e}_\varphi^T = (-\sin\varphi, \cos\varphi, 0)$ and $\hat{e}_\theta^T = (-\cos\theta\cos\varphi, -\cos\theta\sin\varphi, \sin\theta)$ with $\hat{e}_\varphi \times \hat{e}_\theta = \hat{n}$. Clearly, the system can gain energy of order $\rho_s Q^2$ by allowing for a finite gradient for the angles, φ or θ.

In particular, the conical phase obtains by setting $\partial_i\theta = 0$ and $\partial_i\varphi = Q\delta_{i,z}$ so that (2.9) reduces to an effective potential for the polar angle θ

$$\mathcal{V}(\theta) = \frac{\rho_s Q^2}{2}\left[\left(\cos\theta - \frac{\kappa^2}{Q^2}\right)^2 - 1 - \frac{\kappa^4}{Q^4}\right] \tag{2.10}$$

As long as $\kappa^2 < Q^2$ the potential is minimized for an angle θ such that $\cos\theta = \kappa^2/Q^2$ which realizes the conical phase with energy $-\rho_s Q^2(1 + \kappa^4/Q^4)/2$. For $\kappa^2 > Q^2$, on the other hand, the potential is minimized for $\theta = 0$ and one obtains the fully polarized phase with energy $-\rho_s \kappa^2$. The critical value

$$\kappa_{c2}^2 = Q^2 \tag{2.11}$$

identifies the critical field $\mu_0 H_{c2} = J Q^2 \phi/\mu$ separating the two phases.

2.3 Skyrmion Excitation of the Field-Polarized Phase

Interestingly, the theory (2.7) possesses stable soliton solutions in the background of a field-polarized state $\hat{n} = \hat{z}$, see Fig. 2.1b. Introducing cylindrical coordinates

$$\boldsymbol{r}^T = (\rho \cos \chi, \rho \sin \chi, z), \tag{2.12}$$

a vortex solution is generally obtained from a linear relation between the azymuthal angle φ of the magnetization, see (2.8), and χ, i.e., $\varphi = \chi + \varphi_0$. The value of φ_0 can often be chosen arbitrarily for vortex configurations, but not in the present case. The Dzyaloshinskii-Moriya interaction with $D > 0$ is minimized by the choice $\varphi_0 = \pi/2$ that imposes its chirality onto the soliton solution. Furthermore, assuming that the polar angle only depends on the radial distance, $\theta = \theta(\rho)$, the Euler-Lagrange equations deriving from (2.7) reduce to a differential equation for the function $\theta(\rho)$,

$$\theta'' + \frac{\theta'}{\rho} - \frac{\sin\theta\cos\theta}{\rho^2} + \frac{2Q\sin^2\theta}{\rho} - \kappa^2 \sin\theta = 0. \tag{2.13}$$

In the limit of large distances the field-polarized state should be attained, while at the origin the magnetization is assumed to point in the opposite direction, which corresponds to the boundary conditions $\lim_{\rho\to\infty}\theta(\rho) = 0$ and $\theta(0) = \pi$. The saddle point equation (2.13) has been first derived and discussed by Bogdanov and Hubert [15]. With the Ansatz for the soliton, the integrated topological charge density yields

$$W = \int dxdy\, \rho_{\text{top}}\Big|_{\substack{\varphi=\chi+\pi/2,\\ \theta=\theta(\rho)}} = \int_0^\infty d\rho \frac{\theta' \sin\theta}{2} = -\frac{1}{2}\cos\theta|_{\rho=0}^{\infty} = -1. \tag{2.14}$$

The last equality directly follows from the boundary conditions. The soliton solution thus carries a negative unit of topological charge and can be identified as a skyrmion.

From the differential equation (2.13) follows the asymptotics for the solution

$$\theta(\rho) \approx \begin{cases} \pi - c_1 \kappa\rho \text{ for } \rho \to 0 \\ \frac{c_2}{\sqrt{\kappa\rho}} e^{-\kappa\rho} \;\text{ for } \rho \to \infty \end{cases} \tag{2.15}$$

where the constants $c_{1,2}$ however depend on the ratio κ/Q. The skyrmion is thus exponentially confined and decays exponentially at large distances on the length

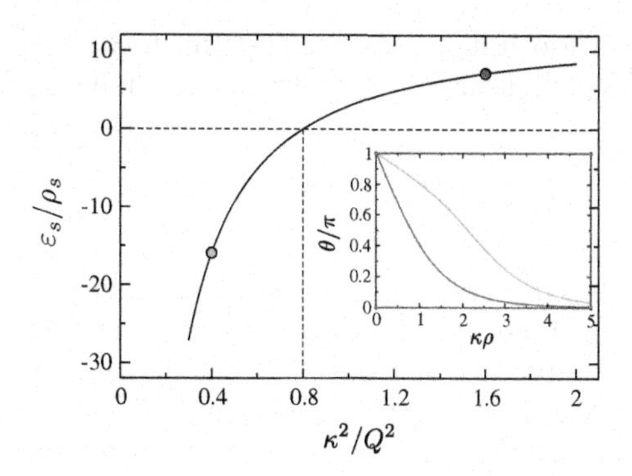

Fig. 2.2 Energy ε_s of the skyrmion excitation as a function of magnetic field as parametrized by κ^2. Inset: skyrmion profiles $\theta(\rho)$ for two values of κ as indicated by the colored dots [28]

scale set by $1/\kappa$. As the magnetic field, i.e., κ^2 increases, the skyrmion size shrinks in order to save magnetic energy. A numerical solution of the profile $\theta(\rho)$ for two values of the κ is shown in the inset of Fig. 2.2. The main panel shows the skyrmion energy ε_s, that is obtained by integrating (2.7) for the skyrmion solution over space, with respect to the energy of the field-polarized phase. The skyrmion is an excitation as long as its energy is positive, $\varepsilon_s > 0$, which requires $\kappa^2 > \kappa_{\text{cr}}^2$ with [15]

$$\kappa_{\text{cr}}^2 \approx 0.8Q^2. \tag{2.16}$$

For smaller values of κ^2, it is energetically advantageous to accomodate many skyrmions in the ground state so that the field-polarized state becomes unstable with respect to the formation of a skyrmion lattice. For a strictly two-dimensional system orthogonal to $\hat{H}$ the conical helix is prevented to develop by geometric constraints, and a skyrmion lattice is expected to materialize at zero temperature as the magnetic field is decreased to smaller values. In thin films and bulk systems, however, the situation is more complicated. Note that the conical-helix phase with a pitch along the field direction is, according to (2.11), already energetically favoured for $\kappa^2 \leq \kappa_{c2}^2 = Q^2$, and thus seems to preempt the skyrmion lattice instability at κ_{cr}^2. This is indeed true for bulk systems where the skyrmion lattice phase is not observed at sufficiently low temperatures. Only for temperatures close to $T_c(H = 0)$ the strong correlations of the chiral paramagnons alluded to in the introduction lead to a modification of the mean-field energetics, and stabilize a skyrmion lattice in a finite temperature range in bulk chiral magnets [11]. In thin films, on the other hand, the skyrmion strings profit energetically by twisting in addition along the z-direction close to the boundary [20]. This boundary contribution stabilizes the skyrmion lattice phase in thin films so that it occupies a much larger region of the phase diagram as compared to bulk systems [21].

The skyrmion in the presence of the Dzyaloshinskii-Moriya interaction D is chiral and possesses the fixed phase difference $\varphi - \chi = \pi/2$. Correspondingly the resulting magnetic configuration is sometimes called a Bloch skyrmion as its profile resembles a Bloch domain wall. This Bloch character of the skyrmion has also qualitative consequences. Due to the chirality, it possesses a finite *toroidal moment* defined by

$$\vec{\mathcal{T}} = \int dxdy\, \boldsymbol{r} \times \hat{n}\Big|_{\substack{\varphi=\chi+\pi/2,\\ \theta=\theta(\rho)}} = \hat{z}\, 2\pi \int_0^\infty d\rho \rho^2 \sin\theta(\rho) \equiv \frac{\hat{z}}{\kappa^3} T(\kappa^2/Q^2) \qquad (2.17)$$

where T is a dimensionless function that depends on the soliton solution $\theta(\rho)$. Toriodal moments have been discussed in the context of the magnetoelectric effect in multiferroic materials [22], and insulating chiral magnets are thus promising candidates for interesting magnetoelectric phenomena. A finite $\vec{\mathcal{T}}$ distinguishes the chiral Bloch skyrmion, in particular, from non-chiral Néel skyrmions for which the cross product $\boldsymbol{r} \times \hat{n}$ vanishes for all $\boldsymbol{r}$. The latter type of skyrmions were recently discovered, for example, in a magnetic monolayer of Fe deposited on a surface of Ir(111) [23–25].

2.3.1 Magnon Spectrum in the Presence of a Skyrmion

The presence of a skyrmion modifies the magnon spectrum, i.e., the small-wavelength excitations. The magnons not only scatter from the skyrmion but also form bound states corresponding to internal excitation modes of the magnetic skyrmion configuration. The problem of magnon excitations in a strictly two-dimensional system perpendicular to the applied field has been studied in [26–29], and the main results are shortly reviewed in the following. A similar analysis has been carried out before but for so-called precessional topological solitons in ferromagnets by Ivanov and collaborators [30].

In order to study the magnon spectrum, the theory (2.7) has to be complemented with a dynamical term,

$$\mathcal{L}_{\text{dyn}} = s\vec{A}(\hat{n})\partial_t \hat{n}, \qquad (2.18)$$

where $s = \hbar M/(g\mu_B)$ is the two-dimensional spin-density with the magnetization $M = \mu\phi$ and the spin-gauge field possesses the property $\varepsilon_{ijk}\partial\vec{A}_j/\partial\hat{n}_i = \hat{n}_k$. It ensures that the Euler-Lagrange equation deriving from $\mathcal{L}_{\text{tot}} = \mathcal{L}_{\text{dyn}} - \mathcal{L}_{\text{pot}}$ with the potential $\mathcal{L}_{\text{pot}} \equiv \mathcal{L}$ just reproduce the Landau-Lifshitz equations

$$\partial_t \hat{n} = -\gamma \hat{n} \times \mathbf{B}_{\text{eff}}, \qquad (2.19)$$

with the gyromagnetic ratio $\gamma = \frac{g\mu_B}{\hbar} > 0$, and the effective magnetic field $\mathbf{B}_{\text{eff}} = -\frac{1}{M}\delta\mathcal{S}/\delta\hat{n}$ is determined by the functional derivative of $\mathcal{S} = \int d^2\mathbf{r}\mathcal{L}$ with the Lagrangian given in (2.7).

Fluctuations around the skyrmion solution $\hat{n}_s \equiv \hat{e}_3$ can be parameterized with the help of the magnon wavefunction ψ, i.e., $\hat{n} = \hat{e}_3\sqrt{1-2|\psi|^2} + \hat{e}_+\psi + \hat{e}_-\psi^*$ where

$\hat{e}_\pm = (\hat{e}_1 \pm i\hat{e}_2)/\sqrt{2}$ with the local orthogonal frame $\hat{e}_\alpha \hat{e}_\beta = \delta_{\alpha\beta}$ for $\alpha, \beta = 1, 2, 3$. Plugging this parametrization into the Landau-Lifshitz equation and expanding in lowest order in ψ, one finds that the spinor $\vec{\psi}^T = (\psi, \psi^*)$ obeys the bosonic Bogoliubov-deGennes equations $i\hbar\tau^z \partial_t \vec{\psi} = \mathcal{H}\vec{\psi}$ with the Hamiltonian [28]

$$\mathcal{H} = \frac{\varepsilon_{\mathrm{DM}}}{Q^2}\left[(-\mathbb{1}i\tau^z\nabla_\perp - \vec{a})^2 + \mathbb{1}v_0 + \tau^x v_x\right], \tag{2.20}$$

where τ^z and τ^x are Pauli matrices, and the energy scale is fixed by the second critical field $\varepsilon_{\mathrm{DM}} = g\mu_B\rho_s Q^2/M = g\mu_0\mu_B H_{c2}$. Here we consider only magnons confined within the plane perpendicular to the field $\hat{H} = \hat{z}$ so that $\nabla_\perp = (\partial_x, \partial_y)$. The vector potential and the potential read explicitly,

$$\vec{a} = \left(\frac{\cos\theta}{\rho} - Q\sin\theta\right)\hat{\chi}, \tag{2.21}$$

$$v_0(\rho) = -\frac{\sin^2\theta}{2\rho^2} - \frac{Q\sin(2\theta)}{2\rho} - Q^2\sin^2\theta + \kappa^2\cos\theta - Q\theta' - \frac{\theta'^2}{2} \tag{2.22}$$

$$v_x(\rho) = \frac{\sin^2\theta}{2\rho^2} + \frac{Q\sin(2\theta)}{2\rho} - Q\theta' - \frac{\theta'^2}{2} \tag{2.23}$$

where $\hat{\chi}^T = (-\sin\chi, \cos\chi)$ and the soliton solution $\theta = \theta(\rho)$.

In order to determine the spectrum of the Hamiltonian (2.20) one has to solve a two-dimensional scattering problem. It is found that besides the magnon scattering states there exist three bound states with eigenenergies below the magnon gap $\varepsilon < \varepsilon_{\mathrm{gap}}$ with $\varepsilon_{\mathrm{gap}} = \varepsilon_{\mathrm{DM}}\kappa^2/Q^2$, which can be labeled by the angular momentum quantum number $m = 0, -2, -3$, see Fig. 2.3. The mode $m = 0$ corresponds to the breathing mode with the skyrmion radius oscillating in time. While the $m = 0$

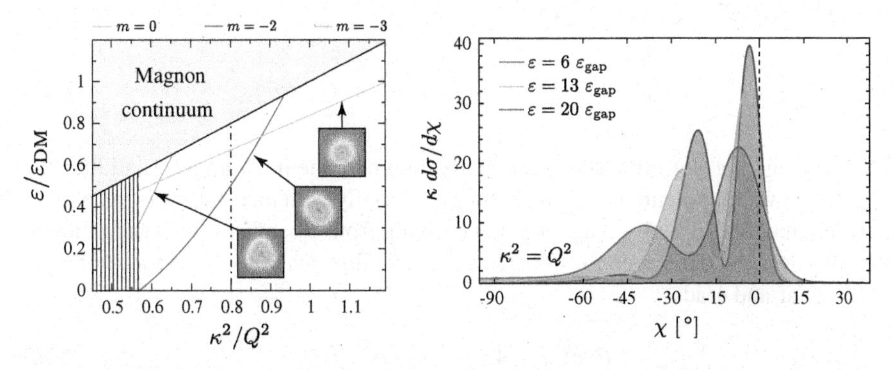

Fig. 2.3 *Left panel* Magnon spectrum in the presence of a skyrmion excitation as a function of magnetic field parametrized by κ^2. Three bound magnon-skyrmion states are found with excitation energies below the magnon continuum. The images represent snapshots of these bound excitation modes. *Right panel* Differential scattering cross section for the scattering of magnons from the skyrmion for three different energies [28]

mode exists also for higher magnetic fields as measured by the magnitude of κ^2, the quadrupolar mode $m = -2$ only materializes for smaller fields just before the global instability at $\kappa^2_{\rm cr} \approx 0.8 Q^2$, see (2.16). In the metastable regime for smaller $\kappa < \kappa_{\rm cr}$, the eigenenergy of this quadrupolar mode decreases and eventually vanishes at $\kappa = \kappa_{\rm bimeron}$ with

$$\kappa^2_{\rm bimeron} \approx 0.56 Q^2. \tag{2.24}$$

The vanishing of this eigenenergy signals a local instability of the skyrmion with respect to a quadrupolar deformation. Such a deformed skyrmion can also be interpreted as a bimeron [31]. Just before the local bimeron instability another mode with angular momentum $m = -3$ appears within the metastable regime.

2.3.2 Magnon Skew and Rainbow Scattering

The resulting differential cross section for the magnon scattering states is shown in Fig. 2.3 for three different energies $\varepsilon > \varepsilon_{\rm gap}$. There are two interesting features: (i) the magnons are scattered preferentially to negative angles χ, i.e., there is *skew scattering* and (ii) there are multiple peaks that arise from interference of classical trajectories, an effect which is known as *rainbow scattering*.

Both effects can be related to the vector potential $\vec{a}$ in the Hamiltonian (2.20). As the magnon traverses the skyrmion its local orthogonal frame $\hat{e}_i$ follows the magnetic texture and this geometric constraint eventually leads to the vector potential $\vec{a}$. It is important to note that this frame rotates with angle $\varphi = \chi + \pi/2$ even for large distances $\rho \gg 1/\kappa$. The asymptotic scattering states, however, will be defined with respect to a fixed laboratory orthogonal frame $\{\hat{x}, \hat{y}, \hat{z}\}$. The transformation from one reference frame to another corresponds to a gauge transformation under which the vector potential transforms as [28]

$$\vec{a}_{\rm lab} = \vec{a} - \nabla\chi = \Big(\frac{\cos\theta - 1}{\rho} - Q\sin\theta\Big)\hat{\chi}. \tag{2.25}$$

As the magnetic skyrmion is exponentially confined, the polar angle θ and thus the scattering vector potential $\vec{a}_{\rm lab}$ vanishes exponentially with increasing distance ρ. As a result, the total effective magnetic flux deriving from $\vec{a}_{\rm lab}$ must vanish according to Stoke's theorem. However, the effective magnetic flux density $\nabla \times (\hbar\vec{a}_{\rm lab}) = \hat{z}\,\mathcal{B}(\mathbf{r})$ is singular and reads

$$\mathcal{B}(\mathbf{r}) = -4\pi\hbar\delta(\mathbf{r}) + \mathcal{B}_{\rm reg}(|\mathbf{r}|). \tag{2.26}$$

Due to the boundary condition $\theta(0) = \pi$, the vector potential diverges as $\vec{a}_{\rm lab} \sim -2\hat{\chi}/\rho$ for small distances $\rho \to 0$ that leads to the singular flux with strength $-4\pi\hbar$. As this singular part is quantized it does not influence physical properties. However, there is an additional smooth regular part $\mathcal{B}_{\text{reg}}(|\mathbf{r}|)$, the integral of which exactly cancels the singular part $\int dxdy\, \mathcal{B}_{\text{reg}}(|\mathbf{r}|) = 4\pi\hbar$.

It is this regular magnetic flux distribution $\mathcal{B}_{\text{reg}}(|\mathbf{r}|)$ that gives rise to an emergent Lorentz force which acts on the magnons and results in skew scattering [27, 28]. Its origin is topological as $\mathcal{B}_{\text{reg}}(|\mathbf{r}|)$ is related to the topological charge density ρ_{top} of the skyrmion,

$$\mathcal{B}_{\text{reg}}(\rho) = 4\pi\hbar\Big(- \rho_{\text{top}}(\rho) - \frac{Q}{4\pi\rho}\partial_\rho(\rho \sin\theta)\Big) \tag{2.27}$$

where $|\mathbf{r}| = \rho$ is the radial distance in two spatial dimension. The second term integrates to zero so that the total flux is just determined by the total winding number (2.14) of the skyrmion texture. At high energy this topological Lorentz force in fact governs the differential cross section [29]. As the flux distribution $\mathcal{B}_{\text{reg}}(|\mathbf{r}|)$ is rotationally symmetric, the classical deflection angle $\Theta(b)$ will be an even function of the impact parameter b in the high-energy limit. A high-energy magnon with impact parameter $|b|$ is thus deflected to the same side irrespective of the sign of b, i.e., whether it passes the skyrmion on the right- or left-hand side. As a result, there exists for a given deflection angle Θ two classical trajectories that interfere leading to oscillations in the differential cross section. A similar phenomenon occurs in the theory of *rainbow scattering*.

2.3.3 Spin-Magnus Force and Magnon Pressure

As the theory (2.7) is translationally invariant, the skyrmion possesses two zero modes corresponding to translations of the skyrmion origin $\mathbf{R}$ along the x and y-directions. The origin of the skyrmionic soliton, $\hat{n}_s(\mathbf{r} - \mathbf{R})$, can be chosen arbitrarly in a homogeneous system without any energy cost. From the dynamical term (2.18) follows in zeroth order in the magnon excitations an effective theory for this skyrmion coordinate

$$L_{\text{eff}} = \mathcal{A}(\mathbf{R})\partial_t \mathbf{R}. \tag{2.28}$$

It describes a massless particle in the presence of a vector potential that is given by the spatial integral over the spin-gauge field

$$\mathcal{A}_i(\mathbf{R}) = -s \int d^2\mathbf{r} \vec{A}(\hat{n}_s(\mathbf{r} - \mathbf{R}))\partial_i \hat{n}_s(\mathbf{r} - \mathbf{R}). \tag{2.29}$$

Importantly, the associated magnetic flux is non-zero and determined by the topological number

$$\mathbf{G} = \nabla_{\mathbf{R}} \times \mathcal{A}(\mathbf{R}) = 4\pi s\hat{z} \int dxdy\, \rho_{\text{top}} = -4\pi s\hat{z}. \tag{2.30}$$

As a result of the non-trivial topology, the skyrmion carries its own magnetic flux. The vector $\mathbf{G}$ was introduce by Thiele [32] and is the so-called gyrocoupling vector. It leads to a spin-Magnus force, i.e., an effective Lorentz force in the equation of motion for the skyrmion. Note that the corresponding emergent magnetic field, $|\mathbf{G}| = \frac{2s}{\hbar}(2\pi\hbar)$ is very large. It amounts to a flux quantum $2\pi\hbar$ per area of a spin-1/2.

How is the skyrmion affected by an incoming flux of magnons? In order to address this question one considers the conservation law for the energy-momentum tensor of the theory specified in (2.7) and (2.18). Integrating this conservation law over space one finds the following effective equation of motion for the skyrmion coordinate [28]

$$\mathbf{G} \times \dot{\mathbf{R}} = \mathbf{F}. \tag{2.31}$$

It just corresponds to the Thiele equation of motion but in the presence of a force $\mathbf{F}$, where we have neglected, for simplicity, the Gilbert damping [32]. Importantly, within this Thiele approximation the skyrmion is massless and behaves similar to the guiding center of electrons in the lowest Landau level: Its velocity $\dot{\mathbf{R}}$ is perpendicular to the force $\mathbf{F}$ so that it moves along equipotential lines.

While scattering, a magnon transfers momentum to the skyrmion resulting in a finite $\mathbf{F}$. An explicitly calculation yields the following expression for a monochromatic magnon current J_ε with energy $\varepsilon = \varepsilon_{\rm DM}(\kappa^2 + k^2)/Q^2$ along the x-direction within the two-dimensional (x, y) plane [28]

$$\mathbf{F} = J_\varepsilon k \begin{pmatrix} \sigma_\parallel(\varepsilon) \\ \sigma_\perp(\varepsilon) \\ 0 \end{pmatrix}. \tag{2.32}$$

This momentum-transfer force is determined by the transport cross sections of the skyrmion

$$\begin{pmatrix} \sigma_\parallel(\varepsilon) \\ \sigma_\perp(\varepsilon) \end{pmatrix} = \int_{-\pi}^{\pi} d\chi \begin{pmatrix} 1 - \cos\chi \\ -\sin\chi \end{pmatrix} \frac{d\sigma}{d\chi}, \tag{2.33}$$

with the differential cross section $d\sigma/d\chi$. Besides the longitudinal force along x there is also a transversal force along y, as shown in Fig. 2.4, because $d\sigma/d\chi$ is not an even function of χ in the presence of the magnon skew scattering so that $\sigma_\perp(\varepsilon) > 0$. The skyrmion reacts to the force $\mathbf{F}$ according to the equation of motion (2.31). The spin-Magnus effect results in a finite skyrmion velocity that is perpendicular to the force $\mathbf{F}$ so that eventually the velocity $\dot{\mathbf{R}}$ possesses a component longitudinal to the applied magnon current that, counterintuitively, points towards the magnon source [27]. The resulting skyrmion Hall angle Φ in Fig. 2.4 is determined by the ratio of transport scattering cross sections $\sigma_\parallel/\sigma_\perp$ [28].

This magnon pressure is at the origin of a series of interesting thermal spin-transport phenomena predicted to occur in insulating chiral magnets [33–35]. A temperature gradient induces magnons to flow from the hot to the cold region of the

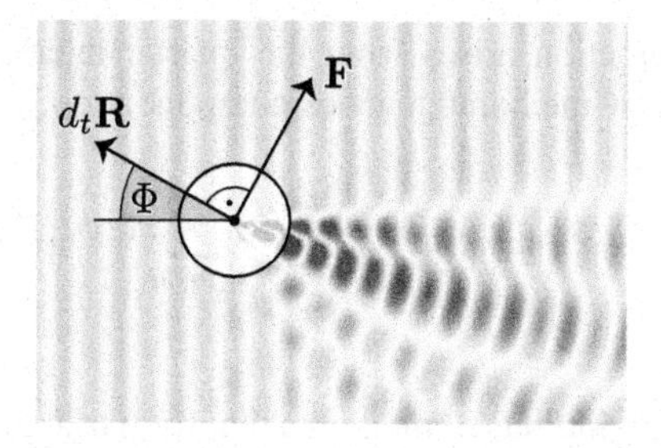

Fig. 2.4 A magnon wave impinging on the skyrmion (represented by the circle) from the left-hand side gives rise to a magnon pressure. The resulting reactive force **F** leads to a finite skyrmion velocity $\partial_t \mathbf{R}$, that as a result of the spin-Magnus force has a longitudinal component pointing towards the magnon source [28]

sample. In the presence of magnetic skyrmion configurations, magnons skew-scatter and generate a thermal topological magnon Hall effect. In turn, the skyrmions experience the reactive counter-force giving rise to a characteristic skyrmion motion with a velocity component towards the hot end of the sample. A unidirectional rotational motion but of a skyrmion crystal induced by a thermal gradient has been already experimentally observed by Mochizuki et al. [36].

2.4 Skyrmion Crystal

We now turn to the discussion of the two-dimensional skyrmion crystal, i.e., a periodic arrangements of skyrmions in the plane perpendicular to the applied field. A description of such a modulated magnetic state in terms of the non-linear sigma model (2.7) is involved [15, 16]. However, it can be rather simply described within the ϕ^4-model of (2.5). The prominent signatures of the skyrmion lattice in neutron scattering experiments [11, 13] are six prominent magnetic Bragg peaks that form a hexagon. A mean-field Ansatz that reflects this experimental observation is the superposition of three magnetic helix solutions with pitches that add to zero, $\mathbf{Q}_1 + \mathbf{Q}_2 + \mathbf{Q}_3 = 0$, which are orthogonal to the applied magnetic field direction $\hat{H} = \hat{z}$, i.e., $\mathbf{Q}_\alpha \hat{z} = 0$,

$$\phi^{\mathrm{mf}}_{\mathrm{SkX}}(\mathbf{r}) = \phi_0 \hat{z} + \sum_{\alpha=1,2,3} \mathrm{Re}\{\phi_\alpha(\hat{e}_{1\alpha} - i\hat{e}_{2\alpha}) e^{i\mathbf{Q}_\alpha \mathbf{r}}\}. \tag{2.34}$$

While the homogeneous component, ϕ_0, is a real variable, the amplitudes ϕ_α are complex. The vectors $\hat{e}_{1\alpha}$ and $\hat{e}_{2\alpha}$ for each $\alpha = 1, 2, 3$ are orthogonal to each other as well as to the pitch $\mathbf{Q}_\alpha = |\mathbf{Q}_\alpha|(\hat{e}_{1\alpha} \times \hat{e}_{2\alpha})$.

The importance of sets of wavevectors that form equilateral triangles is well known from crystallization problems [9]. Such sets of wavevectors are favored by effective cubic interactions due to momentum conservation. In the present problem, an effective cubic interaction arises from the quartic interaction u of (2.5) after replacing a single field by the homogeneous component ϕ_0

$$\frac{u}{4!}\sum_{\mathbf{q}_1,\mathbf{q}_2,\mathbf{q}_3}(\phi_{\mathbf{q}_1}\phi_{\mathbf{q}_2})(\phi_{\mathbf{q}_3}\phi_{-\mathbf{q}_1-\mathbf{q}_2-\mathbf{q}_3}) \rightarrow \frac{u\phi_0}{3!}\sum_{\mathbf{q}_1,\mathbf{q}_2,\mathbf{q}_3}(\hat{z}\phi_{\mathbf{q}_1})(\phi_{\mathbf{q}_2}\phi_{\mathbf{q}_3})\delta_{\mathbf{q}_1+\mathbf{q}_2+\mathbf{q}_3,0} \tag{2.35}$$

A magnetic crystal involving sets of momenta that form triangles can gain additional energy from such a term, thus favouring two-dimensional triangular lattices or even three-dimensional bcc lattices. Whereas three-dimensional chiral magnetic crystals have been discussed theoretically in the literature [37, 38], they have not yet been unambiguously identified experimentally. Possible evidence for a three-dimensional magnetic crystal, however, has been recently presented by Tanigaki et al. [39] for MnGe that possesses a comparatively short pitch length and thus a stronger spin-orbit coupling λ_{SOC}.

Plugging the Ansatz (2.34) into the free energy (2.5) one obtains an effective potential for ϕ_0 and the amplitudes ϕ_α. Minimization then yields a mean-field solution for the two-dimensional skyrmion crystal, that in principle can be improved by allowing for more variational parameters in the Ansatz of (2.34). In order to gain maximal energy from the cubic vertex (2.35) the phases of the helix amplitudes must fulfil the condition $\sum_{\alpha=1,2,3}\arg\phi_\alpha = \pi$. A real space picture of the resulting magnetic configuration is shown in Fig. 2.1c. One recognizes that it amounts to a triangular lattice where the magnetic configuration of each unit cell resembles the single skyrmion solution of Fig. 2.1b. The mean-field configuration of (2.34) is such that its length $|\phi^{\rm mf}_{\rm SkX}(\mathbf{r})|$ varies in space but always remains finite. This allows to define uniquely the unit vector $\hat{n}(\mathbf{r}) = \phi^{\rm mf}_{\rm SkX}(\mathbf{r})/|\phi^{\rm mf}_{\rm SkX}(\mathbf{r})|$ with the help of which one can compute the topological charge (2.2) within the plane perpendicular to the field. Importantly, one finds for each magnetic unit cell $W = -1$ so that each unit cell, as far as its topological properties are concerned, indeed houses a single skyrmion. This topological criterion allows to unambiguously identify the modulated magnetic texture (2.34) as a skyrmion crystal.

While the mean-field Ansatz is very useful to describe the properties of the skyrmion crystal, the resulting mean-field free energy turns out to be still larger than the competing conical helix [11]. The simplified mean-field treatment is thus insufficient to explain the thermodynamic stability of the skyrmion crystal in a bulk material. In bulk materials the skyrmion crystal is only stable within a pocket in the magnetic field and temperature phase diagram where fluctuations are known to be strong. These fluctuations lead to a substantial modification of the mean-field phase diagram, suppress the critical temperature and drive the transition from the paramagnet to the conical helix phase first-order [6, 7]. It has been shown in [11] that these fluctuations are also responsible for the stabilization of the two-dimensional skyrmion crystal. The presence of a stable thermodynamic skyrmion crystal phase in chiral magnets has been also demonstrated theoretically with Monte-Carlo simulations [8].

2.4.1 Excitations of the Skyrmion Crystal

The two-dimensional skyrmion crystal breaks the translational symmetries within the plane perpendicular to the field so that one expects gapless Goldstone modes in the magnon spectrum. The low-energy excitations are described by the displacement vector $\mathbf{u}$ within the two-dimensional plane, that generalizes the description in terms of a single coordinate $\mathbf{R}$ in case of a single skyrmion soliton in Sect. 2.3.3. Correspondingly, it obeys an equation of motion whose dynamics is governed by the spin-Magnus force similar to (2.31) [40, 41]

$$\mathbf{G} \times \dot{\mathbf{u}}(\mathbf{r}, t) = -\frac{\delta E[\mathbf{u}]}{\delta \mathbf{u}(\mathbf{r}, t)}, \tag{2.36}$$

with the gyrocoupling vector $\mathbf{G}$ that reflects the non-trivial topology of the skyrmion crystal. The force is here determined by the functional derivative of the elastic energy $E[\mathbf{u}]$ for the skyrmion crystal, that only depends on $\mathbf{u}$ via the strain field $\varepsilon_{ij} = \frac{1}{2}(\partial_i u_j + \partial_j u_i)$. Solving (2.36) for the low-energy magnon energies one obtains a dispersion that, as a result of the spin-Magnus force, is quadratic in the in-plane momentum, $\omega_{\mathbf{k}} \propto \mathbf{k}^2$ [40, 41], similar to the spinwaves in ferromagnets. This is in contrast however to the usual phonons of atomic crystals that possess a linear dispersion because their dynamics is instead governed by the inertia of atoms.

As the skyrmion crystal is periodic in the plane perpendicular to the field, the magnon excitation energies $\omega_{n\mathbf{k}}$ form a band structure according to Bloch's theorem with band index n. Accordingly, there are plenty of excitation modes even at zero momentum ω_{n0} that are labeled by n. While one of them vanishes due to the Goldstone theorem, the others have a finite excitation energy $\omega_{n0} > 0$. Three of them can be excited in a resonance experiment by the application of a weak ac magnetic field. These magnetic resonances have been theoretically identified by Mochizuki [42] and are illustrated in Fig. 2.5. There is a single breathing mode that can be excited by an

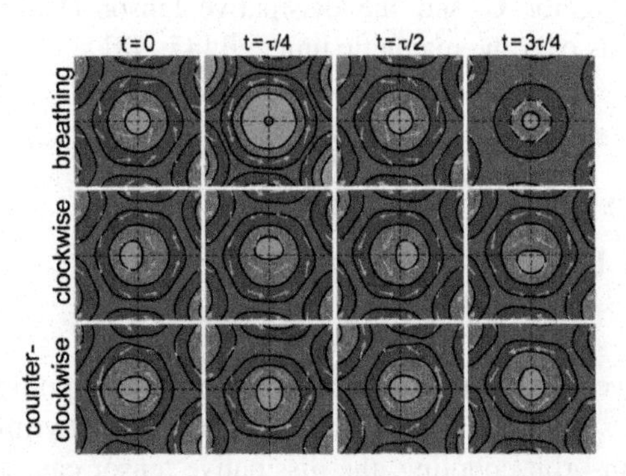

Fig. 2.5 Illustration of the three excitation modes of the skyrmion crystal—a breathing mode and two gyration modes—that can be excited by an homogeneous ac magnetic field [19]. The excitation energy of these modes, $2\pi\hbar/\tau$, is related to their time period τ; the images show snapshots of the magnetic configuration at various times t

out-of-plane ac field and two gyration modes where the skyrmion core gyrates counterclockwise or clockwise and that are excited by in-plane ac fields. These excitation modes have been first observed in Cu_2OSeO_3 [43]. A comprehensive study of these modes on three different materials, MnSi, $Fe_{0.8}Co_{0.2}Si$, and Cu_2OSeO_3, were performed by Schwarze et al. [19] who also showed that the resonance frequencies can be quantitatively explained by theory after taking into account dipolar interactions.

2.5 Spin-Transfer Torques on the Skyrmion Crystal

In the metallic chiral magnets the itinerant electrons can exchange angular momentum with the magnetic texture via the spin-transfer torque, which has two consequences. First, the magnetic texture experiences a force putting it into motion. Second, a counter-force acts on the electrons that will deflect them from their path. The interplay of these two effects has interesting consequences and is at the origin of the topological Hall and skyrmion-flow Hall effect that will be discussed below.

Let us however first discuss the spin-transfer torque that is exerted by a spin-polarized electron current with effective spin velocity $\mathbf{v}_s$. It is described by the generalized Landau-Lifshitz-Gilbert equation [44–46]

$$(\partial_t + (\mathbf{v}_s \nabla))\hat{n} = -\gamma \hat{n} \times \mathbf{B}_{\text{eff}} + \hat{n} \times (\alpha \partial_t + \beta(\mathbf{v}_s \nabla))\hat{n} \tag{2.37}$$

where α and β are damping coefficients. Making an Ansatz of a drifting magnetic texture $\hat{n} = \hat{n}(\mathbf{r} - \mathbf{v}_d t)$ with the drift velocity $\mathbf{v}_d$, projecting (2.37) onto the translational mode and integrating over a magnetic unit cell of the skyrmion lattice one obtains the Thiele equation [32]

$$\mathbf{G} \times (\mathbf{v}_s - \mathbf{v}_d) + \mathbf{D}(\beta \mathbf{v}_s - \alpha \mathbf{v}_d) = \mathbf{F}_{\text{pinning}}. \tag{2.38}$$

The gyrocoupling vector $\mathbf{G}$ and the dissipative tensor $\mathbf{D}$ are defined by two-dimensional integrals over the magnetic unit cell [47, 48]

$$\mathbf{G} = s \int_{\text{unit cell}} d^2\mathbf{r}\, \hat{n}(\partial_x \hat{n} \times \partial_y \hat{n}) = -4\pi s \hat{z}, \tag{2.39}$$

$$\mathbf{D}_{ij} = s \int_{\text{unit cell}} d^2\mathbf{r}\, (\partial_i \hat{n})(\partial_j \hat{n}). \tag{2.40}$$

The gyrocoupling vector is determined by the non-trivial topology of the skyrmion lattice and just possesses the same value as for a single skyrmion, see (2.30). In lowest order in spin-orbit coupling, the dissipative tensor can be approximated to be diagonal [47], $\mathbf{D}_{ij} = D\mathbf{P}_{ij}$ with the projector onto the plane perpendicular to the field $\mathbf{P}_{ij} = \delta_{ij} - \hat{z}_i\hat{z}_j$. The force $\mathbf{F}_{\text{pinning}}$ is here finite due to defects that explicitly break the translational symmetry and pin the magnetic texture.

For a small spin velocity $\mathbf{v}_s$, the pinning force is effectively strong resulting in a vanishing drift velocity $\mathbf{v}_d = 0$. At a critical threshold of the current, however, the pinning forces are overcome so that the drift velocity becomes finite $\mathbf{v}_d \neq 0$. This depinning transition for the skyrmion lattice has been theoretically addressed in [49, 50]. For large $\mathbf{v}_s$ the pinning force $\mathbf{F}_{\text{pinning}}$ can eventually be neglected and the Thiele equation can be explicitly solved for the drift velocity [47, 51]

$$\mathbf{v}_d^{\parallel} = \frac{\beta}{\alpha}\mathbf{v}_s^{\parallel} + \frac{\alpha - \beta}{\alpha^3(D/4\pi)^2 + \alpha}\left(\mathbf{v}_s^{\parallel} + \frac{\alpha D}{4\pi}\hat{z} \times \mathbf{v}_s^{\parallel}\right) \tag{2.41}$$

with $\mathbf{v}_{d,s}^{\parallel} = \mathbf{P}\mathbf{v}_{d,s}$. In the limit of small damping parameters α and β, the drift velocity is approximately equal to the spin-current velocity $\mathbf{v}_d^{\parallel} \approx \mathbf{v}_s^{\parallel}$ which was explicitly confirmed with the help of micromagnetic simulations [50]. This universal current relation arises because the gyrocoupling force $\mathbf{G}$ in (2.38) dominates the dynamics of the skyrmion texture in the limit of small damping.

2.5.1 Gradient in the Effective Spin-Transfer Torques

In the presence of a finite spin velocity $\mathbf{v}_s$ the skyrmion texture experiences a dissipative force due to $\mathbf{D}$ in the Thiele equation (2.38) that is longitudinal to $\mathbf{v}_s$ as well as a spin-Magnus force due to the gyrovector $\mathbf{G}$ that is transversal to $\mathbf{v}_s$ as shown by the green and red arrows, respectively, in Fig. 2.6. Each force is proportional to the local spin-density denoted by s in (2.39) and (2.40). With the help of a thermal or magnetic field gradient one can impose an effective gradient in the spin-density across a macroscopic skyrmion crystal domain. The effective forces will correspondingly vary across the sample as illustrated in Fig. 2.6, which results in a macroscopic torque that might rotate the whole skyrmion crystal domain [48].

Such rotation of the skyrmion crystal due to spin-transfer torques has been experimentally detected in the presence of a thermal gradient with the help of neutron scattering by Jonietz et al. [52]. By applying a relatively small current density of

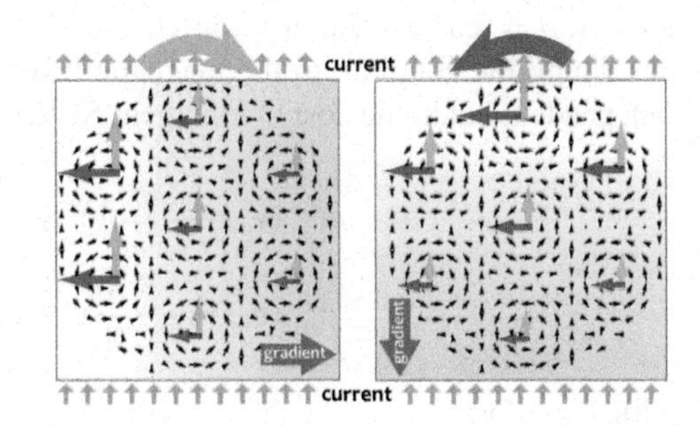

Fig. 2.6 Effective gradients of the spin-density lead to an imbalance of forces across a skyrmion crystal domain [48]. The resulting macroscopic torques were detected in [52] via a rotated neutron scattering pattern of the skyrmion crystal

10^6 A/m^2 the six-fold scattering pattern of the skyrmion lattice was observed to rotate and the rotation angle could be unambiguously related to the orientation of the thermal gradient. This was the first demonstration of a spin-transfer torque phenomenon in skyrmion crystals at current densities that are ultralow compared to the ones usually applied in experiments on current-driven domain-wall motion. The origin of this low critical current density of order $j_c \sim 10^6$ A/m^2 is attributed to the peculiar skyrmion dynamics and the elasticity of the skyrmion crystal. It is the spin-Magnus force that allows the skyrmions to avoid and circumvent pinning potentials efficiently [49, 50, 52].

2.6 Emergent Electrodynamics in Metallic Chiral Magnets

The spin-transfer forces on the magnetic skyrmions are balanced by counter-forces that act on the electrons. Consider the Schrödinger equation of an electron described by the spinor wavefunction $\vec{\Psi}^T = (\Psi_\uparrow, \Psi_\downarrow)$ moving in a magnetic texture $\hat{n}(\mathbf{r}, t)$

$$i\hbar\partial_t \vec{\Psi} = \left(-\frac{\hbar^2}{2m}\mathbb{1}\nabla^2 + J\hat{n}(\mathbf{r}, t)\vec{\sigma}\right)\vec{\Psi}. \tag{2.42}$$

The exchange interaction $J > 0$ favors the electron spin to align antiparallel with the local magnetization field. It is convenient to introduce the local basis that obeys

$$\vec{\sigma}^j\vec{v}^{\pm}(\mathbf{r}, t) = \pm\hat{n}^j(\mathbf{r}, t)\vec{v}^{\pm}(\mathbf{r}, t), \tag{2.43}$$

with $j = x, y, z$ and $\vec{v}^{\alpha\dagger}(\mathbf{r}, t)\vec{v}^{\beta}(\mathbf{r}, t) = \delta_{\alpha\beta}$ and $\sum_{\alpha=\pm1}\vec{v}^{\alpha}(\mathbf{r}, t)\vec{v}^{\alpha\dagger}(\mathbf{r}, t) = \mathbb{1}$. Expanding the spinor wavefunction

$$\vec{\Psi}(\mathbf{r}, t) = \psi_+(\mathbf{r}, t)\vec{v}^+(\mathbf{r}, t) + \psi_-(\mathbf{r}, t)\vec{v}^-(\mathbf{r}, t) \tag{2.44}$$

the expansion coefficients are found to obey the Schrödinger equation

$$\left(\delta_{\alpha\beta}i\hbar\partial_t + eA^0_{\alpha\beta}\right)\psi_\beta = \frac{1}{2m}\Big[-i\hbar\delta_{\alpha\beta}\nabla + e\vec{A}_{\alpha\beta}\Big]\Big[-i\hbar\delta_{\beta\gamma}\nabla + e\vec{A}_{\beta\gamma}\Big]\psi_\gamma - \alpha J\psi_\alpha, \tag{2.45}$$

with $\alpha, \beta = \pm1$ and the electron charge $-e < 0$. In this local basis the last term describing the coupling to the magnetic texture is diagonal, i.e., space- and time-independent but at the cost of emergent SU(2) gauge fields,

$$A^0_{\alpha\beta}(\mathbf{r}, t) = \frac{i\hbar}{e}\vec{v}^{\alpha\dagger}(\mathbf{r}, t)\partial_t\vec{v}^{\beta}(\mathbf{r}, t) \tag{2.46}$$

$$\vec{A}^j_{\alpha\beta}(\mathbf{r}, t) = -\frac{i\hbar}{e}\vec{v}^{\alpha\dagger}(\mathbf{r}, t)\partial_j\vec{v}^{\beta}(\mathbf{r}, t) \tag{2.47}$$

with $j = x, y, z$.

The wavefunctions ψ_+ and ψ_- describe electronic states that the exchange interaction energetically separates by $2J$ while the off-diagonal components of the emergent gauge fields, e.g. $\vec{A}_{+-}$, induce transitions between those states. For magnetic textures that are smooth in space and time these transitions can be treated perturbatively and are neglected in zeroth order. In this adiabatic approximation, the wavefunctions ψ_α are subject to effective U(1) gauge fields; the low-energy state ψ_- that describes the majority spins, for example, obeys

$$\left(i\hbar\partial_t + eA^0_{--}\right)\psi_- \approx \frac{1}{2m}\Big[-i\hbar\nabla + e\vec{A}_{--}\Big]^2\psi_- - J\psi_-. \tag{2.48}$$

The associated emergent electric field and magnetic field are determined from the standard relations

$$\vec{E}_{\text{emergent}} = -\nabla A^0_{--} - \partial_t \vec{A}_{--} \tag{2.49}$$

$$\vec{B}_{\text{emergent}} = \nabla \times \vec{A}_{--}. \tag{2.50}$$

For a magnetic texture that does not vary along the z-direction, the non-vanishing components of the fields read explicitly

$$\vec{E}^{\alpha}_{\text{emergent}} = -\frac{\hbar}{2e}\hat{n}(\partial_\alpha \hat{n} \times \partial_t \hat{n}) = \frac{2\pi\hbar}{e}\varepsilon_{0\alpha\beta} j^{\text{top}}_\beta \tag{2.51}$$

$$\vec{B}^{z}_{\text{emergent}} = -\frac{\hbar}{2e}\hat{n}(\partial_x \hat{n} \times \partial_y \hat{n}) = -\frac{2\pi\hbar}{e}\rho_{\text{top}} \tag{2.52}$$

with $\alpha, \beta = x, y$. Importantly, the emergent fields are determined by the topological charge and current densities of (2.1) and (2.4), respectively. Whereas the emergent magnetic field is perpendicular to the x-y plane and is determined by the topological charge density ρ_{top}, the emergent electric field lies within the plane and is related to the topological current $\vec{j}_{\text{top}}$.

In the absence of singular field configurations the topological charge density obeys the conservation law (2.3), which can be expressed in terms of the emergent fields

$$\partial_t \vec{B}_{\text{emergent}} = -\nabla \times \vec{E}_{\text{emergent}}. \tag{2.53}$$

It can be interpreted as the Maxwell-Faraday equation for the emergent electrodynamics stating that a time-dependent emergent magnetic field is always accompanied by a spatially-varying emergent electric field.

2.6.1 Topological Hall Effect

For the static magnetic skyrmion crystal the emergent magnetic flux per two-dimensional magnetic unit cell just amounts to a single flux quantum,

$$\int_{\text{unit cell}} d^2\mathbf{r}\, \vec{B}^z_{\text{emergent}} = -\frac{2\pi\hbar}{e} \int_{\text{unit cell}} d^2\mathbf{r}\, \rho_{\text{top}} = \frac{2\pi\hbar}{e}. \tag{2.54}$$

In case of MnSi, the size of the magnetic unit cell is approximately 314 nm^2 so that the emergent field for the majority spins corresponds on average to $\langle \vec{B}^z_{\text{emergent}} \rangle \approx -13$ T [54]. It is negative because the itinerant charge carriers in MnSi derive from a hole like Fermi surface, and unlike electrons they possess a positive charge. This large field has physical consequences and, in fact, contributes to the Hall effect. As the skyrmion crystal phase is entered as a function of magnetic field, an additional contribution to the Hall effect is observed in experiments on MnSi [54–56], which can be attributed to a finite emergent magnetic field $\vec{B}_{\text{emergent}}$. As this emergent field is directly related to the non-trivial topology of the magnetic skyrmion texture, this contribution is known as *topological Hall effect* [57].

A recent experimental study [58] has investigated the topological Hall effect in MnSi as a function of pressure, magnetic field and temperature. The results suggest that the topological Hall signal is related to the peculiar metallic properties of MnSi at high pressures, where an unusual temperature dependence of the resistivity, $\delta\rho \sim T^{3/2}$, is observed that is at odds with Fermi-liquid theory. This hints at the exciting possibility that topologically non-trivial magnetic textures are a necessary ingredient for the sought-after explanation of the non-Fermi liquid behavior in MnSi at high-pressures [59, 60].

In a semiclassical picture, the emergent gauge field (2.47) corresponds to a Berry phase that the electron accumulates on its trajectory while adiabatically adjusting its spin to the local magnetic texture, see Fig. 2.7. A full semiclassical analysis of electron dynamics in the presence of a magnetic texture is however rather involved. Besides the Berry phase in real space there exists also a Berry phase in momentum space, that, in particular, gives rise to the anomalous Hall effect [61]. Moreover, the

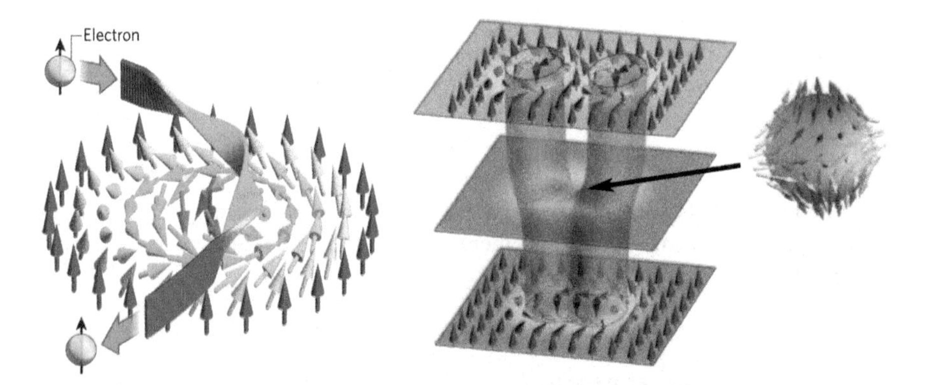

Fig. 2.7 *Left panel* An electron adiabatically adjust its spin-orientation to the local magnetic skyrmion texture thus accumulating a Berry phase [53]. *Right panel* Two skyrmion strings merge into a single skyrmion string. The merging point is identified by a hedgehog defect, which can be interpreted as a magnetic monopole within the emergent electrodynamics [65]

combination of Berry phases in real and momentum space allows for the existence of Berry curvatures in six-dimensional phase space as recently discussed in [62].

2.6.2 Skyrmion-Flow Hall Effect

In the above discussion of the topological Hall effect, we assumed that the electron moves in a static magnetic skyrmion texture, i.e., we assumed the skyrmion crystal to be pinned by defects. If the applied current densities, however, exceed the depinning threshold the skyrmion texture will drift with a velocity $\mathbf{v}_d$ determined by (2.41). For a drifting magnetic texture, $\hat{n}(\mathbf{r} - \mathbf{v}_d t)$, the topological current density of (2.4) is given by the intuitive expression

$$j^{\alpha}_{\text{top}}\Big|_{\hat{n}=\hat{n}(\mathbf{r}-\mathbf{v}_d t)} = \rho_{\text{top}} \mathbf{v}^{\alpha}_d, \tag{2.55}$$

that is, the topological charge density times the drift velocity. This finite topological current translates according to (2.51) to an emergent electric field

$$\vec{E}_{\text{emergent}} = -\mathbf{v}_d \times \vec{B}_{\text{emergent}} \tag{2.56}$$

that is simply related to the emergent magnetic field. The drift of the skyrmion texture results in a drifting emergent magnetic field that in accordance with Faraday's law of induction (2.53) implies a finite emergent electric field. The Lorentz force on the electron ($e > 0$) caused by the emergent fields then simplifies to

$$\vec{F}_{\text{emergent}} = -e\left(\vec{E}_{\text{emergent}} + \mathbf{v} \times \vec{B}_{\text{emergent}}\right) = -e(\mathbf{v} - \mathbf{v}_d) \times \vec{B}_{\text{emergent}} \tag{2.57}$$

where $\mathbf{v}$ is the velocity of the electron. It is only the relative velocity, $\mathbf{v} - \mathbf{v}_d$, of the electron with respect to the moving texture that determines the emergent Lorentz force. Above the depinning transition where $\mathbf{v}_d \neq 0$, this Lorentz force is effectively reduced as $|\mathbf{v} - \mathbf{v}_d| < |\mathbf{v}|$. As a consequence, one expects a reduction of the topological Hall effect when the skyrmion texture starts to flow with the electron liquid. Such a reduction of the topological Hall signal above the critical current density was experimentally demonstrated by Schulz et al. [63]. This skyrmion-flow Hall effect is akin to the flux-flow Hall effect in type II superconductors [64].

2.6.3 Emergent Magnetic Monopoles

For textures that vary in three dimensional space the definition of the emergent magnetic field (2.52) is generalized to

$$\vec{B}^{j}_{\text{emergent}} = \frac{\hbar}{4e} \varepsilon_{jkl} \hat{n}(\partial_k \hat{n} \times \partial_l \hat{n}), \tag{2.58}$$

with $j, k, l = x, y, z$. Similar to the topological conservation law (2.3) in $2+1$ space-time, the conservation law for the topological charge in three spatial dimensions then reads

$$\nabla \vec{B}_{\text{emergent}} = \rho_{\text{monopoles}}. \tag{2.59}$$

It corresponds to Gauss' law for the emergent magnetic field. The right hand side vanishes for a non-singular field configuration, $\rho_{\text{monopoles}} = 0$, i.e., in the absence of hedgehog defects in three spatial dimensions. Such defects effectively correspond to sources and sinks of the emergent magnetic field and can be interpreted as emergent magnetic monopoles. Such monopoles are predicted to be important close to the phase transition where the magnetic skyrmion crystal melts into a magnetic state with a topologically trivial configuration [65], see Fig. 2.7. Experimental evidence for such a hedgehog-mediated melting was provided by Milde et al. who studied the magnetic texture on a surface of a bulk crystal of $Fe_{0.5}Co_{0.5}Si$ by magnetic force microscopy [65].

2.7 Discussion

The topological character of magnetic skyrmion configurations gives rise to a variety of qualitatively novel phenomena and, in particular, influences its dynamical properties in a profound manner. The topological charge carried by the skyrmion, in particular, leads to a finite gyrocoupling vector **G** in its Thiele equation of motion which translates to a spin-Magnus force governing the skyrmion dynamics. The non-trivial topology also characterizes the interaction of skyrmions with magnon excitations as well as electronic degrees of freedom resulting in various topological transport phenomena.

This short review on skyrmions in chiral magnets is not comprehensive and many interesting aspects were only touched upon or not covered at all. For example, the skyrmion might act under certain circumstances like a particle with a finite mass where the mass generation is related to the excitation of its internal degrees of freedom [66, 67]. In insulating compounds skyrmions possess multiferroic properties that allow for a manipulation of skyrmion crystals by electric fields [68] and give rise to interesting microwave magnetoelectric effects [69]. The small critical charge current densities $j_c \sim 10^6$ A/m^2 that are required for spin-transfer torque effects make skyrmion matter interesting for spintronic applications [70]. Their quantized topological charge as well as their topological stability suggest to use skyrmions as basic information units, i.e., bits in future memory devices, for which the controlled creation and destruction of skyrmion configurations [71] and their controlled manipulation [72] is essential. Even skyrmion logic operations have been already discussed in the literature [73].

The remarkable aspect is that this rich physics just arises from complementing an ordinary ferromagnet with a weak Dzyaloshinskii-Moriya interaction $\vec{M}(\nabla \times \vec{M})$ that twists the magnetization $\vec{M}$ on long length scales. This additional interaction albeit very weak gives rise to a plethora of non-perturbative effects and opens the door for topological skyrmion physics. Despite the recent advances in this field there remain many issues to be explored, and it is to be expected that magnetic skyrmion matter will continue to fascinate us for a while.

Acknowledgments I would like to thank the many people with whom I collaborated on the topic of this review, in particular, B. Binz, K. Everschor, P. Krautscheid, S. Schroeter, C. Schütte and the experimental groups of P. Böni, D. Grundler, and, especially, C. Pfleiderer. I am particularly grateful to A. Rosch for the collaboration and support over many years.

References

1. See section §52 Helicoidal magnetic structures in L.D. Landau, E.M. Lifshitz, L.P. Pitaevskii, *Electrodynamics of Continuous Media*, 2nd edn. vol. 8 (Butterworth-Heinemann, 1984)
2. P. Bak, M.H. Jensen, J. Phys. C: Solid State Phys. **13**, L881 (1980)
3. M. Date, K. Okuda, K. Kadowaki, J. Phys. Soc. Jpn. **42**, 1555 (1977)
4. A. Hansen, Ph.D. thesis, Technical University of Denmark (1977). Risø Report Nr. 360. http://www.risoe.dk/rispubl/reports_INIS/RISO360.pdf
5. S.M. Stishov, A.E. Petrova, S. Khasanov, G.K. Panova, A.A. Shikov, J.C. Lashley, D. Wu, T.A. Lograsso, Phys. Rev. B **76**, 52405 (2007)
6. M. Janoschek, M. Garst, A. Bauer, P. Krautscheid, R. Georgii, P. Böni, C. Pfleiderer, Phys. Rev. B **87**, 134407 (2013)
7. A. Bauer, M. Garst, C. Pfleiderer, Phys. Rev. Lett. **110**, 177207 (2013)
8. S. Buhrandt and L. Fritz, Phys. Rev. B **88** (2013)
9. S.A. Brazovskii, I.E. Dzialoshinskii, A.R. Muratov, Sov. Phys. JETP **66**, 625 (1987)
10. S.A. Brazovskii, Sov. Phys. JETP **41**, 85 (1975)
11. S. Mühlbauer, B. Binz, F. Jonietz, C. Pfleiderer, A. Rosch, A. Neubauer, R. Georgii, P. Böni, Science **323**, 915 (2009)
12. X.Z. Yu, Y. Onose, N. Kanazawa, J.H. Park, J.H. Han, Y. Matsui, N. Nagaosa, Y. Tokura, Nature **465**, 901 (2010)
13. T. Adams, S. Mühlbauer, C. Pfleiderer, F. Jonietz, A. Bauer, A. Neubauer, R. Georgii, P. Böni, U. Keiderling, K. Everschor, M. Garst, A. Rosch, Phys. Rev. Lett. **107**, 217206 (2011)
14. A.N. Bogdanov, D.A. Yablonskii, Sov. Phys. JETP **68**, 1 (1989)
15. A.N. Bogdanov, A. Hubert, J. Magn. Magn. Mater. **138**, 255 (1994)
16. U.K. Rößler, A.N. Bogdanov, C. Pfleiderer, Nature **442**, 797 (2006)
17. N. Nagaosa, Y. Tokura, Nat. Nanotechnol. **8**, 899 (2013)
18. K. Everschor, Ph.D. thesis, University of Cologne (2012). http://kups.ub.uni-koeln.de/4811/1/Thesis_zum_Veroeffentlichen.pdf
19. T. Schwarze, J. Waizner, M. Garst, A. Bauer, I. Stasinopoulos, H. Berger, C. Pfleiderer, D. Grundler, Nat. Mater. **14**, 478 (2015)
20. F.N. Rybakov, A.B. Borisov, A.N. Bogdanov, Phys. Rev. B **87**, 094424 (2013)
21. X.Z. Yu, N. Kanazawa, Y. Onose, K. Kimoto, W.Z. Zhang, S. Ishiwata, Y. Matsui, Y. Tokura, Nat. Mater. **10**, 106 (2011)
22. N.A. Spaldin, M. Fiebig, M. Mostovoy, J. Phys.: Condens. Matter **20**, 434203 (2008)

23. S. Heinze, K. von Bergmann, M. Menzel, J. Brede, A. Kubetzka, R. Wiesendanger, G. Bihlmayer, S. Blügel, Nat. Phys. **7**, 713 (2011)
24. N. Romming, C. Hanneken, M. Menzel, J.E. Bickel, B. Wolter, K. von Bergmann, A. Kubetzka, R. Wiesendanger, Science **341**, 636 (2013)
25. K.V. Bergmann, A. Kubetzka, O. Pietzsch, R. Wiesendanger, J. Phys.: Condens. Matter **26**, 394002 (2014)
26. S.-Z. Lin, C.D. Batista, A. Saxena, Phys. Rev. B **89**, 024415 (2014)
27. J. Iwasaki, A.J. Beekman, N. Nagaosa, Phys. Rev. B **89**, 064412 (2014)
28. C. Schütte, M. Garst, Phys. Rev. B **90**, 094423 (2014)
29. S. Schroeter, M. Garst, arXiv:1504.02108
30. A.M. Kosevich, B.A. Ivanov, A.S. Kovalev, Phys. Rep. **194**, 117 (1990); B.A. Ivanov, JETP Lett. **61**, 917 (1995); B.A. Ivanov, H.J. Schnitzer, F.G. Mertens, G.M. Wysin, Phys. Rev. B **58**, 8464 (1998); D.D. Sheka, B.A. Ivanov, F.G. Mertens. Phys. Rev. B **64**, 024432 (2001)
31. M. Ezawa, Phys. Rev. B **83**, 100408 (2011)
32. A.A. Thiele, Phys. Rev. Lett. **30**, 230 (1973)
33. L. Kong, J. Zang, Phys. Rev. Lett. **111**, 067203 (2013)
34. S.-Z. Lin, C.D. Batista, C. Reichhardt, A. Saxena, Phys. Rev. Lett. **112**, 187203 (2014)
35. A.A. Kovalev, Phys. Rev. B **89**, 241101(R) (2014)
36. M. Mochizuki, X.Z. Yu, S. Seki, N. Kanazawa, W. Koshibae, J. Zang, M. Mostovoy, Y. Tokura, N. Nagaosa, Nat. Mater. **13**, 241 (2014)
37. B. Binz, A. Vishwanath, Phys. Rev. B **74**, 214408 (2006)
38. I. Fischer, N. Shah, A. Rosch, Phys. Rev. B **77**, 024415 (2008)
39. T. Tanigaki, K. Shibata, N. Kanazawa, X. Yu, S. Aizawa, Y. Onose, H.S. Park, D. Shindo, Y. Tokura, arXiv:1503.03945
40. O. Petrova, O. Tchernyshyov, Phys. Rev. B **84**, 214433 (2011)
41. J. Zang, M. Mostovoy, J.H. Han, N. Nagaosa, Phys. Rev. Lett. **107**, 136804 (2011)
42. M. Mochizuki, Phys. Rev. Lett. **108**, 017601 (2012)
43. Y. Onose, Y. Okamura, S. Seki, S. Ishiwata, Y. Tokura, Phys. Rev. Lett. **109**, 37603 (2012)
44. J. Slonczewski, J. Magn. Magn. Mater. **159**, L1 (1996)
45. L. Berger, Phys. Rev. B **54**, 9353 (1996)
46. S. Zhang, Z. Li, Phys. Rev. Lett. **93**, 127204 (2004)
47. K. Everschor, M. Garst, R.A. Duine, A. Rosch, Phys. Rev. B **84**, 64401 (2011)
48. K. Everschor, M. Garst, B. Binz, F. Jonietz, S. Mühlbauer, C. Pfleiderer, A. Rosch, Phys. Rev. B **86**, 054432 (2012)
49. S.-Z. Lin, C. Reichhardt, C.D. Batista, A. Saxena, Phys. Rev. B **87**, 214419 (2013)
50. J. Iwasaki, M. Mochizuki, N. Nagaosa, Nat. Comm. **4**, 1463 (2013)
51. J. He, Z. Li, S. Zhang, Phys. Rev. B **73**, 184408 (2006)
52. F. Jonietz, S. Mühlbauer, C. Pfleiderer, A. Neubauer, W. Münzer, A. Bauer, T. Adams, R. Georgii, P. Böni, R.A. Duine, K. Everschor, M. Garst, A. Rosch, Science **330**, 1648 (2010)
53. C. Pfleiderer, A. Rosch, Nature **465**, 880 (2010)
54. R. Ritz, M. Halder, C. Franz, A. Bauer, M. Wagner, R. Bamler, A. Rosch, C. Pfleiderer, Phys. Rev. B **87**, 134424 (2013)
55. A. Neubauer, C. Pfleiderer, B. Binz, A. Rosch, R. Ritz, P.G. Niklowitz, P. Böni, Phys. Rev. Lett. **102**, 186602 (2009)
56. M. Lee, W. Kang, Y. Onose, Y. Tokura, N.P. Ong, Phys. Rev. Lett. **102**, 186601 (2009)
57. P. Bruno, V.K. Dugaev, M. Taillefumier, Phys. Rev. Lett. **93**, 96806 (2004)
58. R. Ritz, M. Halder, M. Wagner, C. Franz, A. Bauer, C. Pfleiderer, Nature **497**, 231 (2013)
59. C. Pfleiderer, S.R. Julian, G.G. Lonzarich, Nature **414**, 427 (2001)
60. C. Pfleiderer, D. Reznik, L. Pintschovius, H.V. Lohneysen, M. Garst, A. Rosch, Nature **427**, 227 (2004)
61. M. Onoda, G. Tatara, N. Nagaosa, J. Phys. Soc. Jpn. **73**, 2624 (2004)
62. F. Freimuth, R. Bamler, Y. Mokrousov, A. Rosch, Phys. Rev. B **88**, 214409 (2013)
63. T. Schulz, R. Ritz, A. Bauer, M. Halder, M. Wagner, C. Franz, C. Pfleiderer, K. Everschor, M. Garst, A. Rosch, Nat. Phys. **8**, 301 (2012)

64. G. Blatter, M.V. Feigel'man, V.B. Geshkenbein, A.I. Larkin, V.M. Vinokur, Rev. Mod. Phys. **66**, 1125 (1994)
65. P. Milde, D. Kohler, J. Seidel, L.M. Eng, A. Bauer, A. Chacon, J. Kindervater, S. Mühlbauer, C. Pfleiderer, S. Buhrandt, C. Schütte, A. Rosch, Science **340**, 1076 (2013)
66. C. Schütte, J. Iwasaki, A. Rosch, N. Nagaosa, Phys. Rev. B **90**, 174434 (2014)
67. F. Büttner, C. Moutafis, M. Schneider, B. Krüger, C.M. Günther, J. Geilhufe, C.V. Korff Schmising, J. Mohanty, B. Pfau, S. Schaffert, A. Bisig, M. Foerster, T. Schulz, C.A.F. Vaz, J.H. Franken, H.J.M. Swagten, M. Kläui, S. Eisebitt, Nat. Phys. **11**, 225 (2015)
68. J.S. White, K. Prsa, P. Huang, A.A. Omrani, I. Zivkovic, M. Bartkowiak, H. Berger, A. Magrez, J.L. Gavilano, G. Nagy, J. Zang, H.M. Ronnow, Phys. Rev. Lett. **113**, 107203 (2014)
69. Y. Okamura, F. Kagawa, M. Mochizuki, M. Kubota, S. Seki, S. Ishiwata, M. Kawasaki, Y. Onose, Y. Tokura, Nat. Commun. **4**, 2391 (2013)
70. J. Sampaio, V. Cros, S. Rohart, A. Thiaville, A. Fert, Nat. Nanotechnol. **8**, 839 (2013)
71. W. Koshibae, N. Nagaosa, Nat. Commun. **5**, 5148 (2014)
72. J. Müller, A. Rosch, Phys. Rev. B **91**, 054410 (2015)
73. X. Zhang, M. Ezawa, Y. Zhou, Sci. Rep. **5**, 9400 (2015)

Chapter 3
Current-Driven Dynamics of Skyrmions

Masahito Mochizuki

Abstract Skyrmion was originally proposed in 1960s by Tony Skyrme as a topological solution of the nonlinear sigma model to account for the stability of hadrons in nuclear physics. Recently realization of skyrmions was indeed discovered in ferromagnets with chiral crystal symmetry as nanometric vortex-like spin textures with a quantized topological invariant. It has turned out that the magnetic skyrmions show intriguing dynamical and transport phenomena through coupling to the electric currents and/or the magnon currents. In this chapter, recent theoretical studies on the current-driven dynamics of magnetic skyrmions are discussed.

3.1 Introduction

Nanometric magnetic whirls, called magnetic skyrmions, realized in ferromagnets without inversion symmetry are recently attracting intensive research interest [1–3]. Skyrmion was theoretically proposed by Tony Skyrme in 1960s to account for stability of the hadrons in the particle physics as a topological solution of the non-linear sigma model in three dimensions [4, 5]. The magnetic skyrmion is composed of spins pointing in all directions to wrap a sphere like a hedgehog as shown in Fig. 3.1a. The number of the wrapping over a sphere corresponds to a topological invariant called skyrmion number. Realization of skyrmions was recently confirmed experimentally in some two-dimensional condensed matter systems, e.g., quantum Hall ferromagnets [6, 7], ferromagnetic monolayers [8, 9], and doped layered antiferromagnets [10]. In these systems, the skyrmions often appear as swirling structures of the magnetizations as shown in Fig. 3.1b. This vortex-like texture corresponds to a projection of the original hedgehog skyrmion onto the two-dimensional plane.

M. Mochizuki (✉)
Department of Physics and Mathematics, Aoyama Gakuin University, PRESTO, Japan Science and Technology Agency, Tokyo, Japan
e-mail: mochizuki@phys.aoyama.ac.jp

J. Seidel (ed.), *Topological Structures in Ferroic Materials*, Springer Series in Materials Science 228, DOI 10.1007/978-3-319-25301-5_3

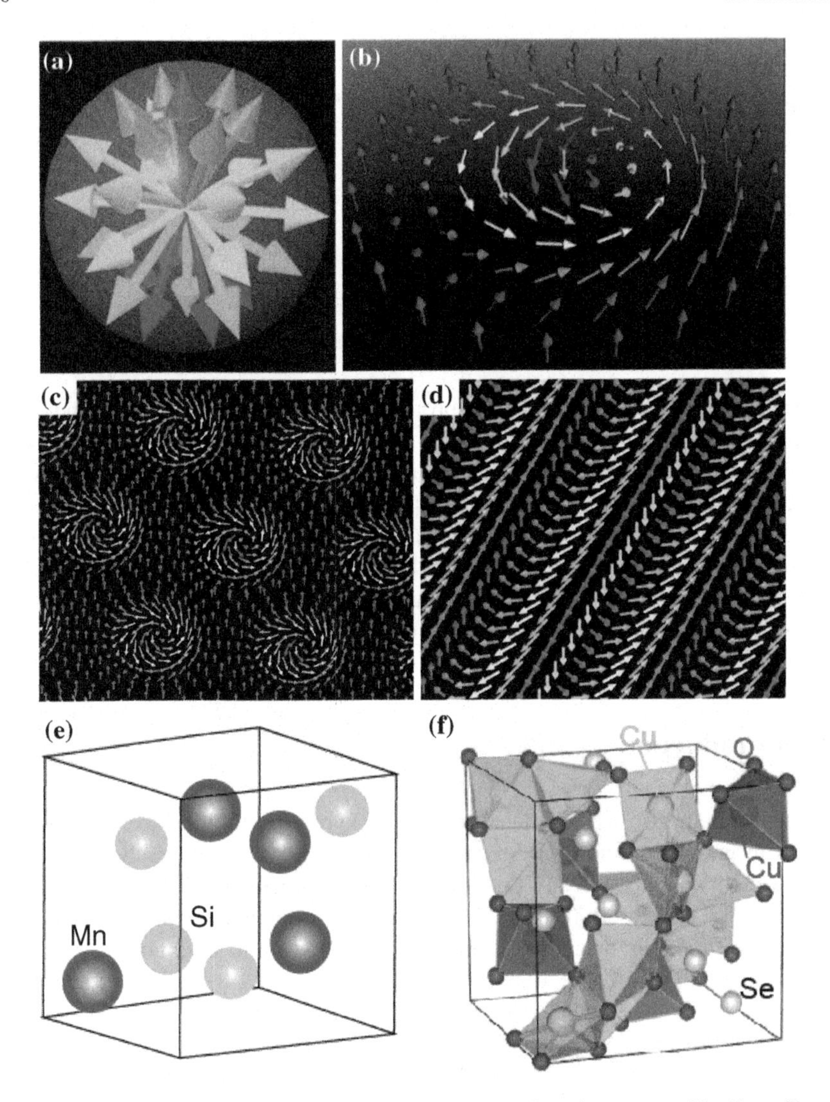

Fig. 3.1 **a** Schematic figure of the original hedgehog type skyrmion proposed by Tony Skyrme in 1960s whose magnetizations point in every direction to wrap a sphere. **b** Schematic figure of a skyrmion recently discovered in chiral-lattice magnets, which corresponds to a projection of the hedgehog type skyrmion onto the two-dimensioanl plane. Its magnetizations also point every direction to wrap a sphere. **c** Schematic figure of the skyrmion crystal realized in the chiral-lattice magnets under an external magnetic field, in which the skyrmions are hexagonally packed to form a triangular lattice. **d** Schematic figure of the helical state realized in the chiral-lattice magnets as a consequence of the competition between the Dzyaloshinskii-Moriya interaction and the ferromagnetic-exchange interaction, which comprises a successive alignment of the Bloch domain walls. **e**, **f** Crystal structures of (**e**) the chiral-lattice metallic magnet MnSi and (**f**) the chiral-lattice insulating magnet Cu_2OSeO_3, both of which belong to the cubic $P2_13$ point group

Bogdanov and his collaborators theoretically predicted crystallization of such vortex-like skyrmions into triangular lattice (so-called skyrmion crystal) as shown in Fig. 3.1c in ferromagnets without inversion symmetry [11–13]. In 2009, the skyrmion crystal was indeed discovered in the so-called A phase of metallic B20 compound MnSi under an external magnetic field $\mathbf{B}$ by the small angle neutron scattering experiment [14]. In each skyrmion constituting the skyrmion crystal, the magnetizations point antiparallel to $\mathbf{B}$ at the vortex core, and rotate upon propagating in the radial directions towards the periphery at which the magnetizations are parallel to $\mathbf{B}$. Such vortex-like skyrmions appear in the planes normal to $\mathbf{B}$, and they are stacked ferromagnetically to form rod-like or tube-like structures in bulk samples. The skyrmion is characterized by the skyrmion number Q, which is defined as,

$$\int \mathrm{d}^2 r \left(\frac{\partial \hat{\mathbf{n}}}{\partial x} \times \frac{\partial \hat{\mathbf{n}}}{\partial y} \right) \cdot \hat{\mathbf{n}} = \pm 4\pi Q. \tag{3.1}$$

Here the unit vector $\hat{\mathbf{n}}$ represents the direction of the local magnetization. The left-hand side of this equation represents a sum of the solid angles spanned by three neighboring magnetizations, and since the magnetizations in one skyrmion point everywhere to wrap a sphere once, its value becomes $+4\pi$ or -4π depending on the sign of the magnetization at the skyrmion core, that is, $Q = +1$ ($Q = -1$) for up (down) core magnetization.

Since the discovery in MnSi, the skyrmion-crystal phase has been discovered successively in several metallic B20 compounds such as $Fe_{1-x}Co_xSi$, FeGe and $Mn_{1-x}Fe_xGe$ by the neutron-scattering experiments [15–19] and the microscopy experiments [20–24]. In addition to the metallic compounds, the skyrmion-crystal phase was discovered also in the insulating copper oxoselenite Cu_2OSeO_3 [25–28]. These compounds commonly have a chiral crystal structure with cubic $P2_13$ symmetry (see Fig. 3.1e, f), and thereby have a finite net component of the Dzyaloshinskii-Moriya interaction [29, 30]. The Dzyaloshinskii-Moriya interaction favors rotating alignment of magnetizations, and thus strongly competes with the ferromagnetic-exchange interaction which favors parallel (collinear) alignment of magnetizations. As a result, the ground state of these chiral-lattice ferromagnets at $\mathbf{B} = 0$ is the helical state (so-called proper screw state), which comprises a successive alignment of the Bloch domain walls. An increase in B at certain temperatures changes the helical phase to the skyrmion-crystal phase and eventually to the field-polarized ferromagnetic phase.

Figure 3.2a, b display the experimentally obtained phase diagrams in plane of temperature T and magnetic field B for bulk samples of MnSi [14] and Cu_2OSeO_3 [25], respectively. Irrespective of the different origin of the magnetism between metal and insulator, both compounds exhibit similar phase diagrams. The skyrmion-crystal phase takes place only as a small pocket (so-called A phase) in the phase diagram at finite T and B on the verge of the boundary between the paramagnetic and the helical (longitudinal conical) phases.

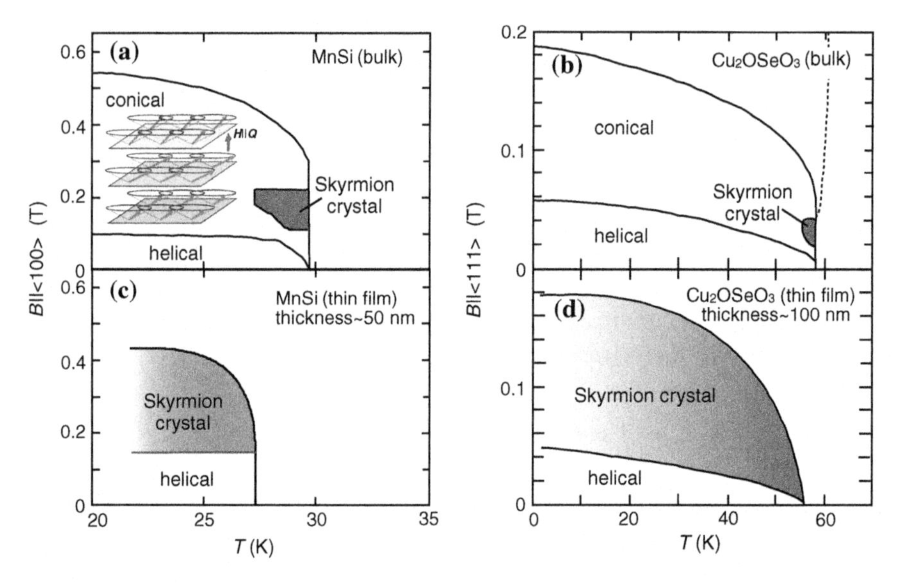

Fig. 3.2 Experimentally obtained T-B phase diagrams for **a** bulk sample of the metallic B20 compound MnSi [14], **b** bulk sample of the insulating oxoselenite Cu_2OSeO_3 [25], **c** thin-film sample of MnSi [22], and **d** thin-film sample of Cu_2OSeO_3 [25]. In spite of the different origin of the magnetism between the metallic and insulating magnets, both MnSi and Cu_2OSeO_3 exhibit similar phase diagrams. For the bulk samples, the skyrmion-crystal phase (so-called A phase) occupies only a tiny region at finite T and B inside the conical phase on the verge of the boundary to the paramagnetic phase. In contrast, the skyrmion-crystal phase spreads over the wide T-B range in the phase diagrams for thin-film samples. The enhanced stability of the skyrmion-crystal phase is attributed to the destabilization of the longitudinal conical phase in thin-film samples (see text)

In 1980, Bak and Jensen proposed a continuum spin model to describe the magnetism in the chiral-lattice ferromagnets [31]:

$$\mathscr{H} = \int \mathrm{d}^3 r \left[\frac{J}{2a}(\nabla \mathbf{M})^2 + \frac{D}{a^2}\mathbf{M}\cdot(\nabla\times\mathbf{M}) - \frac{1}{a^3}\mathbf{B}\cdot\mathbf{M} \right.$$
$$+ \frac{A_1}{a^3}(M_x^4 + M_y^4 + M_z^4)$$
$$\left. - \frac{A_2}{2a}[(\nabla_x M_x)^2 + (\nabla_y M_y)^2 + (\nabla_z M_z)^2] \right]. \tag{3.2}$$

In addition to the ferromagnetic-exchange interaction (the first term), the Dzyaloshinskii-Moriya interaction (the second term), and the Zeeman coupling (the third term), magnetic anisotropies allowed in the cubic crystal symmetry (the fourth and the fifth terms) are considered. Starting from this continuum model, one obtains a lattice spin model, that is, a classical Heisenberg model on the cubic lattice by deviding the space into cubic meshes:

$$
\begin{aligned}
\mathscr{H} = &-J \sum_i \mathbf{m}_i \cdot (\mathbf{m}_{i+\hat{x}} + \mathbf{m}_{i+\hat{y}} + \mathbf{m}_{i+\hat{z}}) \\
&- D \sum_i (\mathbf{m}_i \times \mathbf{m}_{i+\hat{x}} \cdot \hat{x} + \mathbf{m}_i \times \mathbf{m}_{i+\hat{y}} \cdot \hat{y} + \mathbf{m}_i \times \mathbf{m}_{i+\hat{z}} \cdot \hat{z}) \\
&- \mathbf{B} \cdot \sum_i \mathbf{m}_i \\
&+ A_1 \sum_i [(m_i^x)^4 + (m_i^y)^4 + (m_i^z)^4] \\
&- A_2 \sum_i (m_i^x m_{i+\hat{x}}^x + m_i^y m_{i+\hat{y}}^y). \qquad (3.3)
\end{aligned}
$$

As long as slowly varying spin structures such as skyrmion and helix are considered, one can neglect the complex background crystal structure, which justifies the theoretical treatment based on a spin model on the simple cubic lattice.

The skyrmion-crystal phase has turned out to be rather unstable in the bulk samples. It was, however, found that it attains enhanced stability when the sample thickness becomes thinner [21, 32–34]. Figure 3.2c, d display the experimentally obtained phase diagrams for thin-film samples of MnSi [22] and Cu_2OSeO_3 [25], respectively. One finds that the skyrmion-crystal phase spreads over a wide area in the phase diagram, and is realized even at the lowest temperature. This can be understood as follows. In the bulk samples, the longitudinal conical phase propagating parallel to **B** with a uniform magnetization component due to the spin canting towards the **B** direction is stabilized owing to the energy gains from both the Dzyaloshinskii-Moriya interaction and the Zeeman coupling, and the skyrmion-crystal state is usually higher in energy than this conical state. However, when the sample thickness becomes comparable to or thinner than the conical periodicity, the conical state can no longer benefit from the energy gain of the Dzyaloshinskii-Moriya interaction, and thus is destabilized. Instead the skyrmion-crystal state attains the relative stability against the conical state. It was argued that the uniaxial anisotropy, inhomogeneous chiral modulations, and the dipolar interaction can also stabilize skyrmions in thin-film samples [35–40].

The magnetic phases in a thin-film sample of the chiral-lattice ferromagnet have been studied by analyzing the following classical Heisenberg model on the square lattice using the Monte-Carlo technique [32]:

$$
\begin{aligned}
\mathscr{H} = &-J \sum_i \mathbf{m}_i \cdot (\mathbf{m}_{i+\hat{x}} + \mathbf{m}_{i+\hat{y}}) \\
&- D \sum_i (\mathbf{m}_i \times \mathbf{m}_{i+\hat{x}} \cdot \hat{x} + \mathbf{m}_i \times \mathbf{m}_{i+\hat{y}} \cdot \hat{y}) \\
&- \mathbf{B} \cdot \sum_i \mathbf{m}_i \qquad (3.4)
\end{aligned}
$$

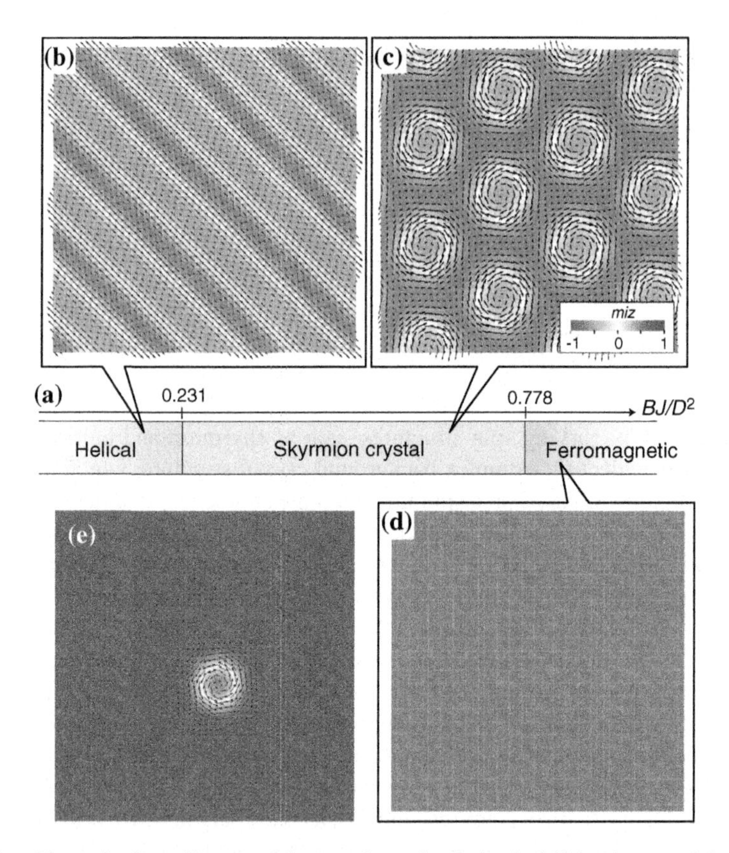

Fig. 3.3 **a** Magnetic phase diagram of the two-dimensional classical Heisenberg model given by (3.4) at $T = 0$ as a function of the magnetic-field strength B. As B increases, the helical phase, the skyrmion-crystal phase, and the field-polarized ferromagnetic phase emerge successively, which reproduces the successive phase transitions in the experimental phase diagrams for thin-film samples of MnSi and Cu_2OSeO_3 at low temperatures (see Fig. 3.2c, d). **b–d** Magnetization configurations in (**b**) the helical phase, **c** the skyrmion-crystal phase, and **d** the field-polarized ferromagnetic phase, respectively. **e** Individual skyrmion as a defect in the ferromagnetic background

Here the magnetic-anisotropy terms are neglected because they turn out to play only minor roles as long as the realistically small coupling constants are considered. The external magnetic field $\mathbf{B} = (0, 0, B)$ is applied perpendicular to the two-dimensional plane. Figure 3.3a shows the magnetic phase diagram of this model at $T = 0$ as a function of the magnetic field B. One finds that three phases, the helical phase, the skyrmion-crystal phase, and the ferromagnetic phase emerge successively as B increases. In the experimental phase diagrams for thin-film samples in Fig. 3.2c, d, one finds corresponding successive phase transitions along the vertical direction at low temperatures. The T-B phase diagrams for bulk samples in Fig. 3.2a, b have also been reproduced by the Monte-Carlo calculation of the spin model in three dimensions given by (3.3) without the anisotropy terms [41]. In this study, the exchange interactions and the Dzyaloshinskii-Moriya interactions on further neighbor bonds on the cubic lattice are taken into account to compensate artificial anisotropies due to the coarse-grained cubic meshing.

It is worth mentioning that skyrmions in chiral-lattice ferromagnets emerge not only in the crystallized form as observed in the skyrmion-crystal phase but also as individual defects in the ferromagnetic state as shown in Fig. 3.3b [20]. Such a skyrmion defect in the ferromagnetic background turns out to be considerably stable because of the topological protection. Because of this robustness, the individual skyrmions are expected for technical application to information carriers in the next-generation magnetic storage devices.

The skyrmions have several advantageous properties for application to high-density and low energy-consuming storage devices. One of the most important properties is that one can drive their motion by applying electric currents via the spin-transfer torque mechanism, and its threshold current density is extremely low as compared to those for other noncollinear magnetic textures. In this article, we review recent theoretical studies on the current-driven dynamics of the magnetic skyrmions. We first discuss the electric-current driven dynamics of the skyrmion crystal in Sect. 3.2 by focusing on the universal relation between the electric current density and the drift velocity of skyrmions as well as the ultralow threshold current density [42, 43]. It is argued that the following three features of skyrmions are key to understanding these issues, that is, (1) nature as a particle, (2) nature as a vortex, and (3) the finite topological number. In Sect. 3.3, the electric-current driven dynamics of individual skyrmions in the ferromagnetic background for several kinds of confined geometries. It is argued that the confinement and the boundary effect dramatically change the electric-current driven dynamics of skyrmions, including the steady-state current-velocity relation and transient phenomena, as well as the creation and annihilation of the skyrmions [44]. In addition to the electric-current driven motion, the magnon-current driven dynamics of skyrmions is also discussed in Sect. 3.4. We discuss that irradiation of light or electron beam to a thin-film specimen induces unidirectional rotations of skyrmion microcrystal through introducing radial temperature gradient. It was uncovered that thermally activated magnons flow in a diffusive way in the presence of temperature gradient, and topological Hall effect of these magnon currents due to fictitious magnetic fields from the skyrmion spin structures works as a driving force of this chiral rotation of skyrmions [45].

3.2 Electric-Current Driven Dynamics of Skyrmions in Non-confined System

It is well known that translational motion of ferromagnetic domain walls can be driven by applying the spin-polarized electric current via the spin-transfer torque effect [46–48]. This phenomenon is attracting intensive research interest because of potential application to next-generation magnetic devices such as race-track memory [49].

However its threshold current density j_c to drive the motion is rather large (typically 10^{10}–10^{12} A/m^2), and hence the Joule heating has been a crucial issue. On the other hand, it was found that the skyrmions in the metallic B20 compounds can also be driven by the electric current [50–55]. Surprisingly its threshold current density j_c turned out to be five or six orders of magnitude smaller than that for the ferromagnetic domain walls, that is, its typical value is 10^5–10^6 A/m^2 [50, 51]. This extremely small j_c indicates that skyrmions are scarcely affected by impurity pinning, and is advantageous for application to low energy-cost storage devices.

The effects of magnetic impurities on the electric-current driven motion of skyrmions were theoretically studied [42] by numerically analyzing the Landau-Lifshitz-Gilbert-Slonczewski equation;

$$\frac{d\mathbf{m}}{dt} = -\gamma \mathbf{m} \times \mathbf{B}^{\rm eff}_{\mathbf{r}} + \frac{\alpha}{m}\mathbf{m} \times \frac{d\mathbf{m}}{dt} + \frac{pa^3}{2em}(\mathbf{j}\cdot\nabla)\mathbf{m} - \frac{pa^3\beta}{2em^2}\left[\mathbf{m}\times(\mathbf{j}\cdot\nabla)\mathbf{m}\right], \tag{3.5}$$

with $\mathbf{m}(\mathbf{r}) = -\mathbf{S}(\mathbf{r})/\hbar$. Here the effective magnetic field $\mathbf{B}^{\rm eff}$ is given by,

$$\mathbf{B}^{\rm eff} = -\frac{1}{\gamma\hbar}\frac{\partial \mathscr{H}}{\partial \mathbf{m}}. \tag{3.6}$$

The first and the second terms describe the gyrotropic motion and the Gilbert damping of the magnetization $\mathbf{m}$ where $\gamma = g\mu_{\rm B}/\hbar(>0)$ and α are the gyromagnetic ratio and the Gilbert damping coefficient, respectively. The third and the fourth terms describe the coupling between $\mathbf{m}$ and the spin-polarized electric current $\mathbf{j}$ where $e(>0)$, p and a are the elementary charge, the spin polarization of electric current, and the lattice constant, respectively. The third term depicts the coupling via the spin-transfer torque, while the fourth via the non-adiabatic effect. The strength of the non-adiabatic effect is represented by the parameter β.

For the Hamiltonian, the classical Heisenberg model given by (3.4) introduced in Sect. 3.1 is employed, but this time the following impurity term is added:

$$\mathscr{H}_{\rm imp.} = -A\sum_{i\in I} m_{iz}^2, \tag{3.7}$$

where I denotes a set of the impurity positions. This term describes magnetic anisotropy at randomly distributed impurity sites where $A > 0$ gives the easy magnetization axis parallel to $\mathbf{z}$, while $A < 0$ gives the easy magnetization plane perpendicular to $\mathbf{z}$. Note that the model (3.4) reproduces successive emergence of the helical, the skyrmion-crystal, and the ferromagnetic phases as a function of magnetic field B. Since the helical state can be regarded as a sequence of Bloch walls in ferromagnets, one can directly compare the electric-current driven motion of skyrmions and that of magnetic domain walls on equal footing without changing any other parameters except B.

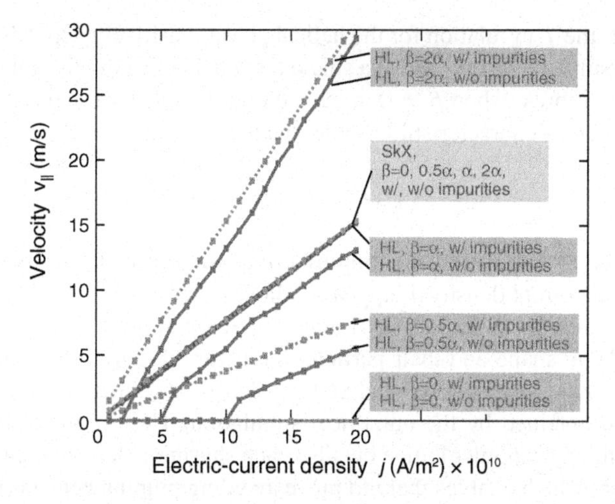

Fig. 3.4 **a** Simulated velocities $v_{\|}$ of the electric-current driven motion for the helical state (HL) and the skyrmion-crystal state (SkX) as functions of the electric current density j for several values of β. In the simulations, the parameters are set to be $J = 1$ meV, $D/J = -0.18$, $A/J = 0.2 \cdot p = 0.2 \cdot \alpha = 0.04$. Both the clean case without impurity ($x = 0$) and the dirty case with impurities ($x = 0.1\,\%$) are examined where x is the impurity concentration. *Red* and *purple points* and *lines* are for the HL state, while *blue* and *lightblue points* and *lines* are for the SkX state. All the *lines* for the SkX are overlapped within the accuracy of the simulations, indicating a universal current-velocity relation insensitive to the nonadiabatic effect and the impurity pinning. In contrast, the current-velocity relation for the HL state sensitively depends on both factors. The velocity becomes faster as β becomes larger, and the threshold current density $j_c \sim 10^{10}$–10^{11} A/m^2 appears upon the impurity doping. (Reproduced from [42].)

Figure 3.4 displays simulated velocities $v_{\|}$ (parallel to $\mathbf{j}$) of spin textures as functions of the electric current density j for the helical and the skyrmion-crystal phases with different values of β ($\beta = 0, 0.5\alpha, \alpha$, and 2α). The impurity concentration is fixed at $x = 0$ for the clean case and at $x = 0.1\,\%$ for the dirty case. The strength of the magnetic anisotropy is fixed at $A = 0.2J\,(>0)$, i.e., the easy-axis anisotropy. The blue and lightblue data points are for the skyrmion-crystal phase, while the red and purple data points are for the helical phase. Remarkably the current-velocity (j-$v_{\|}$) relation for the skyrmion crystal is quite universal, and all the plots overlap within the accuracy of the numerical simulations. They are independent of the nonadiabatic effect β, the Gilbert damping α, and the impurities. As will be proven later, the j-$v_{\|}$ characteristics of the skyrmion crystal nearly obeys the relation:

$$v_{\|} = \frac{pa^3}{2em} j. \tag{3.8}$$

This equation indicates that the electric-current driven motion of skyrmions is insensitive to the value of β and impurities, and thus the skyrmion is an ideal magnetic texture for manipulation via the spin-transfer torque mechanism with a very low current density.

In contrast, the j-$v_\parallel$ relation for the helical phase sensitively depends on all these three factors, similarly to the case of a single ferromagnetic domain wall. The helical structure cannot move when $\beta = 0$, prevented by the intrinsic pinning effect. With a finite β, the j-$v_\parallel$ characteristics in the clean case with $x = 0$ nearly obeys the relation:

$$v_\parallel \propto (\beta/\alpha) j. \tag{3.9}$$

In the presence of impurities, the pinning effect suppresses the velocity $v_\parallel$, and a finite threshold current density j_c appears, whose order is 10^{10}–10^{11} A/m^2.

One of the reasons why the skyrmions are scarcely pinned by impurities are their flexibility in shape and their particle-like nature. Figure 3.5a, b display snapshots of the moving skyrmion crystal and the skyrmions during the electric-current driven motion obtained by the numerical simulation. These figures show that not only the skyrmion triangular lattice but also each skyrmion deform their shapes during the motion, which enables them to move avoiding pinning centers (indicated by green dots). Figure 3.5c displays an example of trajectory of one moving skyrmion, which shows that the skyrmion as a particle-like object winds its trajectory to avoid impurities.

This peculiar motion is attributed to the fact that the X and Y coordinates of the skyrmion core is canonical conjugate due to the spin Berry phase term. The center-of-mass motion of a vortex-like skyrmion texture under the potential U obeys the following equations of motion:

$$\dot{X} = -\frac{a^2}{4\pi m\hbar}\frac{\partial U}{\partial Y}, \tag{3.10}$$

$$\dot{Y} = \frac{a^2}{4\pi m\hbar}\frac{\partial U}{\partial X}. \tag{3.11}$$

These equations are derived from the Lagrangian for spin systems,

$$\mathscr{L} = \frac{m\hbar}{a^2}\int d^2\mathbf{r}(\cos\theta - 1)\dot{\phi} - U(r, B), \tag{3.12}$$

where the first term is referred to as the Berry-phase term. Inserting the following solutions of the skyrmion magnetization configuration,

$$\phi = \tan^{-1}\frac{y-Y}{x-X} - \frac{\pi}{2}, \tag{3.13}$$

$$\theta = f(r) = \begin{cases} 0 & r \to \infty \\ \pi & r = 0 \end{cases} \tag{3.14}$$

the Lagrangian reads,

$$\mathscr{L} = \frac{4\pi m\hbar}{a^2}(X\dot{Y} - Y\dot{X}) - U(r, B). \tag{3.15}$$

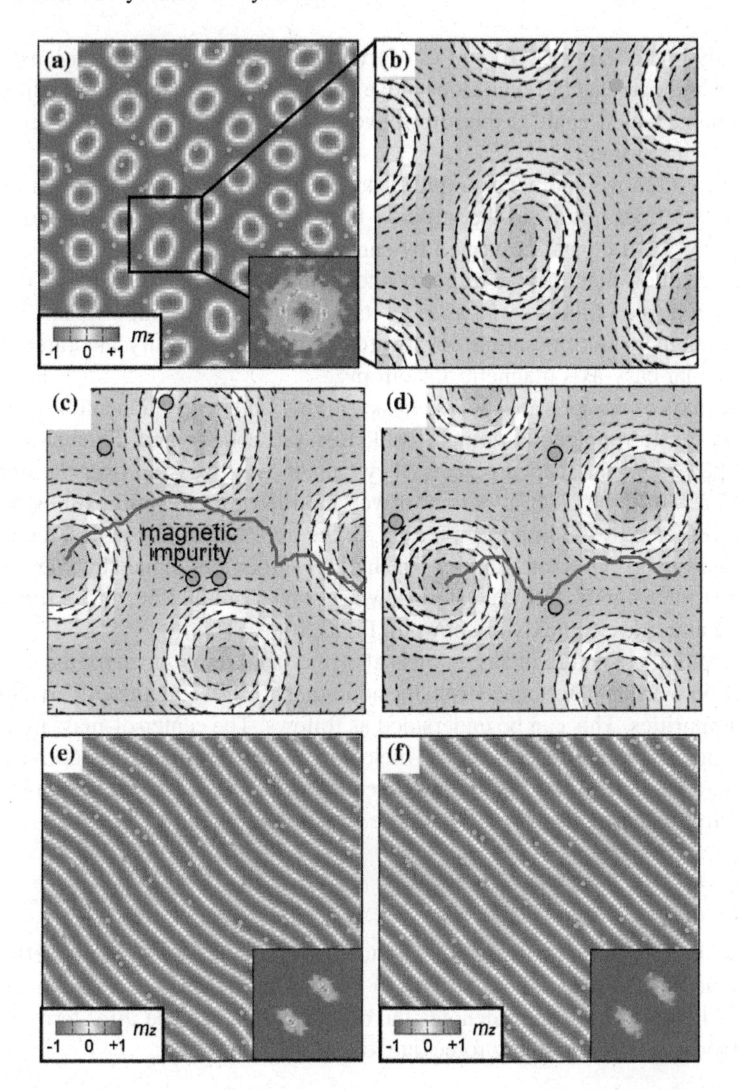

Fig. 3.5 **a** Simulated snapshot of the skyrmion crystal deformed from their original triangular lattice during the electric-current driven motion. **b** Magnified view of (**a**), which shows skyrmions distorted from their original circular shape. **c** Example of a trajectory of one skyrmion in the moving skyrmion crystal during the electric-current driven motion. The skyrmion moves avoiding impurity sites indicated by *green dots*. **d** Another example of skyrmion trajectory. Since a skyrmion in the moving skyrmion crystal is pushed by surrounding other skyrmions, the situation that the skyrmion cannot avoid impurity sites sometimes happens. In such a case, the skyrmion rushes to the impurity site so as to let its core run over the impurity site because the core magnetization pointing downwards is also energetically favorable for the magnetic anisotropy with an easy axis perpendicular to the plane. **e**, **f** Simulated snapshots of the helical state during the electric-current driven motion. The motion is intermittent, that is, sometimes pinned by impurities so that distorted significantly and slow down as shown in (**e**), and sometimes depinned and moves quickly as shown in (**f**). (Reproduced from [42].)

Eventually one obtains (3.10) and (3.11) via the Euler-Lagrange equations. These equations indicate that the skyrmions move in a direction perpendicular to the potential gradient, which enables them to move around a obstacle and to avoid being trapped by impurities.

Figure 3.5d shows another example of skyrmion trajectory, in which one finds that the skyrmion does not avoid the impurity but rushes to it. Since a skyrmion in the moving skyrmion crystal is pushed by surrounding other skyrmions, the situation that the skyrmion cannot avoid impurity sites sometimes happens. In such a case, the skyrmion winds its trajectory so as to let its core run over the impurity site because the core magnetization pointing downwards is also energetically favorable for the perpendicular easy-axis magnetic anisotropy.

Such kinds of motions are specific to skyrmions, and never happen for the helical state and the ferromagnetic domain wall. Because a helix and domain walls are line-shaped or planar-shaped objects, they cannot avoid passing over the impurity sites. As a result, the electric-current driven motion of these spin textures become strongly affected by the impurity pinning. In the presence of impurities, the motion of helical state becomes intermittent. The helix is sometimes pinned by impurities so that is distorted significantly and slow down as shown in Fig. 3.5e, and is sometimes depinned and moves quickly as shown in Fig. 3.5f.

The topological nature of skyrmion is also of crucial importance for the universal j-v relation with small influence from the non-adiabatic effect β, the Gilbert damping α, and impurities. This can be understood as follows. The center-of-mass motion of a rigid spin texture is described by Thiele's equation [56], which is derived from the Landau-Lifshitz-Gilbert-Slonczewski equation by assuming that the spin texture never deforms during its drift motion. The equation is given by [53, 54],

$$\mathbf{G} \times (\mathbf{v}_{\mathrm{s}} - \mathbf{v}_{\mathrm{d}}) + \mathscr{D}(\beta \mathbf{v}_{\mathrm{s}} - \alpha \mathbf{v}_{\mathrm{d}}) + \mathbf{F}_{\mathrm{pin}} - \nabla U = \mathbf{0}, \tag{3.16}$$

where $\mathbf{v}_{\mathrm{d}}$ is the drift velocity of the spin texture and $\mathbf{v}_{\mathrm{s}} = -\frac{pa^3}{2em}\mathbf{j}$ is the velocity of the conduction electrons.The first term in the left-hand side describes the Magnus force, while the second term denotes the dissipative force. The third term denotes the phenomenological pinning force due to impurities [53, 54]:

$$\mathbf{F}_{\mathrm{pin}} \sim -4\pi v_{\mathrm{pin}} f(v_{\mathrm{d}}/v_{\mathrm{pin}})\mathbf{v}_{\mathrm{d}}/|v_{\mathrm{d}}|. \tag{3.17}$$

Here f is a scaling function and $\mathbf{v}_{\mathrm{pin}}$ is a velocity characterizing the pinning strength. The last term represents a force due to the potential from the surrounding environment. The gyromagnetic coupling vector $\mathbf{G} = (0, 0, \mathscr{G})$ is given by

$$\mathscr{G} = \int_{\text{unit cell}} \mathrm{d}^2 r \left(\frac{\partial \hat{\mathbf{n}}}{\partial x} \times \frac{\partial \hat{\mathbf{n}}}{\partial y} \right) \cdot \hat{\mathbf{n}} = 4\pi Q, \tag{3.18}$$

where $Q(= \pm 1)$ is the skyrmion number, and $\hat{\mathbf{n}} = \mathbf{m}(\mathbf{r})/m$. On the other hand, the components of dissipative force tensor $\mathscr{D}$ are given by,

$$\mathscr{D}_{ij} = \int_{\text{unit cell}} d^2r \partial_i \hat{\mathbf{n}} \cdot \partial_j \hat{\mathbf{n}} = \begin{cases} \mathscr{D} & (i, j) = (x, x), (y, y), \\ 0 & \text{otherwise}. \end{cases} \tag{3.19}$$

Note that the first term of (3.16) contains the topological number $\mathscr{G}$. Crucial difference between the skyrmion and the helix is the value of $\mathscr{G}$. It is $\pm 4\pi$ for a single skyrmion, but is zero for helix and domain wall. Because the values of $\alpha(\sim 10^{-2})$ and $\beta(\sim\alpha)$ are much smaller than unity, the second term of (3.16) is negligible if $|\mathscr{G}| = 4\pi$, and the electric-current driven motion becomes governed by the first term. Then the motion of skyrmions is well described by,

$$\mathbf{G} \times (\mathbf{v}_\mathrm{s} - \mathbf{v}_\mathrm{d}) \sim -\mathbf{F}_\mathrm{pin}. \tag{3.20}$$

In this equation, one finds that the skyrmion motion is not affected by the values of β and α. On the other hand, when $\mathscr{G} = 0$, the electric-current driven motion is governed by the second term. Hence the motion of helix and ferromagnetic domain wall is well described by,

$$\mathscr{D}(\beta \mathbf{v}_\mathrm{s} - \alpha \mathbf{v}_\mathrm{d}) \sim -\mathbf{F}_\mathrm{pin}. \tag{3.21}$$

This is the reason why the motion of helix strongly depends on β and α.

Although the numerical simulations show that individual skyrmions as well as the skyrmion crystal are significantly distorted from their original shapes during the motion, the simulation results are reproduced even quantitatively by Thiele's equation (3.16) derived assuming the rigid structure. In the absence of impurities ($\mathbf{F}_\mathrm{pin} = 0$), the drift velocity $\mathbf{v}_\mathrm{d}$ is derived from (3.16) as,

$$\mathbf{v}_\mathrm{d} = \mathbf{v}_\parallel + \mathbf{v}_\perp, \tag{3.22}$$

$$\mathbf{v}_\parallel = \left(\frac{\beta}{\alpha} + \frac{\alpha - \beta}{\alpha^3 (\mathscr{D}/\mathscr{G})^2 + \alpha} \right) \mathbf{v}_\mathrm{s}, \tag{3.23}$$

$$\mathbf{v}_\perp = \frac{(\alpha - \beta)(\mathscr{D}/\mathscr{G})}{\alpha^2 (\mathscr{D}/\mathscr{G})^2 + 1} \left(\hat{\mathbf{z}} \times \mathbf{v}_\mathrm{s} \right). \tag{3.24}$$

Here $\mathbf{v}_\parallel$ and $\mathbf{v}_\perp$ are components of $\mathbf{v}_\mathrm{d}$ parallel and perpendicular to $\mathbf{v}_\mathrm{s}$, respectively. In the case of skyrmions, $\alpha^3 (\mathscr{D}/\mathscr{G})^2$ in the second term of (3.23) is negligible when α is small enough, which gives,

$$\mathbf{v}_\parallel = \mathbf{v}_\mathrm{s}, \tag{3.25}$$

for the current-driven motion of skyrmions in agreement with the β-insensitive universal j-$v_\parallel$ relation obtained in the numerical simulation for $\alpha = 0.04$. Equation (3.23) also suggests that deviation from the universal relation should show up in the extremely dissipative system with much larger α. On the other hand, in the case of helix and domain wall with $\mathscr{G} = 0$, the second term vanishes, which gives,

$$\mathbf{v}_{\parallel} = \frac{\beta}{\alpha}\mathbf{v}_{\mathrm{s}} \propto \frac{\beta}{\alpha}\mathbf{j}. \tag{3.26}$$

This relation is again in agreement with the numerical simulation.

In the presence of impurities with $\mathbf{F}_{\mathrm{pin}} \neq 0$, the Hall angle $R = \mathbf{v}_{\perp}/\mathbf{v}_{\parallel}$ gives important information. Whereas R approaches an asymptotic value in the limit of large j as expected from (3.23) and (3.24), it rapidly increases as j decreases due to the impurity effect. In the limit of small j, the drift velocity v_{d} almost vanishes ($v_{\mathrm{d}} = \sqrt{v_{\parallel}^2 + v_{\perp}^2} \sim 0$), and then the scaling function $f(v_{\mathrm{d}}/v_{\mathrm{pin}})$ in (3.17) becomes unity. In this case we can derive the explicit expression for v_{d}, $v_{\parallel}$ and $v_{\perp}$ of skyrmion,

$$v_{\mathrm{d}} = \frac{\sqrt{(\alpha\mathscr{D}A)^2 + (\alpha^2\mathscr{D}^2 + \mathscr{G}^2)\left[(\beta^2\mathscr{D}^2 + \mathscr{G}^2)v_{\mathrm{s}}^2 - A^2\right]}}{\alpha^2\mathscr{D}^2 + \mathscr{G}^2}, \tag{3.27}$$

$$v_{\parallel} = \frac{v_s}{(\alpha\mathscr{D} + A/v_{\mathrm{d}})^2 + \mathscr{G}^2}\left[(\alpha\mathscr{D} + A/v_{\mathrm{d}})\beta\mathscr{D} + \mathscr{G}^2\right], \tag{3.28}$$

$$v_{\perp} = \frac{v_s}{(\alpha\mathscr{D} + A/v_{\mathrm{d}})^2 + \mathscr{G}^2}\left[(\alpha\mathscr{D} + A/v_{\mathrm{d}})\mathscr{G} - \beta\mathscr{D}\mathscr{G}\right], \tag{3.29}$$

where $A \equiv 4\pi v_{\mathrm{pin}}$. In the limit of $\alpha, \beta \to 0$, we obtain,

$$\frac{v_{\perp}}{v_{\mathrm{d}}} = \frac{R}{\sqrt{1+R^2}} = \frac{j_{\mathrm{c}}}{j}, \tag{3.30}$$

which indicates that $v_{\perp}/v_{\mathrm{d}}$ is proportional to $1/j$. More explicitly, solving (3.16) in the limit of $\alpha, \beta \to 0$, we obtain

$$v_{\mathrm{d}}^2 = \left(\frac{pa^3}{2em}\right)^2 (j^2 - j_{\mathrm{c}}^2). \tag{3.31}$$

3.3 Electric-Current Driven Dynamics of Skyrmions in Confined Geometries

In order to use skyrmions as information carriers in magnetic storage devices, it is essentially important to know their dynamics in nanometric structures and also to create and annihilate them at will. In this section, we discuss results of the theoretical studies on the electric-current induced dynamics of individual skyrmions in several kinds of two-dimensional constricted geometries with ferromagnetic background based on numerical simulations of the Landau-Lifshitz-Gilbert-Slonczewski equation (3.5) with the classical Heisenberg model given by (3.4) [44].

First we argue that the electric-current driven motion of a skyrmion confined in a narrow region is completely different from that in a non-confined plane because of the presence of confining potentials from the boundaries. Figure 3.6 displays simulated

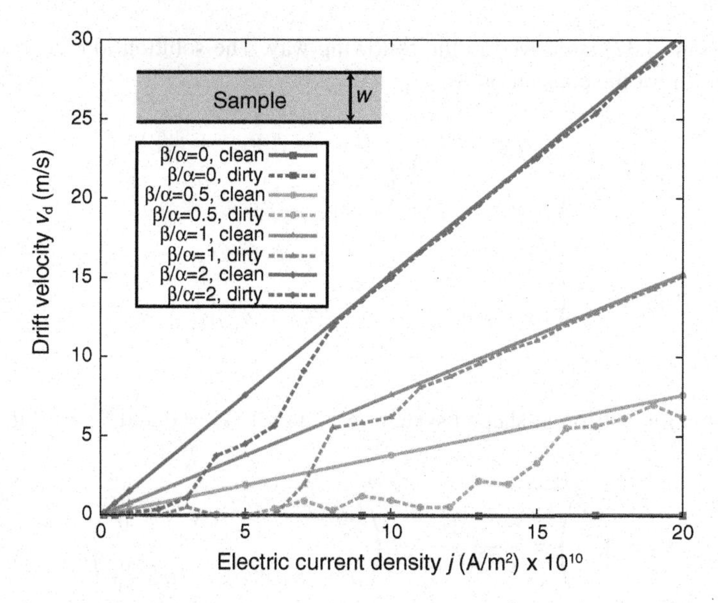

Fig. 3.6 Simulated steady-state velocity of the electric-current induced motion of a skyrmion in a finite-width channel as a function of the electric current density j for several values of β/α. For the dirty case with impurity concentration of $x = 0.1\,\%$, values averaged over eight different patterns of impurity distributions are presented. The parameters used for the simulation are $J = 1\,\text{meV}$, $D/J = 0.18$, $B = 0.0278J$, $A = 0.2J$, and $\alpha = 0.04$. *Inset* shows a schematic figure of the channel of width w. (Reproduced from [44].)

relations between the electric current density j and the drift velocity v_d of a skyrmion in the steady state after the transient time. The sample shape considered here is a long stripline-shaped system along the x-direction, with nanometric width w along the y-direction as shown in the inset of Fig. 3.6. Here the impurity concentration x is fixed at $x = 0$ for the clean case and $x = 0.1\,\%$ for the dirty case.

In this figure, one finds that the j-$v_\parallel$ relation depends strongly on α, β and the impurity effect, in sharp contrast to the universal j-$v_\parallel$ relation of skyrmions without boundary effect as exemplified by (3.25) in Sect. 3.2. When the impurity effect is absent or negligible, the j-$v_\parallel$ relation of a skyrmion for the confined case is derived from Thiele's equation (3.16) as,

$$\mathbf{v}_\parallel = \frac{\beta}{\alpha}\mathbf{v}_s. \tag{3.32}$$

This relation is identical to that of the helical state or ferromagnetic domain walls given in (3.26), which indicates strong dependence on α and β. Note that when $\beta = \alpha$ and $x = 0$, this j-$v_\parallel$ characteristic becomes identical to the universal relation for the skyrmion crystal in a non-confined space.

Equation (3.32) is derived in the following way. The solution to (3.16) without $\mathbf{F}_{\text{pin}}$ is given for the constant $\mathbf{v}_{\text{s}}$ as,

$$\mathbf{v}_{\text{d}} = \frac{1}{\alpha^2\mathcal{D}^2 + \mathcal{G}^2} \begin{pmatrix} -\alpha\mathcal{D} & -\mathcal{G} \\ \mathcal{G} & -\alpha\mathcal{D} \end{pmatrix} \begin{pmatrix} \mathcal{G}v_{\text{s}y} - \beta\mathcal{D}v_{\text{s}x} + \frac{\partial U}{\partial X} \\ -\mathcal{G}v_{\text{s}x} - \beta\mathcal{D}v_{\text{s}y} + \frac{\partial U}{\partial Y} \end{pmatrix}$$
$$= \frac{1}{\alpha^2\mathcal{D}^2 + \mathcal{G}^2} \begin{pmatrix} (\alpha\beta\mathcal{D}^2 + \mathcal{G}^2)v_{\text{s}x} - (\alpha - \beta)\mathcal{D}\mathcal{G}v_{\text{s}y} - \alpha\mathcal{D}\frac{\partial U}{\partial X} - \mathcal{G}\frac{\partial U}{\partial Y} \\ (\alpha - \beta)\mathcal{D}\mathcal{G}v_{\text{s}x} - (\alpha\beta\mathcal{D}^2 + \mathcal{G}^2)v_{\text{s}y} + \mathcal{G}\frac{\partial U}{\partial X} - \alpha\mathcal{D}\frac{\partial U}{\partial Y} \end{pmatrix}. \tag{3.33}$$

In a simple long stripline-shaped system, we can set $v_{\text{s}y} = 0$ and $\frac{\partial U}{\partial X} = 0$ in (3.33) to obtain,

$$\mathbf{v}_{\text{d}} = \begin{pmatrix} \dot{X} \\ \dot{Y} \end{pmatrix} = \frac{1}{\alpha^2\mathcal{D}^2 + \mathcal{G}^2} \begin{pmatrix} (\alpha\beta\mathcal{D}^2 + \mathcal{G}^2)v_{\text{s}x} - \mathcal{G}\frac{\partial U}{\partial Y} \\ (\alpha - \beta)\mathcal{D}\mathcal{G}v_{\text{s}x} - \alpha\mathcal{D}\frac{\partial U}{\partial Y} \end{pmatrix}. \tag{3.34}$$

The y-component of (3.34) determines the Y coordinate in the equilibrium state, that is, the Y coordinate in the limit of $t \to \infty$ converges to a value that satisfies,

$$\frac{\partial U}{\partial Y} = \frac{\alpha - \beta}{\alpha}\mathcal{G}v_{\text{s}x}, \tag{3.35}$$

as long as the skyrmion is confined in the system. Inserting this relation to the x-component of (3.34), we obtain,

$$v_{\text{d}x} = \frac{1}{\alpha^2\mathcal{D}^2 + \mathcal{G}^2}\left(\alpha\beta\mathcal{D}^2 + \frac{\beta}{\alpha}\mathcal{G}^2\right)v_{\text{s}x}$$
$$\simeq \frac{\beta}{\alpha}v_{\text{s}x}. \tag{3.36}$$

When the electric current velocity $v_{\text{s}x}$ (or the electric current density j) is so large that (3.35) cannot be satisfied in the sample, the skyrmion disappears at the longitudinal edge of the sample. The threshold current density j_{c} to delete the skyrmion at the longitudinal edge depends on the ratio β/α. From the simulation, its typical value is evaluated as $j_{\text{c}} \simeq 4.0 \times 10^{11}$ A/m^2 when $\beta/\alpha = 2$.

Next let us consider the skyrmion motion at the junction of a magnetic stripline and a nonmagnetic lead (see inset of Fig. 3.7a). Figure 3.7a, b show two types of skyrmion dynamics near the junction. For a small electric current density, the skyrmion bounces and cannot reach the boundary because it cannot overcome a repulsive potential from the boundary. In this bouncing process, the repulsive potential from the boundary induces a motion transverse to the boundary because of the Magnus force, and eventually the skyrmion stops at a position slightly below the central line of the system.

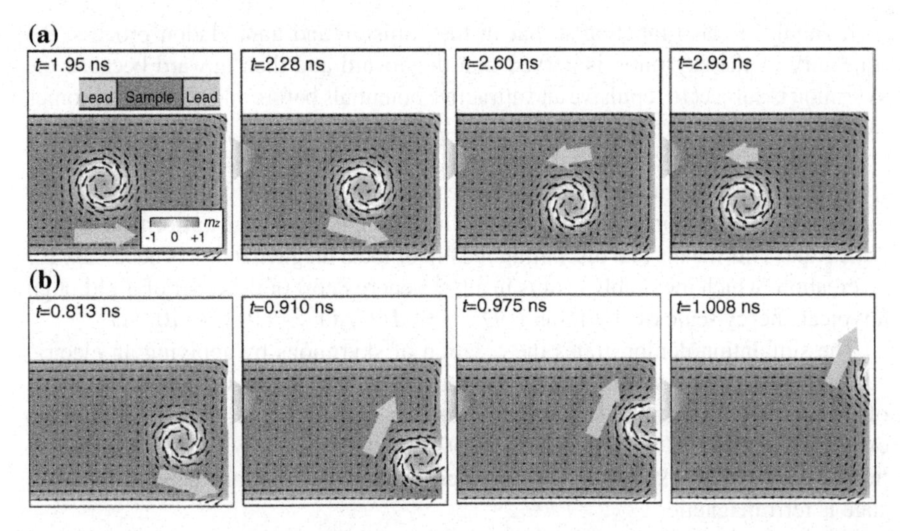

Fig. 3.7 Snapshots of magnetization configurations during the electric-current driven motion of a skyrmion near the edge of the stripline-shaped sample of magnetic material with nonmagnetic leads on both sides (see *inset*) for two different electric current densities **a** $j = 1.0 \times 10^{11}$ A/m^2 and **b** $j = 3.0 \times 10^{11}$ A/m^2. The parameters used for the simulation are equivalent to those for Fig. 3.6, and the clean case without impurities are considered. (Reproduced from [44].)

On the other hand, the larger electric current density enables a skyrmion to overcome the potential barrier as shown in Fig. 3.7b. The electric current further pushes the skyrmion to the sample edge, resulting in the annihilation of the skyrmion.

These kinds of behaviors can also be described by Thiele's equation. Near the vertical sample edge, one can set $v_{sy} = 0$ in (3.33) to obtain,

$$\begin{pmatrix} \dot{X} \\ \dot{Y} \end{pmatrix} = \frac{1}{\alpha^2\mathscr{D}^2 + \mathscr{G}^2} \begin{pmatrix} (\alpha\beta\mathscr{D}^2 + \mathscr{G}^2)v_{sx} - \alpha\mathscr{D}\frac{\partial U}{\partial X} - \mathscr{G}\frac{\partial U}{\partial Y} \\ (\alpha - \beta)\mathscr{D}\mathscr{G}v_{sx} + \mathscr{G}\frac{\partial U}{\partial X} - \alpha\mathscr{D}\frac{\partial U}{\partial Y} \end{pmatrix}.$$

When the current velocity v_{sx} (or the electric current density j) is large enough, the drift velocity of skyrmion v_{sx} is always positive, and the skyrmion goes out of the sample, resulting in annihilation of the skyrmion. On the other hand, if the current velocity v_{sx} is not large enough, bouncing of skyrmion occurs. The skyrmion goes to the boundary and then returns with a small displacement in y-direction. Eventually, the skyrmion stops at a stationary point (X_0, Y_0), which satisfies the following equation:

$$\begin{cases} \frac{\partial U}{\partial X} = \beta\mathscr{D}v_{sx} \\ \frac{\partial U}{\partial Y} = \mathscr{G}v_{sx} \end{cases}. \tag{3.37}$$

It should be also mentioned that in the collision and annihilation process, the trajectory of the skyrmion is curved first downward and then upward because the skyrmion is subject to repulsive and attractive potentials before and after it overcomes the potential barrier, respectively.

The creation of skyrmions is recognized to be very difficult because of the topological stability. Topological spin textures like skyrmion can never be created or annihilated by continuous variation of magnetization configuration from uniform ferromagnetic state. This means that discontinuous flip of local magnetization is necessary for its creation, which inevitably results in a large energy cost of the order of J although a typical energy scale of skyrmion is $D^2/J = J(D/J)^2 \sim (10^{-4} - 10^{-2})J$.

The simulation demonstrates the creation of skyrmions by applying an electric current to a stripline-shaped sample with a rectangular notch structure. Snapshots of the skyrmion creation process are shown in Fig. 3.8. Here the strength of the external magnetic field is fixed at $B = 0.0278J$, which slightly exceeds the critical field strength for the skyrmion-crystal-to-ferromagnetic transition, that is, the ground state is ferromagnetic.

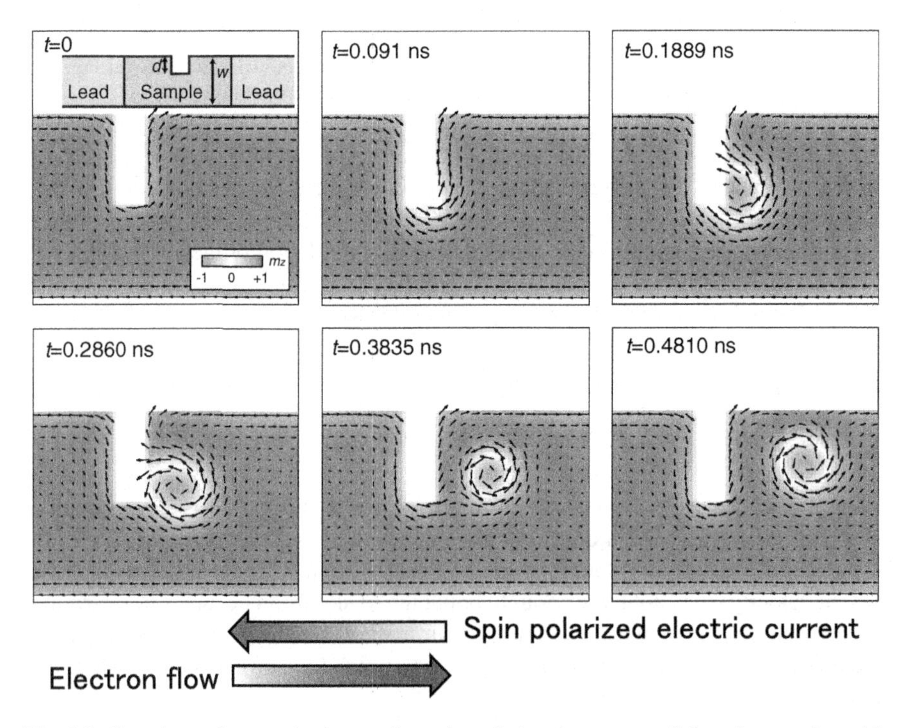

Fig. 3.8 Snapshots of magnetization configurations during the process of skyrmion creation with the electric current density of $j = 3.6 \times 10^{11}$ A/m^2. The parameters used for the simulation are equivalent to those for Fig. 3.6. Here the clean case without impurities are considered. (Reproduced from [44].)

Without electric current, the magnetizations around the notch have large in-plane components due to the Dzyaloshinskii-Moriya interaction. The applied electric current swells out the magnetization texture at the notch via the spin-transfer torque to form a seed of skyrmion. Subsequently, magnetizations behind the seed become twisted via the spin precession and point down due to the Dzyaloshinskii-Moriya interaction, and eventually the skyrmion core is created. The unique direction of the spin precession breaks the reflection symmetry, thereby leading to asymmetry with respect to the sign of **j**. The skyrmion creation occurs only when the electric current flows in a certain direction determined by the sign of **B** irrespective of the sign of the Dzyaloshinskii-Moriya coupling.

The simulations find the following conditions for efficient creation of skyrmions:

- The most suitable angle of the notch corner is 90°.
- Both the notch depth d and the sample width at the notch $w - d$ (w is the width of the stripline) should be comparable to or larger than the size of the skyrmion.
- The sharp edge of the notch corner is not necessarily needed, and the skyrmion creation occurs even with a rounded notch if the curvature radius is comparable to or larger than the size of the skyrmion.

3.4 Magnon-Current Driven Dynamics of Skyrmions

In this section, we discuss the magnon-current driven dynamics of skyrmions [45, 57–60]. Magnon currents couple to noncollinear spin textures such as skyrmions via exchanging the spin-transfer torque. In the presence of temperature gradient, diffusive flows of thermally activated magnons occur in magnets. Reacting force from the diffusive magnon currents turns out to drive motion of skyrmions in a direction opposite to the magnon flow [45, 57, 58]. An intriguing dynamical phenomenon related to this effect has been discovered by the Lorentz transmission electron microscopy [45]. It was observed that micro-scale regions of skyrmion crystal in thin-film samples of MnSi and Cu_2OSeO_3 show persistent rotations in a unique direction in the wide range of temperature and magnetic field. In the Lorentz transmission electron microscope experiments, a static magnetic field is applied in the perpendicular direction from the top down (this direction is referred to as the negative z direction hereafter), and one observes the persistent rotation in a clockwise fashion. This chiral rotation of skyrmion microcrystal is observed both for metallic MnSi and for insulating Cu_2OSeO_3, in spite of their distinct origins of the magnetism between metallic and insulating magnets, as well as differences in skyrmion size, transition temperature, and critical magnetic field, which indicates that this chiral rotation is a generic phenomenon of the skyrmionic system. In the Lorentz microscope, an electron beam is irradiated onto a thin-plate sample of magnet to observe the real-space distribution of magnetizations. One might suspect a circular magnetic field induced by the electron beam as a driving force of this rotation. However, this possibility can be excluded since estimated strength of the induced field is five orders of magnitudes smaller than the geomagnetic field.

Then a question arises: why and how does this phenomenon occur? The electron beam irradiated in the Lorentz microscope is expected to raise a temperature slightly at the beam spot on the sample relative to the outside of the spot, which induces a temperature gradient where the temperature gradually decreases from the center of the beam spot to the periphery. This hypothesis was examined by numerical simulations as a possible driving mechanism.

In the simulation, magnetic configuration of a skyrmion crystal confined in a micrometric circular disk is first prepared by the Monte-Carlo thermalization, and then a radial temperature gradient is introduced to this system as shown in Fig. 3.9b. More concretely a linear temperature gradient with the highest temperature $k_B(T_0 + \Delta T) = 0.106\,J$ and the lowest temperature $k_B T_0 = 0.1\,J$ at the center and the edge of the disk, respectively. Thermally induced dynamics of this confined skyrmion microcrystal is simulated by numerically solving the stochastic Landau-Lifshitz-Gilbert equation using the Heun scheme. The equation is given by,

$$\frac{d\mathbf{m}_i}{dt} = -\frac{1}{1+\alpha^2}\left[\mathbf{m}_i \times \left(\mathbf{B}_i^{\text{eff}} + \boldsymbol{\xi}_i^{\text{fl}}(t)\right) + \frac{\alpha}{m}\mathbf{m}_i \times \left\{\mathbf{m}_i \times \left(\mathbf{B}_i^{\text{eff}} + \boldsymbol{\xi}_i^{\text{fl}}(t)\right)\right\}\right]. \tag{3.38}$$

Here $\mathbf{B}_{\text{eff}}$ is the deterministic effective magnetic field calculated from the Hamiltonian as

$$\mathbf{B}_i^{\text{eff}} = -\frac{1}{\gamma\hbar}\frac{\partial\mathscr{H}}{\partial\mathbf{m}_i}, \tag{3.39}$$

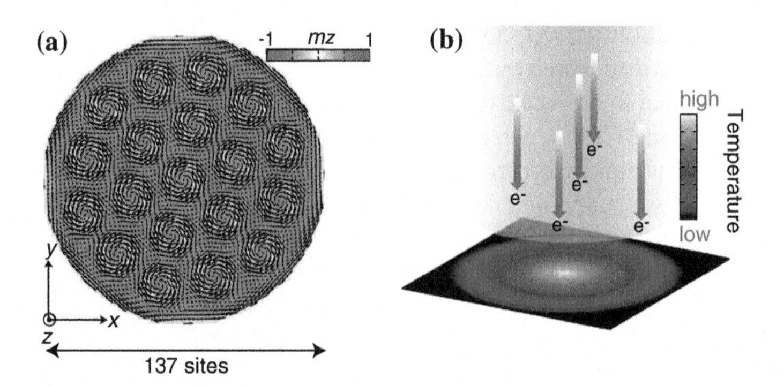

Fig. 3.9 **a** Skyrmion-crystal configuration confined in a circular disk prepared in order to investigate the persistent chiral rotation of skyrmions by numerical simulations. **b** Radial temperature gradient expected to be realized in the Lorentz transmission electron microscopy experiment where the disk center has a high temperature of $T_0 + \Delta T$, while the periphery has a low temperature of T_0 with a temperature difference of $\Delta T(>0)$. In the simulations, a linear interpolation of temperatures between the center and the edge is assumed. (Reproduced from [45].)

while $\boldsymbol{\xi}_i^{\mathrm{fl}}(t)$ is the Gaussian stochastic force. The latter describes the thermally fluctuating environment acting on the magnetizations which satisfies the following relations;

$$\langle \xi_{i,\beta}^{\mathrm{fl}}(t) \rangle = 0, \tag{3.40}$$

$$\langle \xi_{i,\beta}^{\mathrm{fl}}(t) \xi_{j,\lambda}^{\mathrm{fl}}(s) \rangle = 2\kappa_i \delta_{ij} \delta_{\beta\lambda} \delta(t-s), \tag{3.41}$$

where β and λ are the Cartesian coordinates x, y and z. Regarding (3.38) as Langevin's equation, one obtains a relation between κ (variance of the Gaussian distribution) and local temperature T_i from the fluctuation-dissipation theorem as,

$$\kappa_i = \alpha k_{\mathrm{B}} T_i / m \tag{3.42}$$

where α is the Gilbert-damping coefficient and m is the norm of the magnetization vector.

The simulation reproduces the experimentally observed ratchet rotation of the skyrmion crystal. Figure 3.10 shows snapshots of the simulated real-space dynamics of the skyrmion crystal. In this simulation, the magnetic field is applied in the negative z direction in accord with the experimental situation, and one observes the persistent rotation in the clockwise fashion in agreement with the experimental observation. What's important is this chiral rotation is driven purely by thermal fluctuations or the temperature gradient, because no other motive force is considered in this simulation.

A theoretical analysis has uncovered that this phenomenon is induced by the topological Hall effect of magnon currents (see Fig. 3.11a). At finite temperatures, thermal fluctuations activate magnons. In the presence of temperature gradient, such thermally activated magnons flow in the disk from its center with a higher temperature

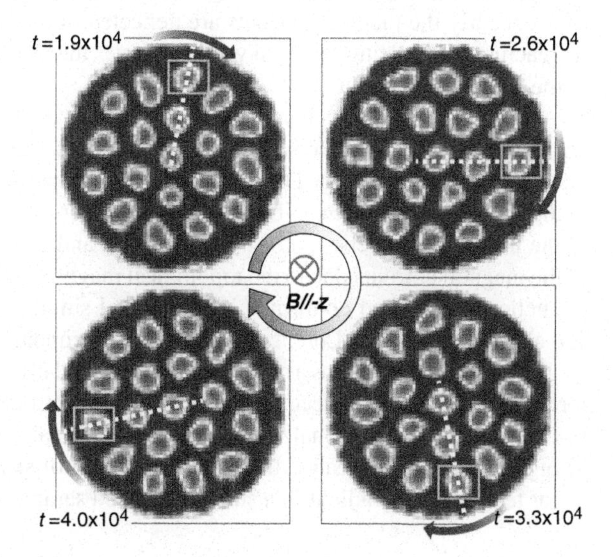

Fig. 3.10 Snapshots of the simulated dynamics of the confined skyrmion crystal with $k_{\mathrm{B}}T_0/J = 0.1$ and $\Delta T/J = 0.006$ where the time unit is $\hbar/J$. The parameters used are $J = 1 \cdot m = 1 \cdot D/J = -0.27 \cdot B_z/J = -0.03$, and $\alpha = 0.01$. The static magnetic field is applied to the negative z direction, and then one observes that the skyrmion crystal is persistently rotating in the clockwise fashion. (Reproduced from [45].)

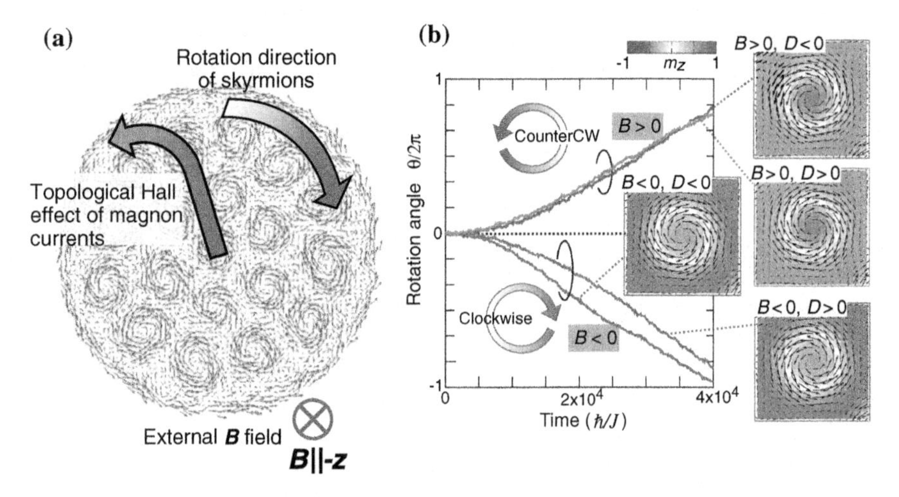

Fig. 3.11 **a** Schematic figure of the persistent unidirectional rotation of skyrmion microcrystal driven by the topological magnon Hall effect in the presence of the temperature gradient. **b** Simulated time profiles of the number of the rotations (or the rotation angle divided by 2π) for four different sets of signs of the Dzyaloshinskii-Moriya parameter D and the external magnetic field $\mathbf{B} = (0, 0, B)$. Note that a positive (negative) slope of the plot indicates the counterclockwise (clockwise) rotation. One finds that the rotation sense is reversed when one reverses the sign of B, whereas is unchanged when one reverses the sign of D. (Reproduced from [45].)

to the edge with a lower temperature. This diffusive magnon currents flowing in the radial direction become deflected by the fictitious magnetic field generated by the topological skyrmion spin textures. When the magnetic field is applied in the negative z direction, each skyrmion has a core magnetization pointing upwards, which gives a skyrmion number $Q = +1$ and a negative quantum magnetic flux to each skyrmion. Consequently the magnon currents are deflected in the counterclockwise direction. Its reacting force acting on the skyrmions drives the rotation of the skyrmion crystal in the opposite direction.

According to this mechanism, one expects that the rotation sense should be changed upon the sign reversal of the external magnetic field, whereas it should not upon the sign reversal of the Dzyaloshinskii-Moriya parameter D or upon the reversal of the crystal chirality. This is because the sign reversal of D does not change the sign of the fictitious magnetic field from skyrmions, and thereby the deflection direction of the magnon currents due to the topological magnon Hall effect is unchanged. This prediction was indeed confirmed by numerical simulations as shown in Fig. 3.11b and also by subsequent Lorentz microscope experiments.

The above finding opens a route to manipulate and drive skyrmions via the reacting force from thermally induced magnon currents by introducing temperature gradient. In addition to the rotational motion under the radial temperature gradient shown in Fig. 3.12a, one can realize translational motion of skyrmions by introducing the linear temperature gradient in a stripline-shaped sample as shown in Fig. 3.12b.

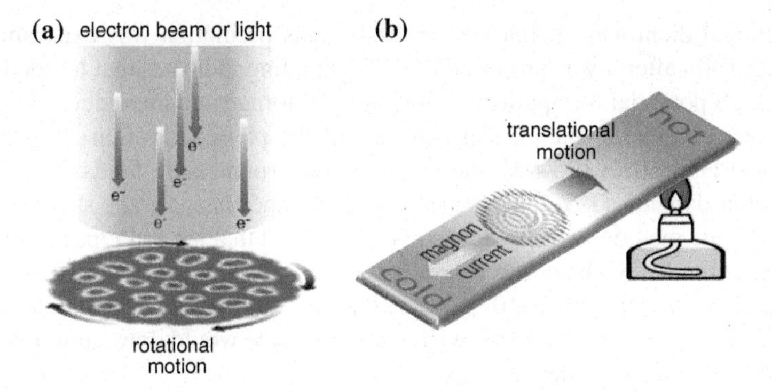

Fig. 3.12 **a** Schematic figure of the rotational motion of skyrmions in the presence of the radial temperature gradient induced by the irradiation of an electron beam or a light. **b** Schematic figure of the driving mechanism of translational motion of skyrmions in a stripline shaped sample with a temperature gradient. Magnons activated thermally at the high-temperature side flow to the low-temperature side. The reaction force from this diffusive magnon current acts on the skyrmion to drive their motion in the opposite direction, that is, in the direction towards the high-temperature side

3.5 Concluding Remarks

We have overviewed recent theoretical studies on the current-driven dynamics of magnetic skyrmions. Three features of skyrmions, that is, the finite topological number, the vortex-like nature, and the particle-like nature have turned out to be key ingredients for their peculiar dynamical properties. The ultralow threshold current density to drive the motion makes skyrmions promising for future technical application. In order to utilize skyrmions as information carriers, one needs to establish efficient ways to write, delete, and read the skyrmion bits. Concerning the writing process, several methods have been theoretically proposed in addition to the method using the electric current and a notch discussed in Sect. 3.3 [61–64]. Also techniques to fabricate thin films, nanowires and nanopatternings of skyrmionic materials should be developed [65–71]. In addition to the current-driven dynamics, the skyrmions show a lot of intriguing phehomena, which have attracted research interest. Peculiar collective modes and response dynamics under the ac electromagnetic fields such as light and microwave were reported recently [72–78]. In particular, the skyrmions in the insulating chiral-lattice magnet Cu_2OSeO_3 exhibit ferroelectricity with magnetism-induced electric polarizations [25–27]. In this multiferroic system, coupling between the electric polarization and the magnetization, so-called magnetoelectric coupling enables us to control the skyrmion spin textures via application of an electric field [79–83]. Furthermore one can activate coupled eigenmodes of magnetizations and polarizations, so-called electromagnons , not only by the ac magnetic-field component of microwave but also by the ac electric-field component of microwave. It was theoretically predicted that interference between these two activation channels causes unprecedentedly large diode effect on microwaves, or directional dichroism of microwaves [84]. This prediction was experimentally confirmed soon after it was proposed [85]. This finding indicates that the skyrmions possess high potential for application not only to storage and logic devices but also to microwave devices. The critical behavior of the phase transitions [86–89], the topological Hall effect [90–93], the emergent electromagnetic fields [54, 94–96], and peculiar dynamics [97–99] are also issues of importance for the skyrmions and the skyrmionic materials. It should also be mentioned that several types of topological spin textures have been discovered or predicted successively in several kinds of materials or systems [100–109]. Researches on the rich and dramatic skyrmion dynamics have just started, and we wish that this article would draw attention to this field and help to develop the research.

Acknowledgments The author thanks N. Nagaosa, J. Iwasaki, M. Mostovoy, J. Zang, and W. Koshibae for collaborations, and Y. Tokura, S. Seki, X.Z. Yu, N. Kanazawa, M. Kubota, Y. Onose, F. Kagawa, T. Arima, M. Kawasaki, A. Rosch, M. Garst, K. Everschor, C. Schütte for enlightening discussions. This work is partly supported by JSPS KAKENHI (Grant Numbers 25870169 and 25287088), and JST PRESTO.

References

1. C. Pfleiderer, Nat. Phys. **7**, 673 (2011)
2. N. Nagaosa, Y. Tokura, Nat. Nanotech. **8**, 899 (2013)
3. A. Fert, V. Cros, J. Sampaio, Nat. Nanotech. **8**, 152 (2013)
4. T.H.R. Skyrme, Proc. R. Soc. A **260**, 127 (1961)
5. T.H.R. Skyrme, Nucl. Phys. **31**, 556 (1962)
6. S.L. Sondhi, A. Karlhede, S.A. Kivelson, E.H. Rezayi, Phys. Rev. B **47**, 16419 (1993)
7. M. Abolfath, J.J. Palacios, H.A. Fertig, S.M. Girvin, A.H. MacDonald, Phys. Rev. B **56**, 6795 (1997)
8. S. Heinze, K. von Bergmann, M. Menzel, J. Brede, A. Kubetzka, R. Wiesendanger, G. Bihlmayer, S. Blügel, Nat. Phys. **7**, 713 (2011)
9. N. Romming, C. Hanneken, M. Menzel, J.E. Bickel, B. Wolter, K. von Bergmann, A. Kubetzka, R. Wiesendanger, Science **341**, 636 (2013)
10. I. Raičević, D. Popović, C. Panagopoulos, L. Benfatto, M.B. Silva Neto, E.S. Choi, T. Sasagawa, Phys. Rev. Lett. **106**, 227206 (2011)
11. A.N. Bogdanov, D.A. Yablonskii, Sov. Phys. JETP **68**, 101 (1989)
12. A. Bogdanov, A. Hubert, J. Mag. Mag. Mat. **138**, 255 (1994)
13. U.K. Rößler, A.N. Bogdanov, C. Pfleiderer, Nature **442**, 797 (2006)
14. S. Mühlbauer, B. Binz, F. Jonietz, C. Pfleiderer, A. Rosch, A. Neubauer, R. Georgii, P. Böni, Science **323**, 915 (2009)
15. C. Pappas, E. Lelievre-Berna, P. Falus, P.M. Bentley, E. Moskvin, S. Grigoriev, P. Fouquet, B. Farago, Phys. Rev. Lett. **102**, 197202 (2009)
16. C. Pfleiderer, T. Adams, A. Bauer, W. Biberacher, B. Binz, F. Birkelbach, P. Böni, C. Franz, R. Georgii, M. Janoschek, F. Jonietz, T. Keller, R. Ritz, S. Mühlbauer, W. Munzer, A. Neubauer, B. Pedersen, A. Rosch, J. Phys. Condens. Matter **22**, 164207 (2010)
17. W. Munzer, A. Neubauer, T. Adams, S. Mühlbauer, C. Franz, F. Jonietz, R. Georgii, P. Böni, B. Pedersen, M. Schmidt, A. Rosch, C. Pfleiderer, Phys. Rev. B **81**, 041203(R) (2010)
18. T. Adams, S. Mühlbauer, C. Pfleiderer, F. Jonietz, A. Bauer, A. Neubauer, R. Georgii, P. Böni, U. Keiderling, K. Everschor, M. Garst, A. Rosch, Phys. Rev. Lett. **107**, 217206 (2011)

19. S.V. Grigoriev, N.M. Potapova, S.-A. Siegfried, V.A. Dyadkin, E.V. Moskvin, V. Dmitriev, D. Menzel, C.D. Dewhurst, D. Chernyshov, R.A. Sadykov, L.N. Fomicheva, A.V. Tsvyashchenko, Phys. Rev. Lett. **110**, 207201 (2013)
20. X.Z. Yu, Y. Onose, N. Kanazawa, J.H. Park, J.H. Han, Y. Matsui, N. Nagaosa, Y. Tokura, Nature **465**, 901 (2010)
21. X.Z. Yu, N. Kanazawa, Y. Onose, K. Kimoto, W.Z. Zhang, S. Ishiwata, Y. Matsui, Y. Tokura, Nat. Mater. **10**, 106 (2011)
22. A. Tonomura, X.Z. Yu, K. Yanagisawa, T. Matsuda, Y. Onose, N. Kanazawa, H.S. Park, Y. Tokura, Nano Lett. **12**, 1673 (2012)
23. K. Shibata, X.Z. Yu, T. Hara, D. Morikawa, N. Kanazawa, K. Kimoto, S. Ishiwata, Y. Matsui, Y. Tokura, Nat. Nanotech. **8**, 723 (2013)
24. D. Morikawa, K. Shibata, N. Kanazawa, X.Z. Yu, Y. Tokura, Phys. Rev. B **88**, 024408 (2013)
25. S. Seki, X.Z. Yu, S. Ishiwata, Y. Tokura, Science **336**, 198 (2012)
26. S. Seki, J.-H. Kim, D.S. Inosov, R. Georgii, B. Keimer, S. Ishiwata, Y. Tokura, Phys. Rev. B **85**, 220406 (2012)
27. S. Seki, S. Ishiwata, Y. Tokura, Phys. Rev. B **86**, 060403 (2012)
28. T. Adams, A. Chacon, M. Wagner, A. Bauer, G. Brandl, B. Pedersen, H. Berger, P. Lemmens, C. Pfleiderer, Phys. Rev. Lett. **108**, 237204 (2012)
29. I. Dzyaloshinskii, J. Phys. Chem. Solids **4**, 241 (1958)
30. T. Moriya, Phys. Rev. **120**, 91 (1960)
31. P. Bak, M.H. Jensen, J. Phys. C **13**, L881 (1980)
32. S.D. Yi, S. Onoda, N. Nagaosa, J.H. Han, Phys. Rev. B **80**, 054416 (2009)
33. J.H. Han, J. Zang, Z. Yang, J.-H. Park, N. Nagaosa, Phys. Rev. B **82**, 094429 (2010)
34. Y.-Q. Li, Y.-H. Liu, Y. Zhou, Phys. Rev. B **84**, 205123 (2011)
35. A.B. Butenko, A.A. Leonov, U.K. Rößler, A.N. Bogdanov, Phys. Rev. B **82**, 052403 (2010)
36. N.S. Kiselev, A.N. Bogdanov, R. Schäfer, U.K. Rößler, J. Phys. D **44**, 392001 (2011)
37. M.N. Wilson, E.A. Karhu, A.S. Quigley, U.K. Rößler, A.B. Butenko, A.N. Bogdanov, M.D. Robertson, T.L. Monchesky, Phys. Rev. B **86**, 144420 (2012)
38. E.A. Karhu, U.K. Rößler, A.N. Bogdanov, S. Kahwaji, B.J. Kirby, H. Fritzsche, M.D. Robertson, C.F. Majkrzak, T.L. Monchesky, Phys. Rev. B **85**, 094429 (2012)
39. F.N. Rybakov, A.B. Borisov, A.N. Bogdanov, Phys. Rev. B **87**, 094424 (2013)
40. H.Y. Kwon, K.M. Bu, Y.Z. Wu, C. Won, J. Magn. Magn. Mater. **324**, 2171 (2012)
41. S. Buhrandt, L. Fritz, Phys. Rev. B **88**, 195137 (2013)
42. J. Iwasaki, M. Mochizuki, N. Nagaosa, Nat. Commun. **4**, 1463 (2013)
43. A. Rosch, Nat. Nanotech. **8**, 160 (2013)
44. J. Iwasaki, M. Mochizuki, N. Nagaosa, Nat. Nanotech. **8**, 742 (2013)
45. M. Mochizuki, X.Z. Yu, S. Seki, N. Kanazawa, W. Koshibae, J. Zang, M. Mostovoy, Y. Tokura, N. Nagaosa, Nat. Mater. **13**, 241 (2014)
46. J.C. Slonczewski, J. Magn. Magn. Mater. **159**, L1 (1996)
47. L. Berger, Phys. Rev. B **54**, 9353 (1996)
48. S.E. Barns, S. Maekawa, Phys. Rev. Lett. **95**, 107204 (2005)
49. S.S.P. Parkin, M. Hayashi, L. Thomas, Science **320**, 190 (2008)
50. F. Jonietz, S. Mühlbauer, C. Pfleiderer, A. Neubauer, W. Münzer, A. Bauer, T. Adams, R. Georgii, P. Böni, R.A. Duine, K. Everschor, M. Garst, A. Rosch, Science **330**, 1648 (2010)
51. X.Z. Yu, N. Kanazawa, W.Z. Zhang, T. Nagai, T. Hara, K. Kimoto, Y. Matsui, Y. Onose, Y. Tokura, Nat. Commun. **3**, 988 (2012)
52. K. Everschor, M. Garst, R.A. Duine, A. Rosch, Phys. Rev. B **84**, 064401 (2011)
53. K. Everschor, M. Garst, B. Binz, F. Jonietz, S. Mühlbauer, C. Pfleiderer, A. Rosch, Phys. Rev. B **86**, 054432 (2012)
54. T. Schulz, R. Ritz, A. Bauer, M. Halder, M. Wagner, C. Franz, C. Pfleiderer, K. Everschor, M. Garst, A. Rosch, Nat. Phys. **8**, 301 (2012)
55. J. Zang, M. Mostovoy, J.H. Han, N. Nagaosa, Phys. Rev. Lett. **107**, 136804 (2011)
56. A.A. Thiele, Phys. Rev. Lett. **30**, 230 (1973)
57. L. Kong, J. Zang, Phys. Rev. Lett. **111**, 067203 (2013)

58. S.-Z. Lin, C.D. Batista, C. Reichhardt, A. Saxena, Phys. Rev. Lett. **112**, 187203 (2014)
59. J. Iwasaki, A.J. Beekman, N. Nagaosa, Phys. Rev. B **89**, 064412 (2014)
60. C. Schutte, M. Garst, Phys. Rev. B **90**, 094423 (2014)
61. T. Ogasawara, N. Iwata, Y. Murakami, H. Okamoto, Y. Tokura, Appl. Phys. Lett. **94**, 162507 (2009)
62. Y. Tchoe, J.H. Han, Phys. Rev. B **85**, 174416 (2012)
63. M. Finazzi, M. Savoini, A.R. Khorsand, A. Tsukamoto, A. Itoh, L. Duo, A. Kirilyuk, Th Rasing, M. Ezawa, Phys. Rev. Lett. **110**, 177205 (2013)
64. J. Sampaio, V. Cros, S. Rohart, A. Thiaville, A. Fert, Nat. Nanotech. **8**, 839 (2013)
65. A.L. Schmitt, J.M. Higgins, J.R. Szczech, S. Jin, J. Mater. Chem. **20**, 223 (2010)
66. S.X. Huang, C.L. Chien, Phys. Rev. Lett. **108**, 267201 (2012)
67. Yufan Li, N. Kanazawa, X. Z. Yu, A. Tsukazaki, M. Kawasaki, M. Ichikawa, X. F. Jin, F. Kagawa, Y. Tokura. Phys. Rev. Lett. **110**, 117202 (2013)
68. X.Z. Yu, J.P. DeGrave, Y. Hara, T. Hara, S. Jin, Y. Tokura, Nano Lett. **13**, 3755 (2013)
69. L. Sun, R.X. Cao, B.F. Miao, Z. Feng, B. You, D. Wu, W. Zhang, A. Hu, H.F. Ding, Phys. Rev. Lett. **110**, 167201 (2013)
70. H. Du, W. Ning, M. Tian, Y. Zhang, EPL **101**, 37001 (2013)
71. H. Du, W. Ning, M. Tian, Y. Zhang, Phys. Rev. B **87**, 014401 (2013)
72. M. Mochizuki, Phys. Rev. Lett. **108**, 017601 (2012)
73. O. Petrova, O. Tchernyshyov, Phys. Rev. B **84**, 214433 (2011)
74. C. Moutafis, S. Komineas, J.A.C. Bland, Phys. Rev. B **79**, 224429 (2009)
75. I. Makhfudz, B. Krueger, O. Tchernyshyov, Phys. Rev. Lett. **109**, 217201 (2012)
76. S.-Z. Lin, C.D. Batista, A. Saxena, Phys. Rev. B **89**, 024415 (2014)
77. G. Tatara, H. Fukuyama, J. Phys. Soc. Jpn. **83**, 104711 (2014)
78. Y. Onose, Y. Okamura, S. Seki, S. Ishiwata, Y. Tokura, Phys. Rev. Lett. **109**, 037603 (2012)
79. J.-W.G. Bos, C.V. Colin, T.T.M. Palstra, Phys. Rev. B **78**, 094416 (2008)
80. M. Belesi, I. Rousochatzakis, M. Abid, U.K. Rößler, H. Berger, J-Ph Ansermet, Phys. Rev. B **85**, 224413 (2012)
81. J.S. White, I. Levatić, A.A. Omrani, N. Egetenmeyer, K. Prša, I. Živković, J.L. Gavilano, J. Kohlbrecher, M. Bartkowiak, H. Berger, H.M. Rønnow, J. Phys. Condens. Matter **24**, 432201 (2012)
82. J.S. White, I. Levatić, A.A. Omrani, N. Egetenmeyer, K. Prša, I. Živković, J.L. Gavilano, J. Kohlbrecher, M. Bartkowiak, H. Berger, H.M. Rønnow, Phys. Rev. Lett. **113**, 107203 (2014)
83. Y.H. Liu, Y.-Q. Li, J.H. Hoon, Phys. Rev. B **87**, 100402 (2013)
84. M. Mochizuki, S. Seki, Phys. Rev. B **87**, 134403 (2013)
85. Y. Okamura, F. Kagawa, M. Mochizuki, M. Kubota, S. Seki, S. Ishiwata, M. Kawasaki, Y. Onose, Y. Tokura, Nat. Commun. **4**, 2391 (2013)
86. A. Bauer, A. Neubauer, C. Franz, W. Munzer, M. Garst, C. Pfleiderer, Phys. Rev. B **82**, 064404 (2010)
87. A. Bauer, M. Garst, C. Pfleiderer, Phys. Rev. Lett. **110**, 177207 (2013)
88. M. Janoschek, M. Garst, A. Bauer, P. Krautscheid, R. Georgii, P. Böni, C. Pfleiderer, Phys. Rev. B **87**, 134407 (2013)
89. H. Wilhelm, M. Baenitz, M. Schmidt, U.K. Rößler, A.A. Leonov, A.N. Bogdanov, Phys. Rev. Lett. **107**, 127203 (2011)
90. B. Binz, A. Vishwanath, Phys. B **403**, 1336 (2008)
91. M. Lee, Y. Onose, Y. Tokura, N.P. Ong, Phys. Rev. B **75**, 172403 (2007)
92. A. Neubauer, C. Pfleiderer, B. Binz, A. Rosch, R. Ritz, P.G. Niklowitz, P. Böni, Phys. Rev. Lett. **102**, 186602 (2009)
93. N. Kanazawa, Y. Onose, T. Arima, D. Okuyama, K. Ohoyama, S. Wakimoto, K. Kakurai, S. Ishiwata, Y. Tokura, Phys. Rev. Lett. **106**, 156603 (2011)
94. N. Nagaosa, Y. Tokura, Phys. Scr. T **146**, 014020 (2012)
95. P. Milde, D. Köhler, J. Seidel, L.M. Eng, A. Bauer, A. Chacon, J. Kindervater, S. Mühlbauer, C. Pfleiderer, S. Buhrandt, C. Schüte, A. Rosch, Science **340**, 1076 (2013)
96. R. Takashima, S. Fujimoto, J. Phys. Soc. Jpn. **83**, 054717 (2014)

97. S.-Z. Lin, C. Reichhardt, C.D. Batista, A. Saxena, Phys. Rev. Lett. **110**, 207202 (2013)
98. S.-Z. Lin, C. Reichhardt, C.D. Batista, A. Saxena, Phys. Rev. B **87**, 214419 (2013)
99. Y.-H. Liu, Y.-Q. Li, J. Phys. Condens. Matter **25**, 076005 (2013)
100. T. Fukumura, H. Sugawara, T. Hasegawa, K. Tanaka, H. Sakaki, T. Kimura, Y. Tokura, Science **284**, 1969 (1999)
101. S. Ishiwata, M. Tokunaga, Y. Kaneko, D. Okuyama, Y. Tokunaga, S. Wakimoto, K. Kakurai, T. Arima, Y. Taguchi, Y. Tokura, Phys. Rev. B **84**, 054427 (2011)
102. N. Kanazawa, J.-H. Kim, D.S. Inosov, J.S. White, N. Egetenmeyer, J.L. Gavilano, S. Ishiwata, Y. Onose, T. Arima, B. Keimer, Y. Tokura, Phys. Rev. B **86**, 134425 (2012)
103. X.Z. Yu, M. Mostovoy, Y. Tokunaga, W. Zhang, K. Kimoto, Y. Matsui, Y. Kaneko, N. Nagaosa, Y. Tokura, Proc. Natl Acad. Sci. USA **109**, 8856 (2012)
104. A. Rosch, Proc. Natl. Acad. Sci. USA **109**, 8793 (2012)
105. X.Z. Yu, Y. Tokunaga, Y. Kaneko, W.Z. Zhang, K. Kimoto, Y. Matsui, Y. Taguchi, Y. Tokura, Nat. Commun. **5**, 4198 (2014)
106. M. Nagao, Y.-G. So, H. Yoshida, M. Isobe, T. Hara, K. Ishizuka, K. Kimoto, Nat. Nanotech. **8**, 325 (2013)
107. J.-H. Park, J.H. Han, Phys. Rev. B **83**, 184406 (2011)
108. T. Okubo, S. Chung, H. Kawamura, Phys. Rev. Lett. **108**, 017206 (2012)
109. A.N. Bogdanov, U.K. Rößler, M. Wolf, K.-H. Müller, Phys. Rev. B **66**, 214410 (2002)

Chapter 4
Functional Topologies in (Multi-) Ferroics: The Ferroelastic Template

E.K.H. Salje, O. Aktas and X. Ding

Abstract Ferroelastic domain boundaries are templates for functional interfaces with superconductivity, ferroelectricity and ferromagnetism constraint to the domain boundary. The topologies of these functional interfaces are described for three, two and one dimension(s), thus showing the basic topological approch to Domain Boundary Engineering.

4.1 Introduction

Bulk ferroics are a well-known class of anisotropic, nonlinear solids that develop a spontaneous order parameter below a symmetry-lowering transition point. These materials typically result from modifications of a high-symmetry structure (prototype). Ferroics are classified according to their primary order parameters, which include strain ε (ferroelastic), polarization P (ferroelectric), and magnetization M (ferromagnetic), where the term "ferro" designates the uniform alignment of the spontaneous moments in neighboring unit cells. Ferri- and anti-ferro-phases can also exist when the order parameter is locally rotated against a crystallographic axis, or when the relevant wavevector (or wavevectors) associated with the structural instability occur(s) at special points at the surface of the Brillouin zone. Incommensurable phase transitions require a complex order parameter with a repetition unit that is not commensurate with the underlying crystal structure. In both of these cases, the translation invariance of the order parameter (incommensurate or commensurate) is

E.K.H. Salje (✉)
Department of Earth Sciences, University of Cambridge,
Downing Street, Cambridge CB2 3EQ, UK
e-mail: es10002@esc.cam.ac.uk

O. Aktas
ETH Zürich, Vladimir-Prelog-Weg 4, 8093 Zurich, Switzerland
e-mail: oktay.aktas@mat.ethz.ch

X. Ding
State Key Laboratory for Mechanical Behavior of Materials,
Xi'an Jiaotong University, Xi'an 710049, People's Republic of China

J. Seidel (ed.), *Topological Structures in Ferroic Materials*, Springer Series in Materials Science 228, DOI 10.1007/978-3-319-25301-5_4

preserved throughout the crystal. Additionally, a novel class of ferroic materials has been garnering increasing attention; these so-called multiferroics feature more than one spontaneous primary order parameter, where the order parameters can be coupled or remain independent depending on the desired application. There are classic textbooks on ferroics [1–13], and numerous recent reviews on multiferroics [14–19].

These bulk ferroic properties have their equivalent in lower dimensional subspaces, such as two-dimensional twin boundaries, one-dimensional Bloch lines and vortex dots. A novel development in ferroic materials begun when it was understood that these low-dimensional ferroic properties are not necessarily related to ferroic bulk properties but that surfaces and interfaces may have ferroic (or superconducting) properties in a more constraint space while the bulk may even not be ferroic at all. Much research has been dedicated to surfaces but any practical applications are unlikely to be confined to surfaces because the total number or particles involved in functionalities will simply be too small (unless the grain size becomes very small). This problem can be overcome when we consider parallel twin boundaries where the number of atoms inside the twin boundaries is much larger than in surface layers and can reach 4 % of the total number of atomes in the sample. This concentration is large enough to think about applications such as memory devices, heat regulators and elastic dampers [19]. In addition, templates such as twin boundaries are mobile under fields, they can form tweed and domain glasses [20, 21]. One may then ask: is it possible to arrange twin boundaries topologically in ways so that the macroscopic use of the material as conductors, switches or heat regulators can be optimized? The answer is affirmative and leads to the development of domain structures, with a high degree of complexity. This field of research is part of 'Domain Boundary Engineering' and has as aim to provide functional domain structures with an optimized topology [14, 19, 28].

Let us illustrate this point with two examples. Consider a twinned martensite: many medical applications require an elastic response which is both reversible and also highly anisotropic [22–25]. This can be achieved by tailoring the twin domains in specific arrays. Such devices are used to open arteries, often used for patients after heart attacks. In another potential application one makes use of twin boundaries being superconducting [26–29]. If these twin boundaries are arranged in comb configurations (see also Fig. 4.6) where they form needle arrays and touch an orthogonal wall, the total system forms an array of Josephson junctions. Such devices would be ideal for the detection of weak magnetic fields useful in astronomy or in brain surgery.

One of the first functional twin boundaries was discovered in tungsten oxide (WO_3) in 1998 [26] with the introduction of Na and oxygen vacancies in twin walls. Through slight modifications of the walls by altering their chemical composition (e.g. from WO_3 to • $WO_{2.993}$), a metal–insulator transition is induced, which at low temperature can lead to the appearance of superconductivity in twin walls (Fig. 4.1) [26–29]. Although the changes are minor and analytically hard to detect, the chemically modified walls become superconducting with a critical field H_{c2} above 15 T and a superconducting transition temperature, T_c, initially near 3 K [26]. The surrounding matrix remains insulating so that this arrangement of superconducting twin boundaries with the formation of needle domains and domain junctions is potentially

Fig. 4.1 Microstructure of WO_3 with highly conducting twin walls and the collapse of piezoelectricity in twin walls **a** topology, **b** conductivity, **c** piezoelectricity. Original data were published in [29]

the key for engineering arrays of Josephson junctions and high-sensitivity magnetic scanners. In addition, it has been suggested that surface layers, presumably similar to the interfacial structures in WO_3, may display superconductivity at temperatures up to $T_c = 91\,K$(Na doping) and 120 K(H doping) [30]. These are extreme values of T_c, which have not been reproduced independently, while the highest recently confirmed value of 13 K has been directly observed by transport measurement.

The fact that a dopant will follow the trajectory of twin walls means that the nano-patterning of the superconducting structure is possible via the patterning of the twin boundaries and subsequent doping. Tungsten oxide, WO_3, and its sub-stoichiometric derivatives, WO_{3-x}, are particularly well suited for this research because they display metal–insulator transitions while remaining thermodynamically highly stable compounds [30–35]. These compounds display a multitude of structural phase transitions—principally linked to shape changes of the WO_6 octahedra and their rotations within an octahedral network. The facility with which oxygen is released under reducing conditions is mainly related to the low energy required to transfer the valence state of localized surplus electrons on the W^{6+} sites to W^{5+} (rather than the chemical bonding to oxygen). This tendency to form W^{5+} states near the surface was directly confirmed by X-ray photoelectron spectroscopy and ultraviolet photoelectron spectroscopy [36] and indirectly by scanning tunnel microscopy. The W^{5+} states are not localized and form bi-polarons in the low-temperature phase [37, 38]. In addition to being superconducting, WO_3 is also a well-known electrochromic, solar cell, and catalytic material [39].

We now discuss the relevant topologies of the ferroelastic templates in increasing order of complexity. First we use simple ferroelastic twin boundaries as templates for the required functionality, we will then use domain walls inside domain walls as a finer tool for such templates.

4.2 Wall Interesections with the Surface

Novak et al. [40] determined the strain profiles of twin walls when they intersect the crystal surface. It was speculated that the twin wall would widen dramatically at the surface so that the wall thicknesses near the surface, as measured by AFM, are

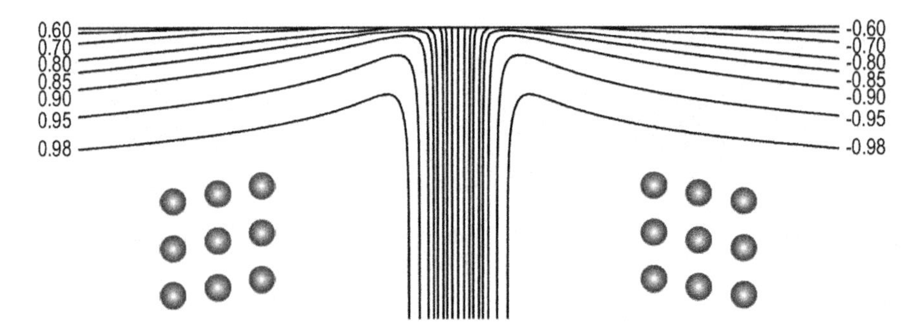

Fig. 4.2 Distribution of the order parameter Q at the surface of the lattice (first 50 layers). *Lines* represent constant Q, with $Q_o = 1$ in the bulk. There are three *lines* in the *middle* of the twin domain wall that are not labeled; they represent the Q values of 0.40, 0.00, and −0.40 respectively. Notice the steepness of the gradient of Q through the twin domain wall. The two structures represent sheared twin atomic configurations in the bulk (far from the twin domain wall and surface intersection)

much wider than the bulk values. Fortunately, this assertion proved to be wrong, the apparently wide twin walls were an artifact of the AFM method resulting from its limited resolution. The surface near strain profiles are shown in Fig. 4.2 and cearly show that the core of the twin wall remains sufficiently narrow to allow the extreme pinning of dopant atoms from the gas phase [41].

When walls intersect the surface under a shallow angle, one finds additional wall bending that was shown by Conti and Weikard [42] and, in a concrete example, by Ishibashi et al. [43, 44]. The intersections become important when the domain boundary concentration becomes large and overlapping domain walls may otherwise exist in the surface [45, 46]. In particular the measurement of the strain profiles by Raman spectroscopy relies on the fact that wall profiles are not much modified near surfaces [47–49].

4.3 Bending of Domain Walls, Needle Domains and 90° Junctions

The internal structure of twin walls is largely determined by the short-range interaction between atoms and the way that the atomic coordinates and occupancies reflect the spatial variations in the thermodynamic order parameter of the ferroelastic or co-elastic phase transition [1, 5]. The structural variations lead to elastic relaxations of the crystal structure, producing long-range strain fields. The strain fields generated by mesoscopic structures need not induce a macroscopic deformation of the sample or a variation in the average lattice parameters. The latter would be expressed by the macroscopic spontaneous strain of the sample and it is usually observed that different mesoscopic structures show almost the same macroscopic lattice parameters and, thus, the same spontaneous strain. When atomic ordering takes place in one part of the crystal it inevitably pulls and pushes neighboring atoms and/or structural units, i.e. it creates a local displacement field. This field then displaces other atoms and hence propagates elastically to distant parts of the crystal via a knock-on effect. The experimental observation of domain boundaries, such as twin walls, shows that their strain profiles corresponds well to that predicted from classic theory of elastic interactions, ignoring any additional short-range interactions and non-local effects [1, 5, 50, 51].

Two energies determine the deviation of wall directions under weak bending: the anisotropy energy $E_{anisotropy} = U(dy/dx)^2$ and the bending energy $E_{bending} = S(d^2y/dx^2)^2$, where the wall trajectory is defined by the coordinates x in the wall and the orthogonal direction y. The wall trajectories follow then from the solution of the Euler Lagrange equations with these energy terms and appropriate boundary conditions.

As example, a macroscopic sample is sheared if domain 1 is stabilized with respect to domain 2. The wall moves if an appropriate constant fore is applied in order to increase the size of domain 1. When the wall hits a defect, a point force is superimposed on the uniform force field and the wall bends around the defect. For thick walls in rather isotropic media, the wall profile is determined by the bending energy with $d^4y/dx^4 = 0$ everywhere except at the locus of the point force (e.g. $x = h/2$) [51]:

$$d^4y/dx^4 = K\delta(x - h/2).$$

The surface of the crystal is allowed to relax so that the boundary conditions are $y = 0$ at $x = 0$ and $x = h$. The solution of the differential equation is a polynomial of third order which is symmetrical with respect to $x = h/2$. No second-order terms exist because of the condition that the wall is flat ($y'' = 0$) without applied force. The required solution is [51]

$$y = (1/h^3)y_{max}x(3h^2 - 4x^2) \quad \text{for } x < h/2$$
$$y = (1/h^3)y_{max}(h - x)(-h^2 + 8xh - 4x^2) \quad \text{for } x < h/2$$

where $y_{max} = Kh^3/48$ is the maximum deviation of the wall centre from the surface. Around the defect the wall is parabolically bent with a straighter shape farther away from the pinning centre. As there is no elastic anisotropy energy present in this example, the wall is never planar along the elastically soft direction but is bent throughout the entire crystal.

Two pinning centres at the surface of the crystal excert point forces are at each end of the wall whereas the rest of the wall is subject to a constant dragging force due to the macroscopic shear of the sample. The differential equation which describes the wall trajectory is [51]

$$d^4y/dx^4 = K[\delta(x) + \delta(x - h)]$$
$$y = (K/24)x(x^3 - 2hx^2 + h^3)$$

where h is again the thickness of the sample in the x direction. The total wall profile is almost identical with that of a single pinning centre. In the case of surface pinning, the wall simply rotates near the centres but does not curve. The maximum curvature is again in the middle of the crystal because the two rotated parts of the wall have to connect in a smooth manner. The important conclusion from the comparison of the trajectories is that it is impossible to distinguish between the possible origins of a bent contour from the experimental observation alone; the contour may be due to one defect in the middle of the bend or several defects at the outside.

We can now add elastic drag to the wall. The wall is again moved by external forces and hits a local defect. In contrast with the previous cases the wall is allowed to relax along the y direction at great distance from the pinning centre. Such relaxation is achieved by a macroscopic deformation of the sample. The restoring force of the relaxation is elastic in nature and increases linearly with the wall displacement K = Py. The trajectory is described by [51]

$$Sd^4y/dx^4 = -PyK'\delta(x).$$

For an infinite crystal, the boundary conditions are $y = 0$ at $x = \infty$ and $x = -\infty$, the bending must be continuous at $y = 0$ in y, y' and y''. The solution of the differential equation is [51]

$$y = y_{max}\exp(-\beta|x|)[\cos(\beta|x|) + \sin(\beta|x|)] \text{ with the characteristic length}$$
$$\lambda = 1/\beta = (4S/P)^{1/4}.$$

The new aspect of this solution, in contrast with the case discussed before, is that there is an intrinsic length scale (λ) of the problem which allows the wall to bend back to the original orientation of the unperturbed crystal far away from the needle tip.

We now discuss one of the most common topology of domain walls in ferroelastics, namely needle domains [51]. We consider Peierls forces which are a linear function of the wall displacement y and simplify the anisotropy energy for small angles to include only the quadratic term in the energy density $E = U(dy/dx)^2 + Py^2$. The wall trajectory is then determined by the Euler–Lagrange equation with $-2Py + 2U\, d^2y/dx^2 = 0$. The solution is an exponential decay with $y = y_{max}\exp(-x/\lambda)$ where y_{max} is the maximum deviation from the unperturbed wall at the needle tip. The length scale λ is given by $\lambda = (U/P)^{1/2}$, namely by the ratio of the anisotropy energy to the Peierls energy. For large pinning forces the needle tip is short, whereas for small pinning forces the tip becomes long and narrow. The profile is smooth and shows a maximum bend near the shaft of the tip (Figs. 4.3 and 4.4). At the tip itself the trajectory is linear. For the actual tip, the trajectory is described by $P = 0$ with the boundary condition $y = 0$ at $x > \lambda$. With $2Ud^2y/dx^2 = 0$ the solution becomes $y = y_{max}(1 - x/\lambda)$ for $x < \lambda$. The trajectory is hence linear over large areas of the needle with exponential corrections near the shaft of the needle tip. The wall profile is shown in Fig. 4.3. Experimental observations confirm the detailed profile analysis [50].

In many cases ferroelastic walls (W and W′) intersect and form corners [1]. It is customary to call these corner configurations ‘right-angled’ domains. This term

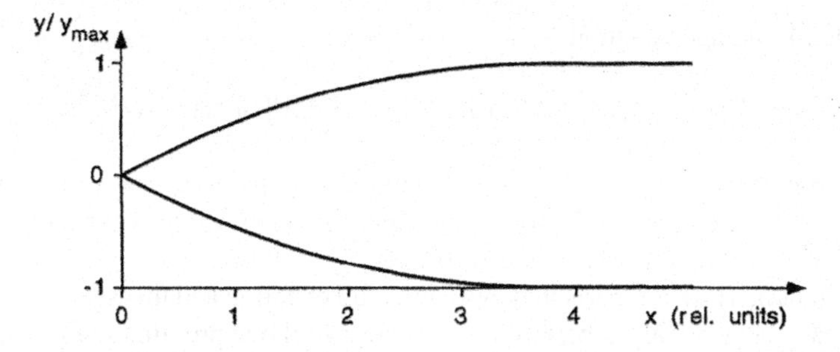

Fig. 4.3 Needle domain with strong Peierls forces and high anisotropy energy

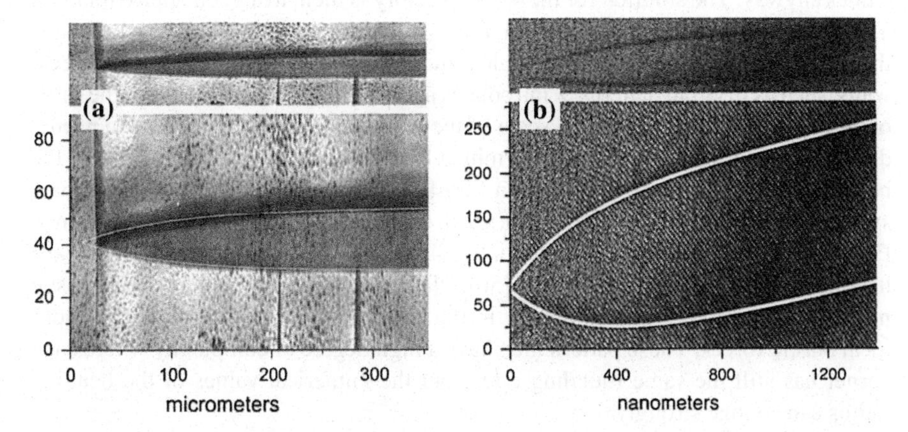

Fig. 4.4 Needle twins in KSCN (**a**) and $BaTiO_3$ (**b**). The trajectories were fit with an exponential function. The upper images show the true aspect ratio, the lower images show the y axis expanded to demonstrate the curvature of the needle tip. The apparent asymmetry in the latter is due to the non-linear expansion. Both axes have the same units (microns in **a** and nanometers in **b**)

is somewhat misleading, however, because the angle between the two walls is not exactly $\pi/2$ but either $\pi/2 + \omega$ or $\pi/2 - \omega$, where ω is the value of the spontaneous strain of the sample ([1] for details). In materials with a small spontaneous strain (10^{-3}, say) the deviation of the angle from $\pi/2$ is too small to be seen in electron microscopy or optical images while for most ferroelastics with spontaneous strains of some 2 % such deviations are clearly recognized. It was shown that this misfit of the wall angle from $\pi/2$ is essential for the understanding of the shape of the corner tip. Elastic strain energies of the bulk may lead to a rounding of the corner or may make the corner bulge out. The role of the Peierls forces domaintes away from the bent and the trajectories have to be asymptotically close to the original straight wall. Using the projection of the length element on the axis we find the energy expression which has to be minimized (Fig. 4.5):

$$\delta \mathrm{E} = \delta \int \left\{ 4/\pi \mathrm{U} \sin^2(2\alpha + \pi/2)[(\pi/4)^2 + \alpha^2]^{1/2} + \mathrm{S}\alpha'^2 \right\} \mathrm{dx} = 0$$

The Euler–Lagrange equation is

$$2S/U\alpha'' = (\delta/\delta\alpha)\{(4/\pi)\sin[2(2\alpha+\pi/2)][(\pi/4)^2+\alpha^2]^{1/2}\}$$

with the boundary conditions $\alpha = -\pi/4$ for $x = -\infty$ and $\alpha = \pi/4$ for $x = \infty$. The solution near the corner can be found for small values of α in a series expansion in α: $2S/U\alpha'' = -A\alpha + B\alpha 2\ldots$ with $A, B > 0$.

In fact, numerical comparison between the full equation and this series expansion shows that for all α-values between $-\pi/4$ and $\pi/4$ the approximation is excellent. Deviations are large outside this interval but the boundary conditions disallow such α-values anyway. The solution for the wall trajectory is then, to a good approximation, $\alpha = (\pi/4)\tanh(x/\lambda)$ where $\lambda \propto (S/U)^{1/2}$ is a measure for the bending radius of the wall around the corner. Mesoscopic structures with a multitude of right-angled walls, such as in tartan patterns, λ introduces again a length scale which is of the same order of magnitude as the length scale of the wall bending near defects or in needle domains if such bending is also determined by the same anisotropy energy and the bending energy. The wall trajectory in Fig. 4.5a is obtained in the (x, y) coordinate system by integration: $y = \lambda\ \ln[\cosh(x/\lambda)]$, this shows the rounding of the corner. Experimental trajectories are shown in Fig. 4.5b. These topologies of the twin walls have been observed experimentally [50]. Note, however, that in most ferroelastic materials several corners will appear in the pattern in order to minimize the non-local elastic forces. These pattern then have a high degree of complexity where each corner has still the same rounding effect but the numerical values of the bending radius can change strongly.

4.4 Adaptive Structure

Energies of the twin boundaries are often in the range of 300 mJ/m^2, which is not dissimilar to surface energies. Much smaller wall energies have been observed, such as in $SrTiO_3$ with very high wall densities [52]. A key observation was made in 1991 by Carpenter [53] who found that the wall energy in anorthite ($CaAl_2Si_2O_8$) depends on the amount of cation ordering between Al and Si. This degree of order is strongly dependent on the annealing temperature that determines the degree of Al, Si order. He could then determine the wall energy as function of temperature and found values between 600 mJm^2 and 300 mJ/m^2 with decreasing temperature between 1675 and 1475 K. Extrapolating to lower temperatures, the wall energy became zero near 1275 K and negative at lower temperatures. Negative wall energies lead to incommensurate structure, which have indeed been found in anorthite. In the context of topologies of ferroelastic domain structures, the key finding is that the wall energy can be extremely small so that the nucleation and destruction of wall requires little or no energy. It was argued by Viehland and Salje [21] that the nucleation and movement of such low energy walls requires little energy itself so that the walls become highly mobile and can hence adapt easily to any change of

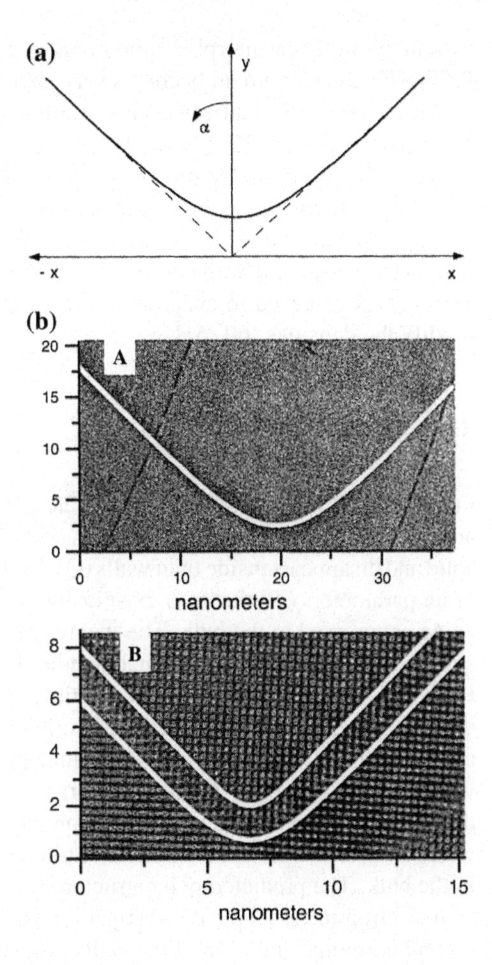

Fig. 4.5 **a** Bending of a domain wall at a right-angled corner [51]. **b** Images of right angle twin walls and the fitted wall trajectories. (**A**) $Gd_2(MoO_4)_3$, (**B**) $YBa_2Cu_3O_7$. Both axes have the same units [50]

boundary conditions. These authors called such topologies 'adaptive structures'. They occur typically near morphotropic boundaries [54, 55] where the structural state is a mix of a multitude of phases, many domain boundaries and the ability to adapt to external forces easily [56]. Adaptive domain structures have been reviewed in detail in [21]. Functional walls with low wall energies are hence expected to adapt to prescribed domain configurations when the driving forces are correctly chosen—although it may not always be possible to design these domain structures in full detail because the scale of the topology may simply be too fine to be handled with current technologies. When adaptive structures are formed in device materials their response to external fields is more akin to that described by fluid dynamics under stress (flow and creep) rather than to elastic responses. This has significant consequences for device materials. The material properties can be optimized using two different parameters, namely firstly the functionality of the domain wall and, second, the position and shape of the domain wall, and hence, implicitly, the domain boundary density [57, 58]. As an example, the domain boundary density can become extremely high near morphotropic boundaries in ferroelectric materials (such as in PZT) where each domain becomes very small with many boundaries in between so that it becomes difficult (or even meaningless) to distinguish between 'bulk' and 'boundary'. Li et al. [59] have shown that this is particularly important when the materials are used in electro-caloric or magneto-caloric heat conductors where the thermal conductivity can be controlled electronically through the domain boundary density and their topology (domain walls perpendicular to the heat flow). The key parameter is again the wall density, which needs to be as high as possible in this case. This is best achieved in adaptive structures whereby finite size effects can greatly modify the densities [60, 61].

4.5 Wall Functionalities

Ferroelastic walls as topological templates for ferroelectric functionalities are characterized by the strain order parameter, which dominates in the bulk of the uniform solid and disappears inside twin walls (Q). Functionalities are described by secondary order parameters (P) that may exist inside twin walls [1, 14, 19, 21, 42, 62–64] but being suppressed in the bulk. The theoretical rationale for such exotic internal wall structures is captured by the Houchmandazeh-Lajzerowicz-Salje [64] coupling term $H_{HLS} \sim Q^2P^2$, which is always compatible with the crystal symmetry [14, 62, 64], or via equivalent gradient coupling decribing flexoelectricity [61, 65–67]. A typical example where the twin walls are uniformly polartized is $CaTiO_3$. Simulations were first performed by Goncalves-Ferreira et al. [68]. These authors showed that the Ti position inside the octahedral complexes is not located in the middle of the wall octahedral if the octahedral is not tilted (as in the bulk) and slightly larger than in the bulk. The predicted polar structure was then observed by extensive transmission electron microscopical investigations [69] and by high resolution studies of the second harmonic emission of the walls [70] (Figs. 4.6 and 4.7).

Polar behaviour of twin walls in $CaTiO_3$ can be further tested using resonant piezoelectric spectroscopy (RPS) [71, 72]. RPS has been very recently introduced as a modification of resonant ultrasound spectroscopy (RUS). Both techniques are used to detect elastic resonances of a sample. RUS does this mechanically, i.e. by generation of elastic waves in the sample by piezoelectric transducers in contact with the sample. In RPS, on the other hand, an electric field is applied across two parallel surfaces of the sample. If the sample is piezoelectric locally or globally, elastic resonances are excited and can be detected by a second piezoelectric transducer (as in the case of RUS). Figure 4.8 shows the temperature evolution of an elastic resonance peak collected by RPS between 10 and 310 K. The squared frequency of this peak reflects the temperature dependence of its associated effective elastic constant, hardening with cooling and then saturating at lower temperatures. The appearance of elastic resonances in the RPS spectra indicates the presence of collective oscillations of dipoles, leading to the piezoelectric response of the $CaTiO_3$ single crystal sample. Such collective behaviour can be attributed to domain walls [71]. The effect of coherently vibrating defect dipoles cannot be completely ruled out as source of

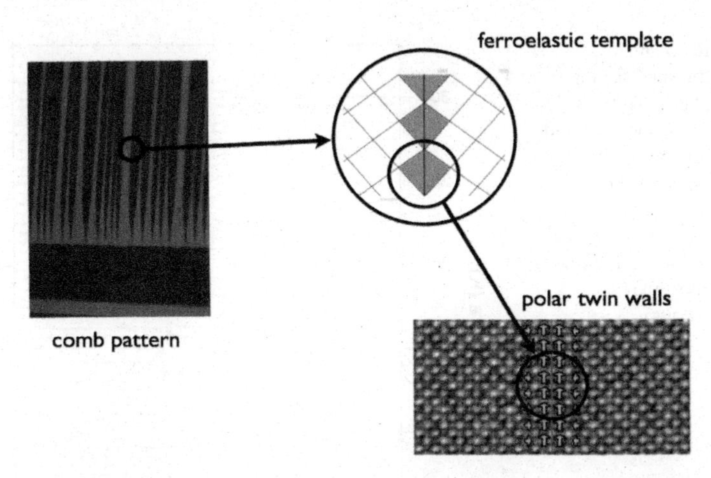

Fig. 4.6 Increasing complexity in twin walls from comb configurations to local strain fields and, finally, to arrays of polar vectors in $CaTiO_3$

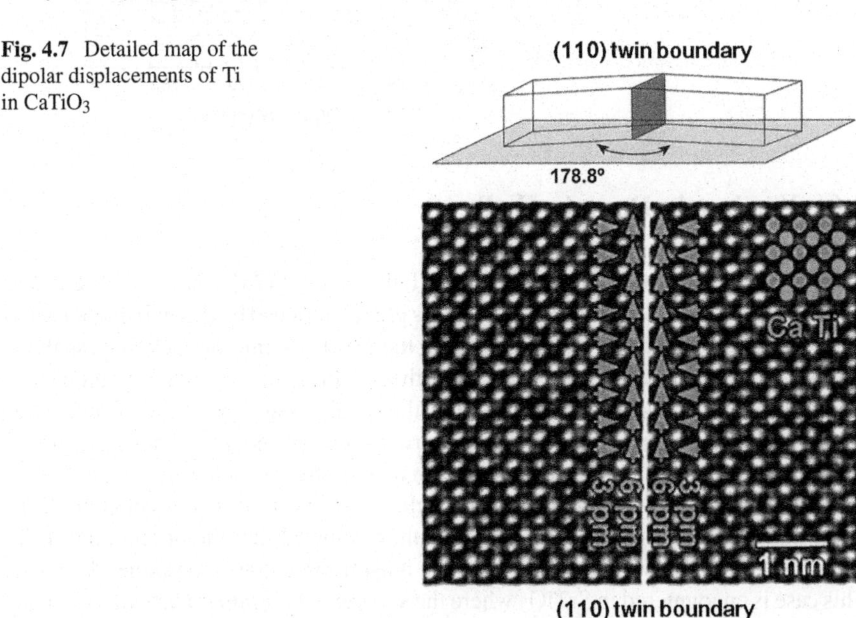

Fig. 4.7 Detailed map of the dipolar displacements of Ti in $CaTiO_3$

the piezoelectric response [72]. Nevertheless, such contribution may be small. Furthermore, coherent defect-related piezoelectric coupling is expected to increase with decreasing temperatures when correlations between defect dipoles increase. Considering that domain walls are pinned in $CaTiO_3$ [69] and no evidence of ferroelectric vortices have been obtained, one may conclude that the direction of the dipole oscillations is probably perpendicular to the direction of dipoles inside the twin walls, hence giving rise to the weak piezoelectric effect by dipolar tilts rather than dipole inversions.

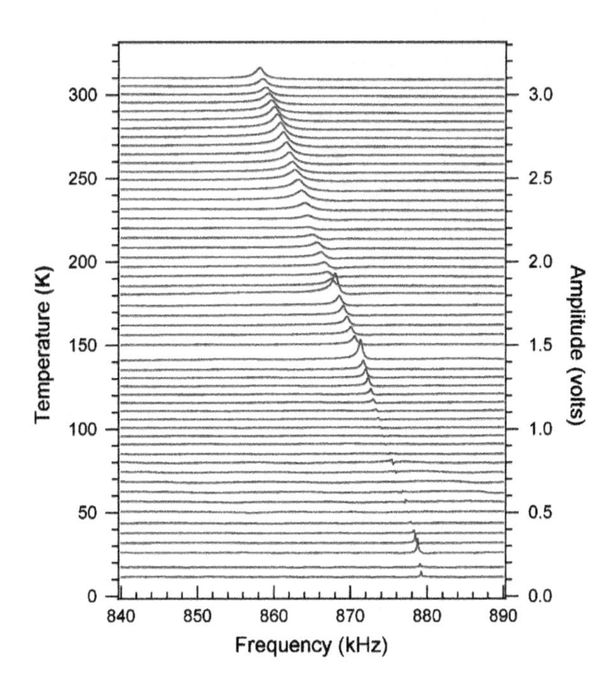

Fig. 4.8 Resonant piezoelectric spectra of a single crystal $CaTiO_3$ sample. An AC electric field of 50 V/mm was applied to obtain the spectra. The spectra are *vertically* translated so that the *left* axis is labelled as temperature. The *right* axis gives amplitude (volts)

4.6 Vortices in Domain Walls

Vortex structures are well known to exist in bulk material [73]. Flicker vortex states, where the lifetime of each individual vortex is greatly reduced by thermal fluctuations, were reported by Zhao et al. [61]. Despite the short lifetime of each vortex, these authors reported that the vortex density of the sample remains constant and can be modulated with weak electric fields. It is the local instability of the vortex state, which makes it possible to easily switch vortices locally and also change the global vortex density. These bulk effects are mirrored by domain wall vortices even if these vortices do not exist in the bulk. The first clear indication of this effect came from molecular dynamics studies [74]. These authors showed that the orientation of the polar vectors in the wall could invert if the anisotropy energy is sufficiently small. This case is encountered in $SrTiO_3$ where the spontaneous strain is particularly small. At low temperatures ferroelectric behavior was found experimentally [71, 75].

The simulated polar pattern in the domain walls is shown in Fig. 4.9, which shows the rotation of the dipoles in the wall. The rotation leads to Bloch like states in the wall where the dipole is oriented perpendicular to the wall. Interestingly, this result is almost identical to previous analytical predictions [42, 64].

Usually polarity in twin walls is ferri-electric, whereas vortex excitations lead to true ferroelectricity on a very local scale. As a result, in-plane electric fields can selectively stabilize one of the vortex polarization states and enhance the ability of the walls to move. This behavior can explain the well-known and uniquely high

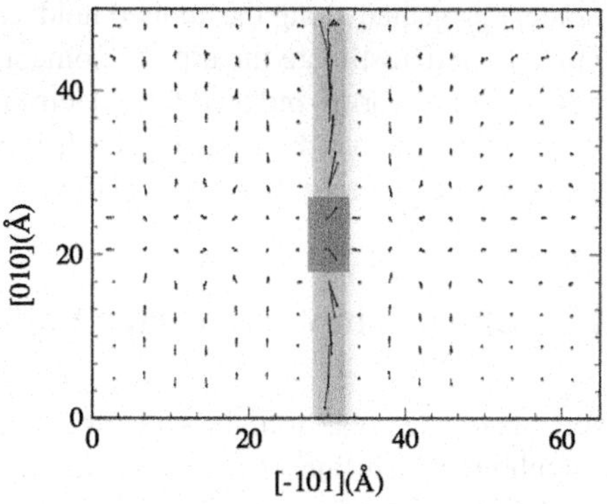

Fig. 4.9 Vortex structure of Ti-displacements at an inversion point inside a twin wall in $SrTiO_3$

Fig. 4.10 Slabs of SrTiO3 with domain boundaries. Each domain boundary contains a twin wall and a polar vortex. Electric fields will switch the vortex state from *left* to *right*, which relates to the electronic states 1 and 0. In addition, the vortex can shift (shown for the two 1 states) which offers an additional degree of freedom for memory functionalities

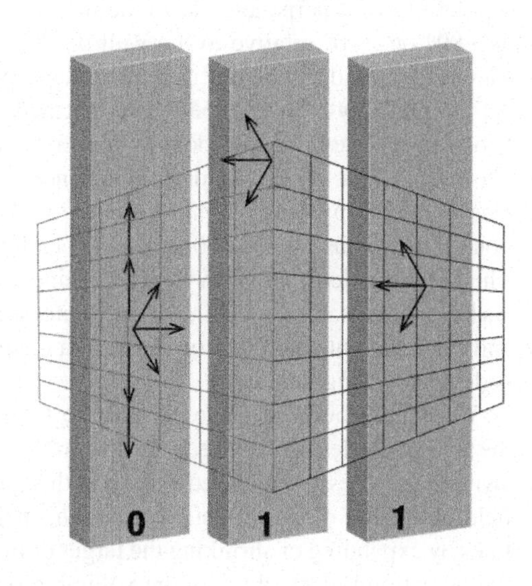

mobility of twin boundaries in $SrTiO_3$. For nanoscale ferroelectric memory devices, one envisages a precisely controllable device, where a desired domain wall pattern is manipulated by shifting the vortex position electrically. Molecular dynamics simulations mimic this situation where a constant electric field acted as an external force individually applied to the charged atoms or dipoles along the direction of the field. To characterize the local response of the vortex polarization to the field, Zykova-Timan and Salje gradually tuned the initial experimental input to the specifications of the atomistic model (e.g. dielectric permittivity of $SrTiO_3$, system size, etc.). The application of high electric fields above 0.007 V/A° along [-101] induced a rotation of the Ti-dipoles near the twin boundary and in the bulk structure. Thus, the vortex deformed and became unstable. In the range from −0.004 to 0.004 V/A° the twin boundary stabilized and a spontaneous switchable polarization of the vortex, aligned with the field, dominated over the initial state. Figure 4.9 shows the microscopic pattern of Ti local displacements inside the oxygen cage in a starting configuration, obtained from steepest descent minimization at 20 K and zero field. The initial dipole moment projected along the applied field axis is estimated as ~0.05 eA°, which may typically underestimate the dipole moment but nevertheless represents the physical mechanism of vortex-induced ferroelectricity inside twin walls. Figure 4.10 shows a conjectured memory device where each domain boundary carries one or more vortex states so that the mixing of these states produces a very high density of logical states.

4.7 Bloch Lines and Vortex Points

Vortices inside polar domain walls in ferroelastic materials can form ordered arrays resembling Bloch-lines in magnets. The Bloch lines are energetically degenerate with dipoles oriented perpendicular to the wall. By symmetry, these dipoles are oriented at $+90^{\circ}$ or -90° relative to the wall dipoles (Fig. 4.9). These two states have the same energy and can be inverted by modest applied electric fields, as argued above. As the majority of wall dipoles are oriented inside the wall, perpendicular to the Bloch line vortex, weak depolarization fields exist for the wall dipoles but not for Bloch lines. The Block line density depends on the density of the twin walls and the elastic anisotropy of the crystal structure. We estimate that distances between twin boundaries are as small as 50 nm and Bloch lines can form with very high densities. The local dipole moment in the Bloch line is similar to the displacement of Ti in $BaTiO_3$. Switchable Bloch lines can be detected by their macroscopic dipole moment and can constitute the functional part of a memory device. The topological situation is depicted in Fig. 4.11.

We depict the Bloch lines inside the twin plane (-101) in Fig. 4.11. We mark the two possible dipole directions in [-101] as red and along [10-1] as blue. Two switching processes can be envisaged in this geometry. The first is to apply a vertical field along the dipole direction up down in Fig. 4.9. This field will move Bloch lines by expanding or shrinking the larger domains. The macroscopic polarization in this direction reflects directly the volume proportion of the up and down domains.

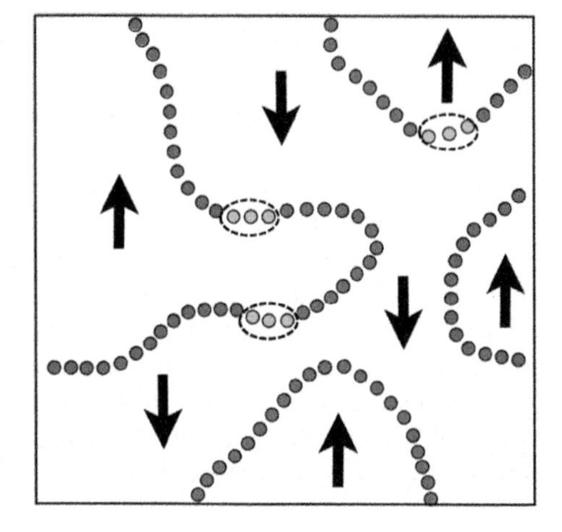

Fig. 4.11 Twin plane with local dipole moments in the *up* and *down* direction. Bloch lines limit the domains within the twin wall where the dipole direction is either out of the plane (*red*) or into the plane (*blue*). Bloch vortex points, were two Bloch *lines* with different polarity join, are indicated in *green*

Depending on the geometrical distribution of Bloch lines, we expect the changes of the macroscopic polarization to be continuous or abrupt (the latter happens when Bloch lines jam [57] similar to jamming of ferroelastic domains).

The second switching mechanism relates to fields perpendicular to the twin plane. In this case there is little interaction with the dipoles in each subdomain besides some weak dipole canting. The field couples directly with the dipoles which are orthogonal to the twin plane (red and blue in Fig. 4.11). This may happen by switching dipoles in segments of Bloch lines into the opposite directions. This local switching does not involve the shift of Bloch points, as indicated are green segments in Fig. 4.11 and is hence fast. Simulations of complex ferroelastic structures found switching times only slightly longer than the phonon times [76, 77]. A much slower movement can be envisaged when the Bloch points shift and thereby change the ratio of red and blue domains inside the Bloch walls. This sliding of Bloch points is similar to the sliding of kink excitations in ferroelastic materials and requires ca. 50 phonon times to pass a diameter of 10 nm [78]. The total bit density of such a Bloch-line ferroelectric vertex memory can be very high: We estimate that distances between twin boundaries are as small as 50 nm and that Bloch lines can form with some with densities of 100 Bloch lines in an area of 100×100 nm^2, giving a bit density of 1016/m^2. This may be compared with a maximum bit density of 1.0 Tbit/sq.in. (the unusual mixed units favored in the magnetic memory industry), and it offers potentially one order of magnitude higher density. The problems of access remain; however, these can be addressed by nano-lithography with e-beam writing. A separate question is one of write speed, but in the present case this is not limited by domain wall mobility, which in the low-field regime is ca. 1 nm/s [79, 80]. Here only local polarization switches with no transport; the situation is analogous to that of 90-degree ferroelectric switching without conventional domain wall motion.

4.8 Conclusion

Ferroelastic twin walls are—in specific materials—templates for functional properties such as superconductivity, ferroelectricity, and magnetism. These properties can be distributed homogeneously throughout the twin wall producing two-dimensional device devices. Additional singularities of these dipolar arrays can lead to lower dimensional subspaces, which may also contain ferroic functionalities. Examples are Bloch lines of vortices inside walls which contain vectors perpendicular to the wall. The two orientations, positive and negative, are symmetry-equivalent. Conjugated fields can switch these vectors, either electric dipoles or spins, leading potentially to a ferroic hysteresis. Vortices inside the twin walls will usually form Bloch lines with positive and negative polarity. These Bloch lines represent one-dimensional topological elements whereby each Bloch line is ferroic. When two Bloch lines meet they form vortex points where the polar ferroic properties disappear and where large strain fields are generated.

Acknowledgments EKHS is grateful for support by EPSRC (EP/K009702/1) and the Leverhulme Trust (RPG-2012-564).

References

1. E.K.H. Salje, *Phase Transitions in Ferroelastic and Co-elastic Crystals* (Cambridge University Press, Cambridge, 1993)
2. K. Bhattacharya, *Microstructure of Martensite: Why It Forms and How It Gives Rise to Shape Memory Effects* (Oxford University Press, Oxford, 2004)
3. A.S. Sidorkin, *Domain Structure in Ferroelectrics and Related Materials* (Cambridge International Science Publishing, Cambridge, 2006)
4. M. Lines, A. Glass, *Principles and Applications of Ferroelectrics and Related Materials* (Clarendon Press, Oxford, 1979)
5. A.G. Khachaturyan, *The Theory of Structural Transformations in Solids* (Wiley, New York, 1983)
6. J.F. Scott, *Ferroelectric Memories* (Springer, Berlin, 2000)
7. K.M. Rabe, J.M. Triscone, C.H. Ahn, *Physics of Ferroelectrics: A Modern Perspective* (Springer, Berlin, 2007)
8. D. Jiles, *Introduction to Magnetism and Magnetic Materials* (Chapman & Hall, London, 1998)
9. B.D. Cullity, C.D. Graham, *Introduction to Magnetic Materials*, 2nd edn. (Wiley, New York, 2008)
10. E. Pavarini, E. Koch, U. Schollwock, *Emergent Phenomena in Correlated Matter* (Forschungszentrum Julich, Julich, 2013)
11. G. Catalan, J. Seidel, R. Ramesh, J.F. Scott, Domain wall nanoelectronics. Rev. Mod. Phys. **84**, 119 (2012)
12. M. Dawber, K.M. Rabe, J.F. Scott, Physics of thin-film ferroelectric oxides. Rev. Mod. Phys. **77**, 1083 (2005)
13. P. Weiss, The variation of ferromagnetism with temperature. Comptes Rendus **143**, 1136 (1906)
14. E.K.H. Salje, Multiferroic domain boundaries as active memory devices: trajectories towards domain boundary engineering. Chemphyschem **11**, 940 (2010)
15. W. Eerenstein, N.D. Mathur, J.F. Scott, Multiferroic and magnetoelectric materials. Nature **442**, 759 (2006)
16. N. Spaldin, M. Fiebig, The renaissance of magnetoelectric multiferroics. Science **309**, 391 (2005)
17. M. Fiebig, Revival of the magnetoelectric effect. J. Phys. D Appl. Phys. **38**, R123 (2005)
18. A. Kadomtseva, Y. Popov, A. Pyatakov, G. Vorob'ev, A.K. Zvezdin, D. Viehland, Phase transitions in multiferroic BiFeO3 crystals, thin-layers, and ceramics: enduring potential for a single phase, room-temperature magnetoelectric 'holy grail'. Phase Transit. **79**, 1019 (2006)
19. E.K.H. Salje, Ferroelastic materials. Annu. Rev. Mater. Res. **42**, 265 (2012)
20. E.K.H. Salje, X. Ding, O. Aktas, Domain Glass. Physica Status Solidi B **251**, 2061 (2014)
21. D.D. Viehland, E.K.H. Salje, Domain boundary-dominated systems: adaptive structures and functional twin boundaries. Adv. Phys. **63**(4) (2014)
22. G.S., Brady, H.R., Clauser, J.A., Vaccari, Materials Handbook 15th (ed.). McGraw-Hill Professional, p. 633 (2002). ISBN 978-0-07-136076-0
23. A. Pelton, S. Russell, J. DiCello, The physical metallurgy of nitinol for medical applications. JOM **55**, 33 (2003)
24. T.M. Mereau, T.C. Ford, Nitinol compression staples for bone fixation in foot surgery. J. Am. Podiatr. Med. Assoc. **96**, 102 (2006)
25. K. Otsuka, C.M. Wayman, *Shape Memory Materials* (Cambridge University Press, Cambridge, 1999)

26. A. Aird, E.K.H. Salje, Sheet superconductivity in twin walls: experimental evidence of WO3-x. J. Phys. Condens. Matter **10**, L377 (1998)
27. A. Aird, M.C. Domeneghetti, F. Mazzi, V. Tazzoli, E.K.H. Salje, Sheet superconductivity in WO3-x: crystal structure of the tetragonal matrix. J. Phys. Condens. Matter **10**, L569 (1998)
28. E. Salje, H. Zhang, Domain boundary engineering. Ph. Transit. **82**, 452 (2009)
29. Y. Kim, M. Alexe, E.K.H. Salje, Nanoscale properties of thin twin walls and surface layers in piezoelectric WO3-x. Appl. Phys. Lett. **96**, 032904 (2010)
30. S. Reich, G. Leitus, R. Popovitz-Biro, A. Goldbourt, S. Vega, J. Supercond. Nov. Magn. **22**, 343 (2009)
31. B.O. Loopstra, P. Boldrini, Neutron diffraction investigation of WO3. Acta Crystallogr. **21**, 15 (1966)
32. E.K.H. Salje, S. Rehmann, F. Pobell, D. Morris, K.S. Knight, R. Herrmanndorfer, M. Dove, Crystal structure and paramagnetic behaviour of epsilon-WO3-x. J. Phys. Condens. Matter **9**, 6563 (1997)
33. R. Diehl, G. Brandt, E.K.H. Salje, Crystal Structure of triclinic WO3. Acta Crystallogr. B **34**, 1105 (1978)
34. E. Salje, K. Viswanathan, Physical properties and phase transitions in WO3. Acta Crystallogr. A **31**, 356 (1975)
35. K.R. Locherer, I.P. Swainson, E.K.H. Salje, Phase transitions in tungsten trioxide at high temperatures—a new look. J. Phys. Condens. Matter **11**, 6737 (1999)
36. E. Salje, A.F. Carley, M. Roberts, Effect of reduction and temperature on the electronic core levels of tungsten and molybdenum in WO3 and WxMo1-xO3—photoelectron spectroscopic study. J. Solid State Chem. **29**, 237 (1979)
37. O.F. Schimer, E. Salje, Conduction bipolarons in low-temperature crystalline WO3-x. J. Phys. Condens. Matter **13**, 1067 (1980)
38. O.F. Schirmer, E. Salje, W5+ polarons in crystalline low-temperature WO3 electron-spin-resonance and optical absorption. Solid State Commun. **33**, 333 (1980)
39. H. Al-Kandari, E. Al-Kharafi, N. Al-Awadi, O.M. El-Dusouqui, A. Katrib, Surface electronic structure—catalytic activity relationship of partially reduced WO3 bulk or deposited on TiO2. J. Electron Spectrosc. Relat. Phenom. **151**, 128 (2006)
40. J. Novak, E.K.H. Salje, Surface structure of domain walls. J. Phys. Condens. Matter **10**, L359 (1998)
41. J. Novak, E.K.H. Salje, Simulated mesoscopic structures of a domain wall in a ferroelastic lattice. Eur. Phys. J. B **4**, 279 (1998)
42. S. Conti, U. Weikard, Interaction between free boundaries and domain walls in ferroelastics. Eur. Phys. J. B **41**, 413 (2004)
43. Y. Ishibashi, M. Iwata, E. Salje, Polarization reversals in the presence of 90 degrees domain walls. Jpn. J. Appl. Phys. Part 1 **44**, 7512 (2005)
44. Y. Ishibashi, E. Salje, A theory of ferroelectric 90 degree domain wall. J. Phys. Soc. Jpn. **71**, 2800 (2002)
45. D.C. Palmer, E.K.H. Salje, W.W. Schmahl, Phase transitions in leucite- x-ray diffraction studies. Phys. Chem. Miner. **16**, 714 (1989)
46. E. Salje, Thermodynamics of plagioclases. 1: theory of the I bar 1—P bar 1 phase transition in anorthite and ca-rich palgioclases. Phys. Chem. Miner. **14**, 181 (1987)
47. E. Salje, V. Devarajan, U. Bismayer, Phase transitions in Pb3(P1-xAsxO4)2—influence of the central peak and flip mode on the Raman-scattering of hard modes. J. Phys. C **16**, 5233 (1983)
48. E.K.H. Salje, U. Bismayer, Hard mode spectroscopy: the concept and applications. Conference: Workshop on the State-of-the-Art of Hard Mode Spectroscopy, Lyon, France, Sept 26–27 1996. Ph. Transit. **63**, 1 (1997)
49. E.K.H. Salje, Hard mode spectroscopy—experimental studies of structural phase–transitions. Ph. Transit. **37**, 83 (1992)
50. E.K.H. Salje, A. Buckley, G. Van Tendeloo, Y. Ishibashi, G.L. Nord, Needle twins and right-angled twins in minerals: comparison between experiment and theory. Am. Mineral. **83**, 811 (1998)

51. E.K.H. Salje, Y. Ishibashi, Mesoscopic structures in ferroelastic crystals: needle twins and right-angled domains. J. Phys. Condens. Matter **8**, 8477 (1996)
52. J. Chrosch, E.K.H. Salje, Near-surface domain structures in uniaxially stressed SrTiO3. J. Phys. Condens. Matter **10**, 2817 (1998)
53. M.A. Carpenter, Mechanisms and kinetics of Al-Si ordering in anorthite: 1. incommensurate structure and domain coarsening. Am. Mineral. **76**, 1120 (1991)
54. B. Noheda, D.E. Cox, G. Shirane, J.A. Gonzalo, L.E. Cross, S.-E. Park, A monoclinic ferroelectric phase in the Pb(Zr1-xTix)O-3 solid solution. Appl. Phys. Lett. **74**, 2059 (1999)
55. B. Noheda, D.E. Cox, G. Shirane, J. Gao, Z.-G. Ye, Phase diagram of the ferroelectric relaxor (1-x)PbMg1/3Nb2/3O3-xPbTiO(3). Phys. Rev. B **66**, 054104 (2002)
56. B. Noheda, Structure and high-piezoelectricity in lead oxide solid solutions. Curr. Opin. Solid State Mater. Sci. **6**, 27 (2002)
57. E.K.H. Salje, X. Ding, Z. Zhao, T. Lookman, A. Saxena, Thermally activated avalanches: jamming and the progression of needle domains. Phys. Rev. B **83**, 104109 (2011)
58. X. Ding, Z. Zhao, T. Lookman, E.K.H. Salje, High junction and twin boundary densities in driven dynamical systems. Adv. Mater. **24**, 5385 (2012)
59. S. Li, X. Ding, J. Ren, X. Moya, J. Li, J. Sun, E.K.H. Salje, Strain-controlled thermal conductivity in ferroic twinned films. Sci. Rep. **4**, 6375 (2014)
60. E.K.H. Salje, X. Ding, Z. Zhao, Noise and finite size effects in multiferroics with strong elastic interactions. Appl. Phys. Lett. **102**, 152909 (2013)
61. Z. Zhao, X. Ding, E.K.H. Salje, Flicker vortex structures in multiferroic materials. Appl. Phys. Lett. **105**, 112906 (2014)
62. W.T. Lee, E.K.H. Salje, U. Bismayer, Domain-wall structure and domain-wall strain. J. Appl. Phys. **93**, 9890 (2003)
63. W.T. Lee, E.K.H. Salje, U. Bismayer, Domain wall diffusion and domain wall softening. J. Phys.: Condens. Matter **15**, 1353 (2003)
64. B. Houchmandzadeh, J. Lajzerowicz, E.K.H. Salje, Order prameter coupling and chirality of domain-walls. J. Phys. Condens. Matter **4**, 9779 (1992)
65. J. Hong, D. Vanderbilt, First-principles theory and calculation of flexoelectricity. Phys. Rev. B **88**, 174107 (2013)
66. A.K. Tagantsev, A.S. Yurkov, Flexoelectric effect in finite samples. J. Appl. Phys. **112**, 044103 (2012)
67. E.A. Eliseev, A.N. Morozovska, Y. Gu, A. Borisevich, L.-Q. Chen, V. Gopalan, S.V. Kalinin, Conductivity of twin-domain-wall/surface junctions in ferroelastics: interplay of deformation potential, octahedral rotations, improper ferroelectricity, and flexoelectric coupling. Phys. Rev. B **86**, 085416 (2012)
68. L. Goncalves-Ferreira, S.A.T. Redfern, E. Artacho, E.K.H. Salje, Ferrielectric twin walls in CaTiO(3). Phys. Rev. Lett. **101**, 097602 (2008)
69. S. Van Aert, S. Turner, R. Delville, D. Schryvers, G. Van Tendeloo, E.K.H. Salje, Direct observation of ferrielectricity at ferroelastic domain boundaries in CaTiO3 by electron microscopy. Adv. Mater. **24**, 523 (2012)
70. H. Yokota, H. Usami, R. Haumont, P. Hicher, J. Kaneshiro, E.K.H. Salje, Y. Uesu, Direct evidence of polar nature of ferroelastic twin boundaries in CaTiO3 obtained by second harmonic generation microscope. Phys. Rev. B **89**, 144109 (2014)
71. E.K.H. Salje, O. Aktas, M.A. Carpenter, J.F. Scott, Domains within domains and walls within walls: evidence for polar domains in cryogenic SrTiO3. Phys. Rev. Lett. **111**, 247603 (2013)
72. O. Aktas, S. Crossley, M.A. Carpenter, E.K.H. Salje, Polar correlations and defect-induced ferroelectricity in cryogenic KTaO3. Phys. Rev. B **90**, 165309 (2014)
73. S. Prosandeev, A. Malashevich, Z. Gui, L. Louis, R. Walter, I. Souza, L. Bellaiche, Natural optical activity and its control by electric field in electrotoroidic systems. Phys. Rev. B **87**, 195111 (2013)
74. T. Zykova-Timan, E.K.H. Salje, Highly mobile vortex structures inside polar twin boundaries in SrTiO3. Appl. Phys. Lett. **104**, 082907 (2014)

75. J.F. Scott, E.K.H. Salje, M.A. Carpenter, Domain wall damping and elastic softening in SrTiO3: evidence for polar twin walls. Phys. Rev. Lett. **109**, 187601 (2012)
76. E.K.H. Salje, Z. Zhao, X. Ding, J. Sun, Mechanical spectroscopy in twinned minerals: simulation of resonance patterns at high frequencies. Am. Mineral. **98**, 1449 (2013)
77. L. Zhang, E.K.H. Salje, X. Ding, J. Sun, Strain rate dependence of twinning avalanches at high speed impact. Appl. Phys. Lett. **104**, 162906 (2014)
78. E.K.H. Salje, X. Wang, X. Ding, J. Sun, Simulating acoustic emission: The noise of collapsing domains. Phys. Rev. B **90**, 064103 (2014)
79. T. Tybell, P. Paruch, T. Giamarchi, J.M. Triscone, Domain wall creep in epitaxial ferroelectric Pb(Zr0.2Ti0.8)O_3 thin films. Phys. Rev. Lett. **89**, 097601 (2002)
80. R. McQuaid, A. Gruverman, J.F. Scott, J.M. Gregg, Self-similar nested flux closure structures in a tetragonal ferroelectric. Nano Lett. **14**, 4230 (2014)

Chapter 5
Charged Domain Walls in Ferroelectrics

Tomas Sluka, Petr Bednyakov, Petr Yudin, Arnaud Crassous and Alexander Tagantsev

Abstract Charged Domain Walls (CDWs) in ferroelectrics are compositionally homogeneous interfaces, some of which display metallic-like conductivity and can be created, displaced and erased inside a monolith of nominally insulating materials. Such CDWs are promising electronic elements for reconfigurable nanoelectronics. This chapter introduces types of CDWs, their theoretically predicted and experimentally observed properties, and methods of their artificial engineering.

5.1 Introduction

Charged Domain Walls (CDWs) are a subset of compositionally homogeneous interfaces in materials with the ferroelectric order parameter. Unlike electrically Neutral Domain Walls (NDWs), which are more extensively studied, CDWs contain substantial amount of bound charge due to Head-to-head convergence or Tail-to-Tail divergence of spontaneous polarization at the wall. The peculiar feature of CDWs, predicted in theory already in 1970's [1], is that it is possible to compensate the bound charge with free electrons or holes. This makes CDWs the first example of a movable electronically active homo-interfaces inside a monolith of nominally nonconducting material. However, the presence of bound charge makes CDWs energetically costly and unstable objects (with exceptions) which explains why they are rarely found in ferroelectric crystals, ceramics and thin films. Although CDWs were occasionally documented in the past [2–5] and their elevated conductivity was reported in 1970's [6] signatures of free carrier gas-like conductivity of CDWs were not reported until 2012 [7]. Indeed, electronically compensated CDWs were often assumed as impossible for several reasons. CDWs in typical ferroelectrics cannot exist

T. Sluka (✉) · P. Bednyakov · P. Yudin · A. Crassous · A. Tagantsev
Ceramics Laboratory, Swiss Federal Institute of Technology (EPFL),
CH-1015 Lausanne, Switzerland
e-mail: tomas.sluka@epfl.ch

T. Sluka
Department of Quantum Matter Physics, University of Geneva, CH-1211
Geneva, Switzerland

J. Seidel (ed.), *Topological Structures in Ferroic Materials*, Springer Series
in Materials Science 228, DOI 10.1007/978-3-319-25301-5_5

without almost complete compensation. It means that a CDW requires delivery or prior presence of compensating charge during its formation. Dielectrics however contain almost no free carriers and domain walls have mostly an option to avoid charging—acquire a neutral configuration. Additionally, bound charge can be compensated by mobile charged defects instead of free carriers.

The flood of research in 2000's focused on oxide interfaces [8, 9] revived the interest in electronically compensated CDWs as potential hardware reconfigurable conducting paths. First direct observation of CDWs with high resolution transmission electron microscopy (TEM) was documented in ultra thin film of $Pb(Zr_{0.2}Ti_{0.8})O_3$ [10] and later in thin film of $BiFeO_3$ [11, 12]. As of 2015, additional optical, electron and piezoresponse force microscopy (PFM) evidence of CDW is available for $Pb(Zr_{0.2}Ti_{0.8})O_3$ [7, 13, 14], $PbTiO_3$ [15] and $BiFeO_3$ [12, 16–18] thin films, $BaTiO_3$ [19–21] and $LiNbO_3$ [22–24] crystals, and also in crystals of improper ferroelectric manganites [25–27], hybrid improper ferroelectric $(Ca,Sr)_3Ti_2O_7$ [28] and in organic ferroelectrics [29].

Revived analytical models, phase field simulations and first principle calculations detailed the electronic properties of CDWs and continued insisting on the possible presence of highly conducting electron or hole gas at CDWs. First models dealt with 180° CDWs [30–33] which regularly appear e.g. during domain nucleation. Indeed, the predicted metallic type conductance of CDWs was first experimentally observed at transient domain walls of nucleating nanodomains in $Pb(Zr_{0.2}Ti_{0.8})O_3$ thin film [7]. The mechanism of CDW stabilization by mechanical (ferroelastic) clamping was later proposed [34] primarily as an explanation of enhanced electromechanical response in $BaTiO_3$ crystals [19] but, later, it was shown that an artificially engineered positively charged CDWs display steady metallic-type conductivity which exceeds the thermally activated bulk conductivity by up to nine orders of magnitude [20]. It was suggested [7, 20] that the free electrons which compensate the positively charged CDWs originate from shallow levels of oxygen vacancies. The ionised oxygen vacancies, on the other hand, become attracted to the negatively charged CDWs which appear non-conducting. Engineering of CDW patterns with controlled density was later developed in $BaTiO_3$ crystals and mechanisms of charge compensation were discussed [21]. Creation and nanometer scale manipulation with individual stable CDWs was demonstrated in technologically more viable $BiFeO_3$ thin films [18]. Here, the CDWs with metallic-type conductivity were compensated by electrons injected from an AFM tip. This work also suggested that CDWs might be exploitable as a reconfigurable quasi-dopant in wide band-gap ferroelectrics. Several other specific properties of CDWs such as their low mobility [35] and possibly anomalously large thickness in the proximity of a phase transition of the parent material [36] were theoretically predicted. Interestingly, nominally neutral DWs that are partly charged due to film $(Pb(Zr_{0.1}Ti_{0.9})O_3)$/substrate ($DyScO_3$) mechanical interaction also display a non-thermally activated conductivity in the range from room temperature down to 4 K [37].

A specific category of fundamentally stable CDWs was found in improper ferroelectric manganites where their elevated conductivity was documented [26, 27]. Similarly, conductive CDWs were found in a hybrid improper ferroelectric

$(Ca,Sr)_3Ti_2O_7$ [28]. First principle calculation predicted a substantial bandgap narrowing at CDW in ferroelectric halides [38].

This chapter introduces in detail the theoretically predicted and experimentally observed properties of CDWs, and methods of their artificial engineering. Note, that the term ferroelectric in the following text refers to proper ferroelectrics unless introduced otherwise.

5.2 Classification of Charged Domain Walls

To define CDWs and to understand their uniqueness among Domain Walls (DWs) let us start with a "naive picture" of ferroelectrics. Ferroelectric materials consist of domains that are spontaneously polarized in one of their symmetry-permitted directions, i.e. different domain states have differently displaced ions from the centrosymmetric position as illustrated with the example of the perovskite tetragonal unit cell in Fig. 5.1a. The switchable part of the ionic displacements is described by the vector of spontaneous polarization **P** and is exhibited by the appearance of polarization charge at the polar surfaces of each domain as illustrated in Fig. 5.1b. If a domain is terminated by the surface of the ferroelectric material, the polarization charge is usually compensated with free electrons of an electrode, or by adsorbates from the outer environment together with accumulated intrinsic mobile defects and free carriers from the domain interior (illustrated by the full circles in Fig. 5.1).

However, when two domains border directly with each other they are separated by a DW, i.e. typically 1–10 nm thick compositionally homogeneous and movable transition region. As most ferroelectrics are wide band-gap materials they contain only negligible amount of intrinsic free carriers (unless they are heavily doped) and possibly slow charged defects which cannot rapidly compensate the polarization charge. DWs are therefore forced by the electrostatic forces to acquire an orientation which minimizes—ideally to zero—the total polarization charge at their location. This occurs either when polarization projection to the DW plane is zero (like in the case of anti-parallel 180° DW Fig. 5.1c) or when polarization vector keeps Head-to-Tail continuity across the DW, i.e. when the normal component of the polarization with respect to DW does not change. The latter means that the polarization charge "peeping" from one domain is almost perfectly compensated by the polarization charge at the surface of the adjacent domain (Fig. 5.1d). The neutrality of a DW is called the *condition of electrostatic compatibility*. It is met by the vast majority of DWs in proper[1] ferroelectrics which are, thus, called Neutral Domain Walls (NDW).

Violation of the condition of the electrostatic compatibility, when a DW deviates from its neutral orientation, creates a nonzero net bound charge at the DW, which is the result of converging or diverging spontaneous polarization. In mathematical

[1] Proper ferroelectrics are ferroic materials where polarization is the primary order parameter and whose dielectric response follows the Curie-Weiss law. There exist also improper ferroelectrics, which is discussed in Sect. 5.2.2, where polarization is a secondary order parameter and CDWs are naturally locked objects.

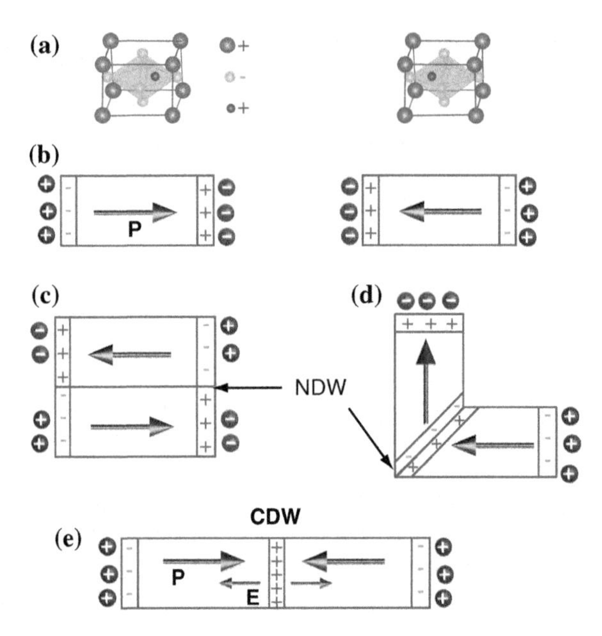

Fig. 5.1 Ferroelectric domains are spontaneously polarized in one of the symmetry-permitted directions. **a** The example of two (out of six) domain states in a perovskite tetragonal unit cell. **b** The *purple* vector of spontaneous polarization **P** represents the switchable part of the ionic displacement which results in the appearance of the polarization charge (indicated by *red* + and *green* −) on the polar surfaces of each domain. This charge is usually compensated by mobile charged species collected from the outer environment (indicated by ⊕ and ⊖). **c** Two antiparallel domains separated by an electrically neutral 180° domain wall. **d** Two domains separated by neutral non-180° ferroelastic domain wall where polarization charge is perfectly compensated. **e** Charged Head-to-Head domain wall which produces depolarizing field E

terms the volume density of this bound charge, let us call it the polarization charge ρ_P, is:

$$\nabla \cdot \mathbf{P} = -\rho_P.$$

Note that in this chapter we assume the definition of dielectric displacement $\mathbf{D} = \varepsilon_0 \varepsilon_b \mathbf{E} + \mathbf{P}$ where ε_0 and ε_b are the vacuum and background permittivity [39], respectively, $\mathbf{E}$ is the electric field and $\mathbf{P}$ is the ferroelectric part of polarization.

The more handy definition of surface density of polarization charge σ_P is the change of the polarization components normal to the DW:

$$\Delta \mathbf{P} \cdot \mathbf{n} = -\sigma_P, \tag{5.1}$$

where $\Delta \mathbf{P} = \mathbf{P}_2 - \mathbf{P}_1$ is difference between the polarizations of adjacent domains and $\mathbf{n}$ is the DW normal as illustrated in Fig. 5.2a. The basic distinction between NDWs (Fig. 5.2b) and CDWs (Fig. 5.2c) is therefore given the change of their polarization component normal to the DW:

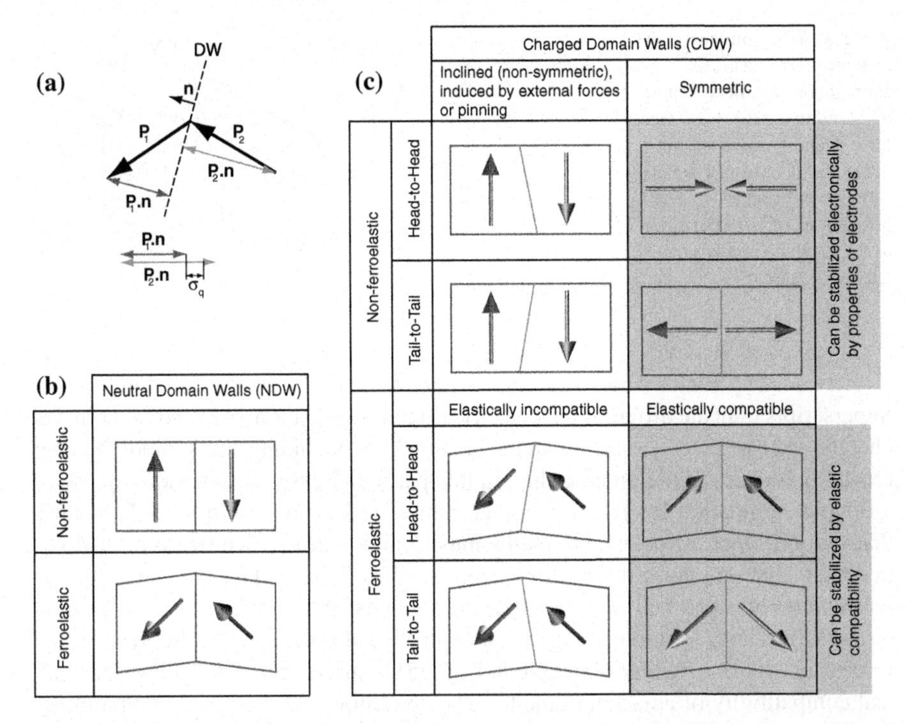

Fig. 5.2 Basic classification of neutral and charged domain walls. **a** Graphical representation of polarization charge produced by polarization divergence on a ferroelectric domain wall (DW). The difference between normal components of polarization $\mathbf{P_1} \cdot \mathbf{n}$ and $\mathbf{P_2} \cdot \mathbf{n}$, where $\mathbf{n}$ is DW normal vector, represents the surface charge density σ_P which gives the basic distinction between neutral and charged domain walls. **b** Ferroelastic and non-ferroelastic neutral domain walls. **c**, Ferroelastic and non-ferroelastic charged domains

NDW: No change of normal polarization component $\Delta\mathbf{P} \cdot \mathbf{n} = 0$,

CDW: Change of normal polarization component $\Delta\mathbf{P} \cdot \mathbf{n} \neq 0$;

Let us further distinguish between the ferroelastic and non-ferroelastic CDWs (Fig. 5.2c). The spontaneous ionic displacement in proper ferroelectrics is a source of spontaneous strain **e** which is usually identical in mutually 180° rotated domains (in proper ferroelectrics which have no piezoelectric effect above T_C [39]) but different in all other combinations. Two domains create mechanically compatible stress-free CDW when [40]:

$$(e_{ij}^{(1)} - e_{ij}^{(2)})s_i s_j = 0. \tag{5.2}$$

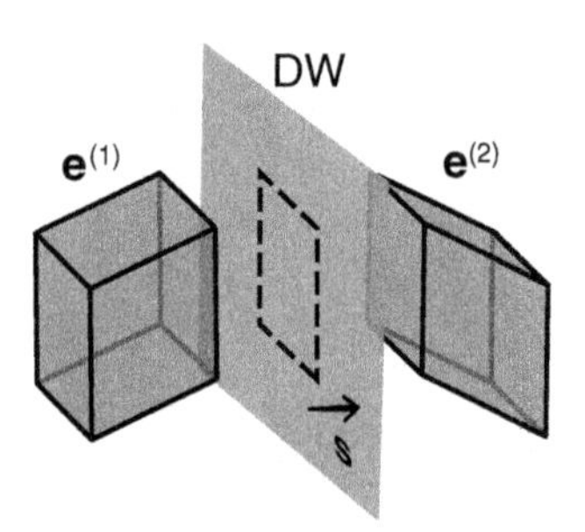

Fig. 5.3 Two domains with different ferroelastic spontaneous strains $\mathbf{e}^{(1)}$ and $\mathbf{e}^{(2)}$ tend to create stress-free ferroelastic domain wall in the plane of vectors **s** which satisfy condition (5.2). The projections of both domains to the "s-plane" have zero lattice mismatch

Here **s** is any vector which lies in the plane of the CDW as illustrated in Fig. 5.3. Superscripts indicate different ferroelectric domains. Equation (5.2) and other in this chapter are expressed in the Einstein notation (i.e. excluding the symbol $\sum$ over repeating indexes) which means that, in this particular case, the left side represents a double summation for i and j going from 1 to 3. The condition (5.2) dictates that two different ferroelastic domains must join at surfaces with zero mutual lattice mismatch otherwise the system is penalized by increased elastic energy due to tension/compression between the adjacent domains. While orientation of the non-ferroelastic CDW (example in Fig. 5.4) is controlled almost solely by the electrostatic forces, a ferroelastic CDW (example in Fig. 5.5) is strictly subjected to the mechanical compatibility of adjacent domains. The condition of mechanical compatibility therefore represents a very important factor in CDW stabilization.

$$\begin{aligned} \text{mechanically compatible CDW:} \quad & (e_{ij}^{(1)} - e_{ij}^{(2)})s_i s_j = 0, \\ \text{mechanically incopatible CDW:} \quad & (e_{ij}^{(1)} - e_{ij}^{(2)})s_i s_j \neq 0. \end{aligned}$$

Not all DWs are however ferroelastic. The non-ferroelastic DWs (like 180° DWs in perovskites) always satisfy the mechanical compatibility condition and, therefore, such CDW cannot be mechanically stabilised. The DW ferroelasticity can be identified as follows:

$$\begin{aligned} \text{non-ferroelastic CDW:} \quad & \text{identical strain in adjacent domains } \Delta\mathbf{e} = \mathbf{0}, \\ \text{ferroelastic CDW:} \quad & \text{different strain in adjacent domains } \Delta\mathbf{e} \neq \mathbf{0}. \end{aligned}$$

Another important classification of CDWs is based on their symmetry. CDWs can be inclined (non-symmetric) or mirror symmetric with respect to the spontaneous polarization (Fig. 5.2c). NDWs become CDWs when they are misaligned from their neutral orientation by inhomogeneous electric or elastic fields, defect pinning or

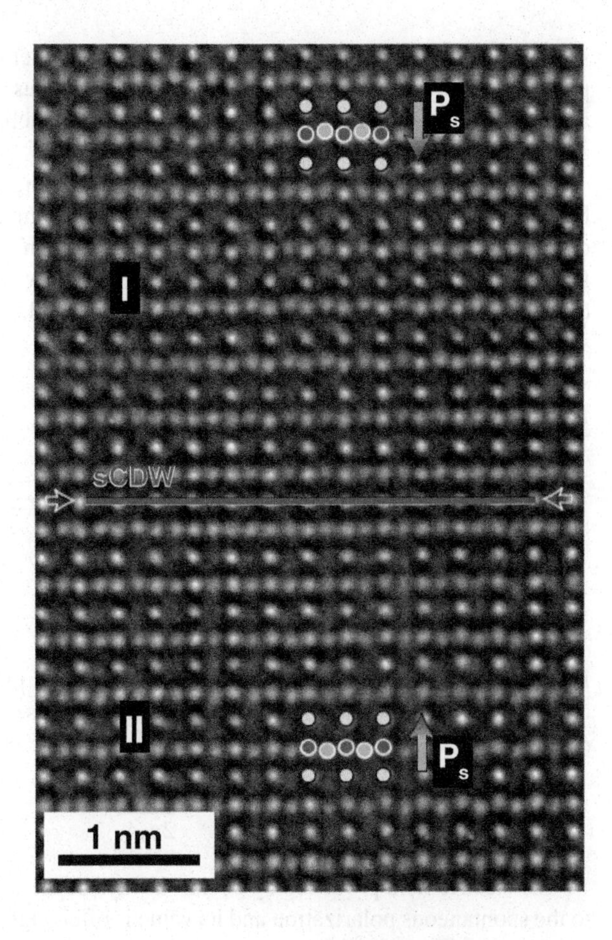

Fig. 5.4 High resolution TEM image of a symmetric non-ferroelastic Head-to-Head charged domain wall (indicated by *red line*) in an ultra thin $Pb(Zr_{0.2}Ti_{0.8})O_3$ film sandwiched between $SrTiO_3$ layers [10]

Fig. 5.5 High resolution TEM image of a symmetric ferroelastic Tail-to-Tail charged domain wall (indicated by the *red arrows*) in a $BiFeO_3$ film (BFO) on $GdScO_3$ substrate (GSO) [11]

substrate clamping. Such CDWs are formed, at least temporarily, during each 180° polarization switching because DWs at a domain nucleus are inevitably inclined from a neutral orientation (Fig. 5.6) [7, 13, 14]. The DW inclination can be stabilised by a specific film/substrate clamping [37].

The symmetric CDWs represent the most intriguing category because they can be stabilized without any external forces or defect pinning while they are strongly charged. The importance of CDW stability will become clear when we will look at ferroelectrics as ideal dielectrics and realize that the DW charging is strictly penalized or entirely prohibited due to depolarizing electric field.

5.2.1 Depolarizing Electric Field

The polarization charge ρ_P, like any charge, is a source of an electric field $\mathbf{E}$ (red arrows in Fig. 5.1e) according to:

$$\nabla \cdot \mathbf{E} = \frac{\rho_P}{\varepsilon_0 \varepsilon_b}.$$

By applying the Gauss law on an infinite flat CDW plane (like in Fig. 5.2a) one readily finds an expression for the electric field in terms of the surface charge density σ_P:

$$\mathbf{E} = \frac{\sigma_P}{2\varepsilon_0 \varepsilon_b},$$

The electric field produced by σ_P has always depolarizing direction with respect to the spontaneous polarization and its value is often higher than the thermodynamic coercive field, e.g. in typical perovskite ferroelectrics. For example, the polarization charge at symmetrical Head-to-Head CDW in materials like $Pb(Zr_xTi_{1-x})O_3$ or $BaTiO_3$ is $\sigma_P = 2P_s \sim 10^1\ \mu C/cm^2$ which produces a depolarizing field $E_{dep} = \sigma_{P_s}/(2\varepsilon_0\varepsilon_b) \sim 10^0$ MV/cm while their thermodynamic coercive field is $E_{crit} \sim 10^{-1}$ MV/cm. Note that the E_{crit} is a theoretical parameter while experimentally measured coercive field is $10^{-3} - 10^{-2}$ MV/cm [39] which is orders of magnitude lower than the estimated E_{dep}.

An even more severe argument against the existence of CDWs comes from the fact that, in the dielectric approximation, the depolarizing field induces a shift of the Curie temperature T_0 by

$$\Delta T_0 = \frac{C}{\varepsilon_b}, \tag{5.3}$$

where C is the Curie-Weiss constant. According to (5.3) in $BaTiO_3$ (with $C = 1.7 \times 10^5$ K and $\varepsilon_b = 7$) [39], T_0 would be shifted by 24,000 K which is impossible. Similar estimates apply to other perovskite ferroelectrics. It means that the

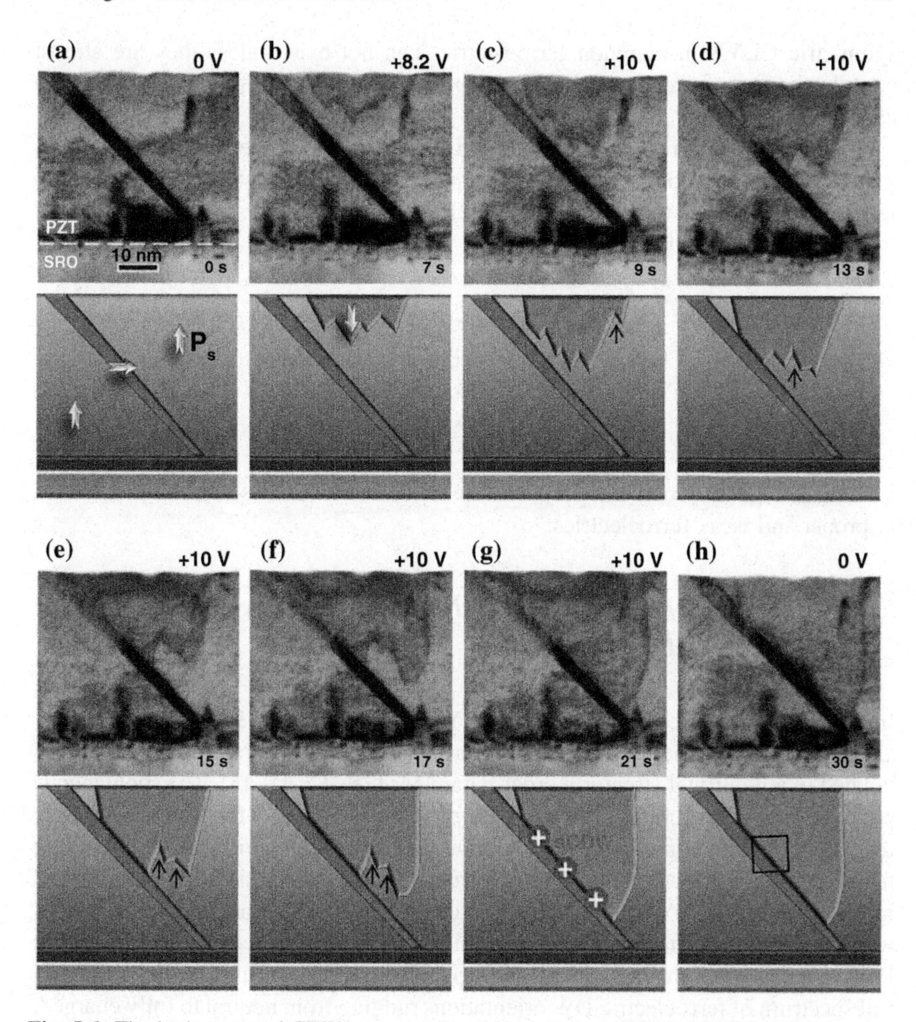

Fig. 5.6 The basic types of CDWs are seen in a chronological TEM bright field image series showing the evolution of a domain nucleus in $Pb(Zr_{0.2}Ti_{0.8})O_3$ film on $SrRuO_3$ electrode and $DyScO_3$ substrate. The domain nucleus (*blue*) is created under an AFM tip biased gradually from 0→10→ 0 V. *White arrows* indicate the spontaneous polarization. **a** The virgin structure shows a ~ 50 nm thick narrow a-domain separated from the up polarized c-domain with ferroelastic domain walls that are slightly inclined from their neutral orientation (note that the DWs are not parallel due to the film/substrate mismatch). **b–f** The down oriented domain nucleus propagates from the AFM tip into the original c-domain. The domain walls of the nucleus are non-ferroelastic, and locally almost symmetrical at its front (**f**), but affected by defect pinning (the *thin black arrows* indicate arguable pinning centers), and inclined from an antiparallel orientation on its sides. **g** 90° symmetrical ferroelastic CDW is created when nucleus reaches the a/c-domain boundary and remains stable (**h**) [14]

symmetric CDWs in common ferroelectrics can not exist unless they are almost completely compensated. At the same time, dielectrics have practically zero free carrier concentration.

Qualitatively different situation is found at the slightly inclined domain walls or in materials with exceptionally small spontaneous polarization which produces depolarizing field smaller than the coercive field. Therefore, CDWs can be categorized into additional two types:

Weakly CDW (wCDW):	can exist without any charge compensation
Strongly CDW (sCDW):	cannot exist uncompensated.

In this context special category must be dedicated to CDWs in improper, hybrid improper and weak ferroelectrics.

5.2.2 *Charged Domain Walls in Improper and Hybrid Improper Ferroelectrics*

Improper ferroelectrics like $Gd_2(MoO_4)_3$, $YMnO_3$, $ErMnO_3$ or $HoMnO_3$ are materials where the spontaneous polarization is a secondary order parameter that is subordinated to a nontrivial structural order parameter [25, 27, 41–43]. The spontaneous polarization is therefore governed by different thermodynamic forces than in proper ferroelectrics which implies two important effects on CDWs. First, the depolarizing field does not induce any shift of the ferroelectric phase transition of an improper ferroelectric. It leads only to a reduction of the spontaneous polarization modulus which, consequently, reduces the depolarizing field [39, 44]. Second, the ferroelectric DWs in manganites are interlocked with structural antiphase boundaries resulting in a full spectrum of ferroelectric DW orientations ranging from neutral to fully charged, Fig. 5.7, [25–27]. The manganites are relatively narrow band-gap materials (1.5–2 eV) with considerable intrinsic free carrier concentration at room temperature. The introduced factors contribute to the regular existence of CDWs in these materials, but also do not guarantee their electronic activity as they can exist entirely uncompensated.

Similarly, stable conducting CDWs were reported in relatively narrow bandgap so called hybrid improper ferroelectric $(Ca,Sr)_3Ti_2O_7$ [28] shown in Fig. 5.8.

5.2.3 *Charged Domain Walls in Weak Ferroelectrics*

Weak ferroelectrics are proper ferroelectrics which exhibit exceptionally small values of the Curie-Weiss constants, in the range $C \sim 3$–30 K, due to small soft-mode effective charge [45, 46]. It is clear from (5.3) that for such small C, the presence of bound charge at CDWs leads merely to a small reduction of the Curie temperature.

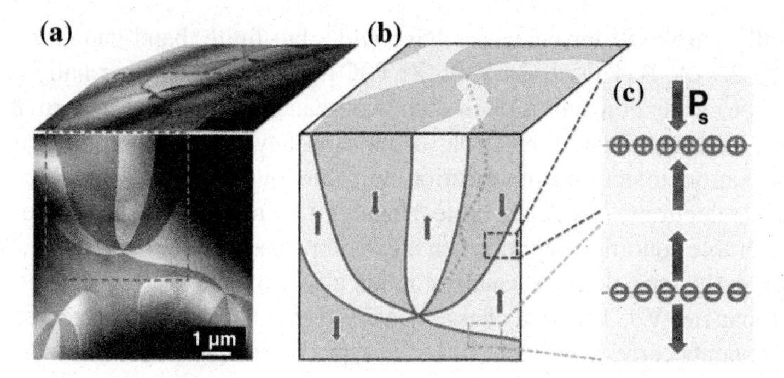

Fig. 5.7 **a** TEM dark field images of the *top* and side views of vortex domains in hexagonal improper ferroelectric $HoMnO_3$. **b** A cartoon sketch of the 3D profile of a *curved* vortex in the *boxed area* in (**a**). **c** Zoom-in cartoon of Head-to-Head and Tail-to-Tail charged domain walls [26]

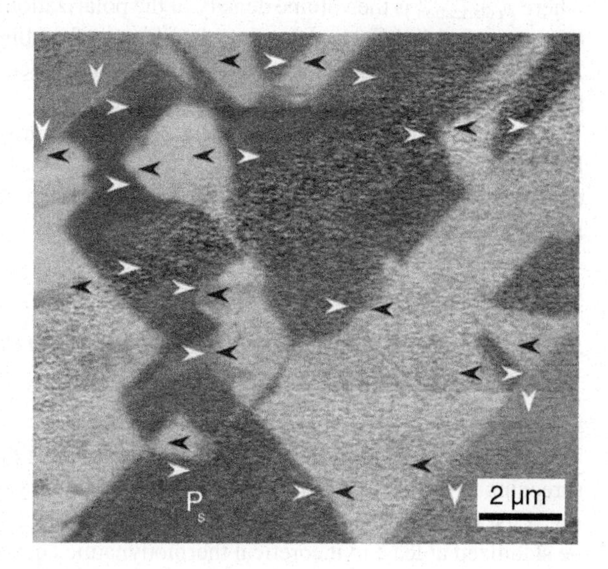

Fig. 5.8 Domain structure in a hybrid improper ferroelectric: in-plane PFM image of the (001) surface of an oxygen-deficient $Ca_{2.46}Sr_{0.54}Ti_2O_7$ crystal at room temperature. The long axis of the AFM cantilever is oriented along the vertical direction of the image. The *black* and *white arrows* indicate the local in-plane component of the polarization [28]

Thus, ferroelectricity in the domains adjacent to a CDW in a weak ferroelectric is not suppressed, except in the close vicinity of the transition temperature. However, the depolarizing field may still destabilize a CDW if its magnitude is higher than the effective coercive field.

5.3 Charged Domain Wall Screening

As it was explained above, fundamental existence of so-called sCDWs in proper ferroelectrics requires almost complete compensation of their polarization charge. Typical ferroelectrics are nominally nonconducting materials, one may however find several sources of compensating charge each associated with different phenomenon.

First of all, real world ferroelectrics have wide, but finite, band-gap (for example $BiFeO_3$: ~2.7 eV, $BaTiO_3$: ~3 eV, $Pb(Zr,Ti)O_3$: ~3.5 eV). The thermally activated intrinsic free carrier concentration in such wide-band-gap materials is virtually zero, but it does not mean that an electron (or hole) of any origin (interband excitation, defect ionisation, injection from exterior etc.) can not occupy the available states in conduction (electrons) or valence (holes) band if it is energetically favorable. Indeed, such a favorable condition occurs when they compensate a bound charge [1, 30, 34].

Additionally, ferroelectrics usually contain also movable charged defects like the oxygen vacancies $V\ddot{o}$. The total charge density ρ can therefore be conveniently split into components corresponding to different types of charge:

$$\rho = \rho_{\text{polarization}} + \rho_{\text{free}} + \rho_{\text{mobile dopants}} + \rho_{\text{immobile dopants}} + \dots, \tag{5.4}$$

where $\rho_{\text{polarization}}$ is the volume density of the polarization charge, ρ_{free} the free charge (electrons and holes), $\rho_{\text{mobile dopants}}$ the charge of mobile dopants and defects, and $\rho_{\text{imobile dopants}}$ the charge of immobile dopants and defects, etc. An analogical expression can be written for surface charge densities.

The mobile charge carriers ρ_{free} and $\rho_{\text{mobile dopants}}$ are attracted and therefore collected by the polarization charge $\rho_{\text{polarization}}$ which locally reduces the total charge density ρ and, thus, the depolarizing electric field. Let us therefore explore the screening scenarios in detail.

5.3.1 Electron-Hole Screening in the Thermodynamic Equilibrium

The most intriguing option of sCDW compensation is the pure free electron or hole screening as it results in a quasi 2D free electron/hole gas inside an insulating material. First of all let us explore whether sCDWs screened solely by free carriers can be stabilized at least in theoretical thermodynamic equilibrium. The early analytical models [1, 30, 47, 48] dealt with one dimensional systems which show clearly the possibility of free carrier compensation of polarization charge at sCDW. These models were however not fully self-consistent and dealt with one-dimensional system in which a sCDW has only one degree of freedom of motion. It means, for example, that it can not be destabilised by rotation to a neutral orientation like in multidimensional models. A general mechanism of sCDW stabilization induced by a low or a high electrode work-function was however proposed [47, 48]. Self-consistent multidimensional models require a numerical solution such as those in phase field simulations in [7, 31–33] for non-ferroelastic sCDW and in [20, 34] for ferroelastic sCDWs. The non-ferroelastic case does not include any strong inherent stabilization mechanism (except defect pinning or the suggested massive doping [33]), but it shows [7] the presence of free carriers at sCDWs. The ferroelastic CDWs, on the other hand, can be entirely stabilized by mechanical compatibility condition as proven in [18, 20, 21].

The following text introduces a self-consistent model of ferroelastic 90° sCDWs in tetragonal perovskite. The readers interested only in the conclusions may skip the model description to the last paragraphs of this section.

Ferroelectric domain structure is metastable in any local minimum of its free energy. Model equations are therefore obtained by Lagrange principle from Helmholtz free energy density [49]:

$$f[\{P_i, P_{i,j}, e_{ij}, D_i\}] = f^{(e)}_{\text{bulk}} + f_{\text{ela}} + f_{\text{es}} + f_{\text{grad}} + f_{\text{ele}}, \tag{5.5}$$

where P_i is the ferroelectric part of polarization, $P_{i,j}$ its derivatives (the subscript ‘, i’ represents the operator of spatial derivatives $\partial/\partial x_i$), D_i the electric displacement and $e_{ij} = 1/2(u_{i,j} + u_{j,i})$ is the elastic strain where u_i is a displacement vector.

The bulk free energy density

$$\begin{aligned} f^{(e)}_{\text{bulk}}[\{P_i\}] = \\ \alpha_1 \sum_i P_i^2 + \alpha^{(e)}_{11} \sum_i P_i^4 + \alpha^{(e)}_{12} \sum_{i>j} P_i^2 P_j^2 + \alpha_{111} \sum_i P_i^6 \\ +\alpha_{112} \sum_{i>j} (P_i^4 P_j^2 + P_j^4 P_i^2) + \alpha_{123} \prod_i P_i^2 \end{aligned} \tag{5.6}$$

is expressed for a zero strain as a six-order polynomial expansion [50], where α_i, $\alpha^{(e)}_{ij}$, α_{ijk} are parameters fitted to the single crystal properties. The remaining contributions represent bilinear forms of densities of the elastic energy $f_{\text{ela}}[\{e_{ij}\}] = 1/2 c_{ijkl} e_{ij} e_{kl}$, where c_{ijkl} is the elastic stiffness, electrostriction energy $f_{\text{es}}[\{P_i, e_{ij}\}] = -q_{ijkl} e_{ij} P_k P_l$, where q_{ijkl} are the electrostriction coefficients, gradient energy $f_{\text{grad}}[\{P_{i,j}\}] = 1/2 G_{ijkl} P_{i,j} P_{k,l}$, where G_{ijkl} are the gradient energy coefficients, and electrostatic energy $f_{\text{ele}}[\{P_i, D_i\}] = 1/(2\varepsilon_0\varepsilon_b)(D_i - P_i)^2$. Here it is clear that $D_i = \varepsilon_0\varepsilon_b E_i + P_i$. The zero-strain coefficients $\alpha^{(e)}_{ij}$ can be expressed in terms of usually introduced stress-free coefficients α_{ij} as follows:

$$\begin{aligned} \alpha^{(e)}_{11} &= \alpha_{11} + \frac{1}{6}\left(\frac{2(q_{11} - q_{12})^2}{c_{11} - c_{12}} + \frac{(q_{11} + 2q_{12})^2}{c_{11} + 2c_{12}}\right), \\ \alpha^{(e)}_{12} &= \alpha_{12} + \frac{1}{6}\left(\frac{2(q_{11} + 2q_{12})^2}{c_{11} + 2c_{12}} - \frac{2(q_{11} - q_{12})^2}{c_{11} - c_{12}} + \frac{3q_{44}^2}{4c_{44}}\right). \end{aligned}$$

By using the Legendre transformation to the electric enthalpy

$$h[\{P_i, P_{i,j}, u_{i,j}, \varphi_{,i}\}] = f[\{P_i, P_{i,j}, e_{ij}, D_i\}] - D_i E_i,$$

where $E_i = -\varphi_{,i}$ is the electric field and φ the electric potential, and using Lagrange principle, we can uniformly express the set of field equations which govern the kinetics of ferroelectrics:

$$\left(\frac{\partial h}{\partial e_{ij}}\right)_{,j} = 0, \tag{5.7}$$

$$\left(\frac{\partial h}{\partial E_i}\right)_{,i} = q(p-n), \tag{5.8}$$

$$\frac{1}{\Gamma}\frac{\partial P_i}{\partial t} - \left(\frac{\partial h}{\partial P_{i,j}}\right)_{,j} = -\frac{\partial h}{\partial P_i}. \tag{5.9}$$

Equation (5.7) defines the mechanical equilibrium while inertia is neglected. The Poisson's equation (5.8) represents the Gauss's law for the charge and the electric field in a dielectric including a nonzero concentration of free electrons n and holes p. Equation (5.9) is the time dependent Landau-Ginzburg-Devonshire equation [51] which governs the spatiotemporal evolution of the spontaneous polarization with kinetics given by damping Γ.

The coupling between the ferroelectric/ferroelastic system with its semiconductor properties is introduced by considering a nonzero density of free carriers in the electrostatic equation (5.8). The distribution of free carriers is governed by continuity equations:

$$q\frac{\partial n}{\partial t} + J_{i,i}^{(n)} = qR_n, \tag{5.10}$$

$$q\frac{\partial p}{\partial t} + J_{i,i}^{(p)} = qR_p, \tag{5.11}$$

where electron and hole currents $J_i^{(n)}$ and $J_i^{(p)}$, respectively, are given by the drift and diffusion as follows: $J_i^{(n)} = \mu_n(qnE_i + k_B T n_{,i})$ and $J_i^{(p)} = \mu_p(qpE_i - k_B T p_{,i})$. Here μ_n and μ_p are the electron and hole mobilities, respectively. If we are interested only in the stationary solution at the thermodynamic equilibrium we can introduce the computationally convenient form of recombination rates R_n and R_p as follows: $R_n = -(n-n_0)/\tau$ and $R_p = -(p-p_0)/\tau$, where τ is the life-time constant and n_0 and p_0 are the electron and hole concentrations at the thermodynamic equilibrium:

$$n_0 = NF_{1/2}\left(-\frac{E_{\mathrm{C}} - E_{\mathrm{F}} - q\varphi}{k_{\mathrm{B}}T}\right),$$

$$p_0 = NF_{1/2}\left(-\frac{E_{\mathrm{F}} - E_{\mathrm{V}} + q\varphi}{k_{\mathrm{B}}T}\right).$$

Here $F_{1/2}$ is the Fermi-Dirac integral. The density of states is given by the effective mass approximation:

$$N \simeq 2\left(\frac{m_{\mathrm{eff}}k_{\mathrm{B}}T}{2\pi\hbar^2}\right)^{\frac{3}{2}},$$

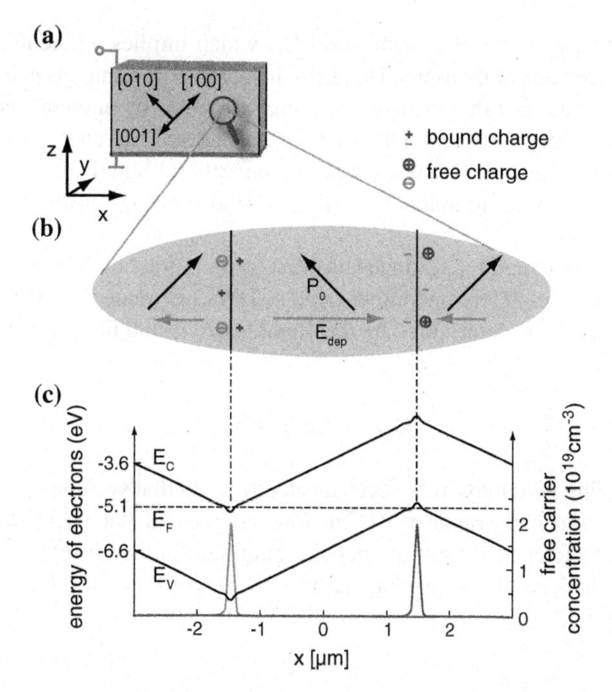

Fig. 5.9 **a** An electroded $(110)_c$ plate of tetragonal $BaTiO_3$. **b** A periodic structure of charged 90° domain walls where bound polarization charge (+, −) induced by the divergence of polarization, P_0, is almost perfectly compensated by free charge (⊕, ⊖) whereas the imperfection of compensation creates built-in depolarizing field $\mathbf{E}_{dep}$. **c** Phase-field simulation-calculated band bending induced by $\mathbf{E}_{dep}$. The bending causes the edges of the conduction E_C or valence E_V bands (*solid black lines*) to approach the Fermi level E_F (*dashed black line*) where high concentration of free electrons (*red line*) or holes (*blue line*) are generated and become available for compensation of the bound charge. The bandgap between the conduction and valence bands is assumed to be 3 eV. The competing band bending and charge compensation are in equilibrium when conduction and valence bands cross the Fermi level about 0.22 eV at Head-to-Head and Tail-to-Tail domain walls, respectively [34]

where m_{eff} is the effective mass of electrons or holes.

When a ferroelastic Head-to-Head and Tail-to-Tail polarization arrangement is artificially introduced into the initial conditions of the problem given by (5.7)–(5.11), the free carrier compensated CDWs are stabilized as shown for the case of 90° CDWs in tetragonal $BaTiO_3$ in Fig. 5.9.

Two intriguing features of the stable periodic sCDW structure may be seen in Fig. 5.9, (i) the presence of the free carriers reaching concentrations of doped silicon at CDWs and (ii) the band bending with constant potential change between Head-to-Head and Tail-to-Tail sCDWs. This situation represents a self-maintaining equilibrium due to the coupling of two mechanisms: (i) The non-thermally activated free electrons (holes) can be present when the conduction band $\mathbf{E}_C$ (valence band $\mathbf{E}_V$)

drops below (raise above) the Fermi level $\mathbf{E}_{\mathrm{F}}$, which implies a potential difference between the electron compensated Head-to-Head and hole compensated Tail-to-Tail sCDWs and, (ii) the fact that the band bending is induced by uncompensated charge at sCDWs. In other words, the CDWs on one hand require charge compensation but, on the other hand, the sCDWs must remain partly (although almost completely) uncompensated in order to maintain sufficient band bending for the stabilization of free charge carriers.

The depolarizing field $\mathbf{E}_{\mathrm{dep}}$ induced by the sCDWs is drastically reduced compared to the case with completely uncompensated sCDWs (estimated in Sect. 5.2.1) but it is still present and it is determined by the domain size w and the bandgap energy $\mathbf{E}_{\mathrm{G}}$ as:

$$\mathbf{E}_{\mathrm{dep}} = \mathbf{E}_{\mathrm{G}}/(qw). \tag{5.12}$$

This depolarizing field may still exceed an electric field that switches polarization by 90° for domain size in the range of 10^{-6} m. It was suggested that when the domain size is approaching the critical limit the depolarizing field induces polarization rotation that causes enhanced electromechanical response by mimicking the approach to a phase transition [34].

5.3.1.1 Free Carriers at Charged Domain Walls by First Principle Calculations

First-principle density functional theory was used to calculate local density of states at the Head-to-Head and Tail-to-Tail CDWs in improper ferroelectric $YMnO_3$ [27]. The results shown in Fig. 5.10 introduce qualitatively identical features of electronically compensated CDWs as in the phase field calculated results for proper ferroelectric in Fig. 5.9. In both cases, the Fermi level enters the conduction and valence bands at the Head-to-Head and Tail-to-Tail CDWs, respectively, where electrons and holes can occupy the available states.

Note, as discussed in Sect. 5.2.2, that the stabilization of CDWs in improper ferroelectrics is provided by their interlocking with anti-phase boundaries. A few nanometer spacing of CDWs might therefore be possible.

First-principle models of sCDWs in proper ferroelectrics suffer from additional complications. The limited size of simulation domain (determining the computational complexity) and use of periodic boundary conditions in standard simulation packages results in the necessity to consider periodic domain structures with extremely small CDW spacing. The CDWs period in proper ferroelectrics is however limited due to the depolarizing field that may exceed the coercive fields, according to (5.12).

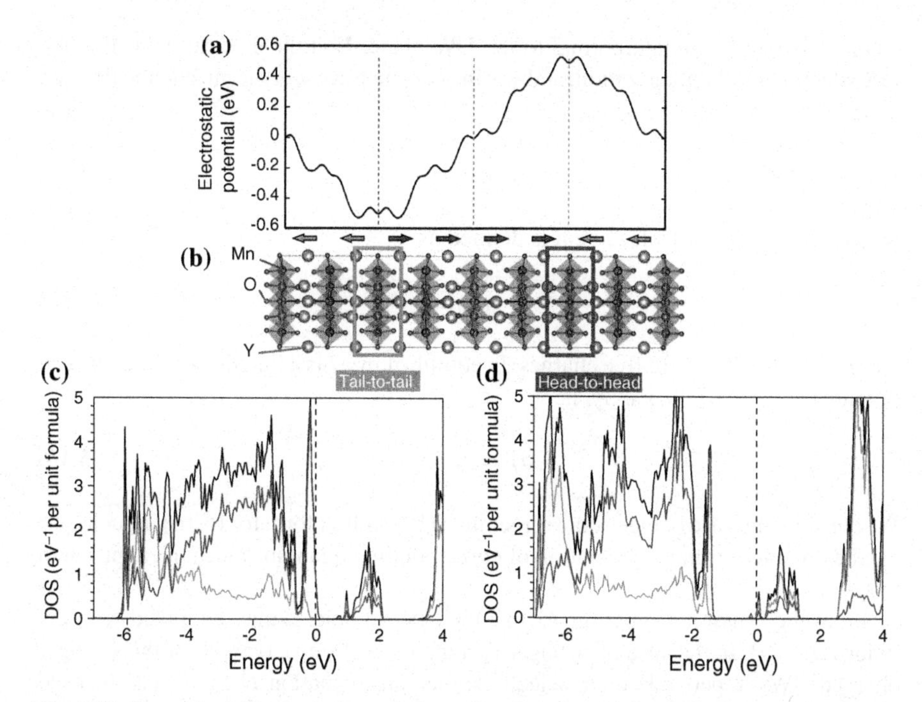

Fig. 5.10 First-principle calculations of charged and uncharged ferroelectric domain walls in improper ferroelectric $YMnO_3$ show the potential profile (**a**) across CDWs and surrounding domains (similar—but horizontally flipped—to electron energy profile in Fig. 5.9c), the simulation supercell (**b**), and local density of states in the Tail-to-Tail (**c**) and the Head-to-Head (**d**) domain walls. The *black lines* in the density-of-states plots indicate the sum of the local density of states, while the *blue* and *red lines* show the oxygen and manganese contributions respectively. Note the shift up in the energy of the bands from the Head-to-Head to the Tail-to-Tail configuration is caused by the gradient in the electrostatic potential [27]

5.3.1.2 The Anomalously Thick Charged Domain Walls

The domain wall thickness is an important parameter which correlates with its intrinsic energy, mobility [39], and internal structure.

Unlike in ferromagnets, where DWs are typicaly 100 nm thick DWs in ferroelectrics are typically extremely thin, mostly <2 nm [39]. However sCDWs are typically thicker: TEM observation of 180° sCDW in $Pb(Zr,Ti)O_3$ revealed a thickness of $\simeq$7 nm [10]. Here we discuss phenomenological reasons for this and indicate the case where sCDWs may be yet thicker, namely anomalously large thickness is predicted to occur in 90° sCDW in morphotropic phase boundary (MPB) systems.

The simple Landau theory description for 180° sCDWs by Gureev et al. [30] is built on the constitutive equation for polarization

$$E = \alpha P + \beta_1 P^3 - g\frac{\partial^2 P}{\partial x^2}, \tag{5.13}$$

where x is the coordinate normal to the DW plane, $\mathbf{E}$ is the electric field, $\mathbf{P}$ is the ferroelectric part of polarization, and g is the correlation energy term, and the Poisson equation:

$$\nabla \mathbf{D} = \rho_f, \tag{5.14}$$

where the electrical displacement is defined as

$$\mathbf{D} = \varepsilon_0 \varepsilon_b \mathbf{E} + \mathbf{P}. \tag{5.15}$$

Here, ρ_f is the density of free charges. In equilibrium ρ_f is a function of the electrical potential $\varphi = -\int \mathbf{E} \cdot d\mathbf{x}$ only, i.e.

$$\rho_f = \rho_f(\varphi). \tag{5.16}$$

We consider the case of electronic screening. The explicit form for (5.16) in such case can be obtained from consideration of corresponding quantum-mechanical problem [30].

From the Poisson's equation, (5.14), the electric field E may be linked with the polarization P. In the case of strong screening ($\rho_f \simeq -\rho_P$), typical for the strongly charged DWs in perovskite ferroelectrics, the approximation $D_x \simeq P_x$ can be used [30] for the electric displacement, (5.15). Its substitution to (5.14) yields

$$dP/dx \simeq \rho_f, \tag{5.17}$$

which is consistent with $\rho_P = -\nabla \mathbf{P}$, in the case where $\rho_f \simeq -\rho_P$. The derivative of (5.17) with respect to x using (5.16) yields:

$$E = -\frac{d\varphi}{dx} = -\left(\frac{\partial \rho_f}{\partial \varphi}\right)^{-1} \frac{d^2 P}{dx^2}. \tag{5.18}$$

When substituted into (5.13), (5.18) determines an additional effective correlation energy term.[2] Gureev et. al. [30] demonstrated that this additional correlation term is—in case of ordinary perovskite ferroelectrics—typically two orders of magnitude larger than the initial one: $| E | \gg g\frac{d^2 P}{dx^2}$. As a consequence, sCDWs are a few times thicker than neutral DWs and an approximation can be applied where the initial correlation terms are neglected. Under this approximation in [30] it has been shown that (5.13) can be solved analytically for a number of limiting cases, where an explicit analytical form for (5.16) is available. These cases correspond to the so-called screening regimes where the electronic gas in the wall can be classified as either degenerate or classical and the screening itself as either linear or non-linear.

---

[2]Note that in general case $\left(\frac{\partial \rho}{\partial \varphi}\right)^{-1}$ is a functional of $P(x)$, [30].

The sCDW half-width δ_{180} of the polarization profiles obtained for these screening regimes can be covered with a generic formula:

$$\delta_{180} = \frac{a}{|\alpha|^b P_0^c}, \tag{5.19}$$

where $1/2 \leq b \leq 1$ and $0 \leq c \leq 1$ while a is a material dependent but temperature independent coefficient. The coefficients a, b, and c depend upon the screening regime. In perovskite ferroelectrics interesting for applications a so-called nonlinear regime of the bound charge screening takes place, and an approximation may be used where the electron gas becomes degenerate in the sCDW-region. For this regime (hereafter nonlinear degenerate screening regime), (5.19) becomes:

$$\delta_{180}^{(ND)} = \left(\frac{9\pi^4 \hbar^6}{q^5 m^3 |\alpha|^3 P_0} \right)^{1/5}. \tag{5.20}$$

Here $\hbar = 1.05 \cdot 10^{-34}J\cdot s$ is the Planck constant, m is the effective electron mass, and $q = 1.6 \cdot 10^{-19}$ C is the elementary charge. For $PbTiO_3$, using free electron mass $m = 9.1 \cdot 10^{31}$ kg, $P_0 = 0.75\,C/m^2$ and $|\alpha| = 0.05$, the domain-wall width ($2\delta_{180}^{(ND)}$), (5.20), was calculated to be 5.4 nm which is in good agreement with observations in [10].

The theory was further developed for 90° sCDW [36] where, in a simple approximation neglecting elastic effects around the DW, the result for the DW width was found to be equivalent to (5.19) for 180° sCDW within the substitution

$$\alpha \rightarrow \alpha\theta,\ \beta_1 \rightarrow 2\beta_1\theta,\ P_0 \rightarrow P_0/\sqrt{2}. \tag{5.21}$$

$$\theta = \frac{2(\beta_2 - \beta_1)}{\beta_2 + \beta_1}. \tag{5.22}$$

Applying substitution (5.21) to (5.19) one readily obtains an expression for the half-width of 90° sCDWs:

$$\delta_{90} = \delta_{180}\theta^{-b}2^{c/2}. \tag{5.23}$$

The parameter θ from (5.22) tends to zero as the morphotropic boundary ($\beta_2 = \beta_1$) is approached while far away from MPB it is of order of unity. This leads to considerable difference between widths of 90° and 180° sCDWs near MPB. The thickness of 90° sCDW, scaling as $|\theta|^{-b}$, ($1/2 \leq b \leq 1$) as controlled by (5.23), diverges as one approaches the MPB. For the nonlinear degenerate screening regime (5.23) becomes:

$$\delta_{90}^{(ND)} = \left(\frac{9\sqrt{2}\pi^4 \hbar^6}{q^5 m^3 |\alpha\theta|^3 P_0} \right)^{1/5}. \tag{5.24}$$

At the same time the thickness of 180° sCDW does not undergo considerable changes near MPB.

The above analysis using a simple framework demonstrates the reasons for the thickening of 90° charged domain walls near morphotropic boundary. In the considered approximation, the thickness of 90° sCDW grows unlimitedly when approaching MPB. However, in realistic systems, this growth is expected to be limited because of the anisotropy of Landau potential, given by higher-order polarization powers, not vanishing at MPB, and because of constraints related to electromechanical coupling.

Quantitative description of the effect of 90° sCDW thickening near MPB with these two factors included in the model was further done in [36]. The 90° sCDW thickness was calculated numerically for the parameters of $Pb(Zr_{1-x}Ti_x)O_3$, from [52, 53], a 4-fold DW thickness growth from 5.4 nm far from MPB to 22 nm at the MPB was predicted.

It was also found in [36] that in 90° sCDW polarization rotates with almost constant modulus, like magnetisation in ferromagnetic Neel walls. The reason for the unusual behaviour of polarization in 90° ferroelectric sCDWs, is the ease of polarization rotation in the proximity of the MPB. The Landau potential for polarization becomes nearly isotropic near the morphotropic boundary (within the approximation of expansion up to 4th polarization power). Thus from the point of view of Landau potential, the trajectory of the sCDW, where the polarization rotates with constant modulus is optimal. The situation is, however, different in electrically neutral 90° DWs, where the polarization rotation is constrained by the depolarizing field and occurs with simultaneous reduction of the polarization modulus. For an electrically neutral 90° DW the polarization component normal to this wall remains constant in first approximation, regardless of the Landau potential anisotropy (see e.g. [54]). In contrast, in charged 90° domain walls, the electrostatic constraints are released due to the availability of free charge carriers. The free charges redistribute and screen the depolarizing field, enabling the polarization rotation.

The scenario described above for MPB materials may also be applied to tetragonal $BaTiO_3$ where polarization rotation becomes easy near transition to orthorhombic phase and large dielectric susceptibility anisotropy is documented at room temperatures [55]. Using the parameters of $BaTiO_3$, (given in [50]), it was calculated that near the transition to the orthorhombic phase, sCDW thickness reaches 70 nm, which is about 2.5 times larger than in the middle of the tetragonal phase.

In summary, sCDW in ferroelectrics are typically one order of magnitude thicker than neutral DWs. This is due to the large effective correlation energy, resulting from kinetic energy of the gas of screening electrons. In addition to this thickening effect, in 90° sCDWs, there exists thickness anomaly related to the ease of polarisation rotation. The thickness anomaly is predicted for free-carrier compensated 90° sCDWs near the MPB and for $BaTiO_3$ at room temperature, near transition from tetragonal to orthorhombic phase; in the both cases sCDW thicknesses of the order of 10–100 nm are expected. The internal structure of the anomalously thick ferroelectric domain walls is analogical to Neel walls in ferromagnets.

5.3.2 *Combined Free Carrier and Defect Screening in the Thermodynamic Equilibrium*

Real world materials always contain a certain fraction of defects and dopants. We should therefore investigate what happens when defects, especially the mobile ones, participate on sCDW screening. It was calculated that a massive concentration of fixed dopants ($\sim 5 \times 10^{23}\mathrm{m}^{-3}$) can theoretically stabilize one of the sCDWs [33], e.g. donors stabilize Head-to-Head sCDWs. However a sCDW in so massively doped materials becomes technologically less interesting due to associated substantial bulk leakage. The bulk conductivity is mediated by doping impurities which provide additional electron states usually inside the band-gap and shift the Fermi level towards the conduction (in case of donors) or valence (in case of acceptors) bands. The Fermi level in the proximity to the conduction or valence bands allows the presence of thermally activated free electrons or holes, respectively. In other words the defect states can be easily ionized, which provides bulk conduction and allows immediate charge redistribution which compensates the polarization charge.

Very different situation occurs with smaller concentrations of mobile dopants such as the oxygen vacancies $V\ddot{o}$ in perovskites. $V\ddot{o}$ are relatively mobile defects (with mobility $\sim 10^{-8} - 10^{-11}$ $\mathrm{cm}^2\mathrm{V}^{-1}\mathrm{s}^{-1}$ [56] and references therein) in perovskites due to the fact that there are three oxygen ions in each complete unite cell consisting of five atoms. Oxygen ions can therefore easily jump to the neighbouring $V\ddot{o}$ which represents a virtual object migrating in the opposite direction analogically to holes. $V\ddot{o}$ are effectively charged and can trap up to two electrons in relatively shallow levels (e.g. 0.28–0.66 eV below the bottom of the conduction band in $BaTiO_3$ [57]). In order to maintain the total electrostatic neutrality, each $V\ddot{o}$ is associated either with oppositely charge cation defect (this occurs e.g. during processing of non-stoichiometric compounds) or with free or trapped electrons [58]. In the latter, $V\ddot{o}$ are basically mobile shallow electron traps from which the electrons can be easily liberated.

As described in [20] it was surprising that sCDWs formed slowly, but regularly within several hours of frustrative poling (discussed below) and that only Head-to-Head sCDWs displayed giant conductivity while Tail-to-Tail sCDWs remained indistinguishable from the bulk. It was even contrary to the expectations because the Pt electrodes used in this experiment should theoretically create ohmic contact with the Tail-to-Tail sCDW and a large barrier with the Head-to-Head sCDWs.

It was therefore suggested that the sCDWs are compensated with the assistance of positively charged $V\ddot{o}$ that are attracted to the Tail-to-Tail sCDWs and the electrons that are liberated from $V\ddot{o}$ and are attracted by the Head-to-Head sCDWs. A phase field simulation then showed, that the depolarizing field, which appears as soon as the sCDWs start forming, separates electrons from $V\ddot{o}$ and, consequently, drags them towards Head-to-Head and Tail-to-Tail sCDWs, respectively. Obviously, the much more mobile electrons are able to compensate the Head-to-Head sCDWs orders of magnitude faster than $V\ddot{o}$ compensate the Tail-to-Tail ones. It was shown in [21] that Tail-to-Tail sCDWs more often acquire a zig-zag shape with a smaller charge density and that the sCDW density is directly proportional to the initial V$\ddot{o}$ concentration (provided by high temperature annealing in vacuum).

The model in Sect. 5.3.1 can therefore be upgraded in (5.8) by adding the volume density of charged donors n_D as follows:

$$\left(\frac{\partial h}{\partial E_i}\right)_{,i} = q(p-n) + n_\mathrm{D}, \tag{5.25}$$

The concentration of ionised donors is obtained as $n_\mathrm{D} = qzf(\varphi)N_\mathrm{D}$, where z is the donor valency and

$$f(\varphi) = 1 - \left(1 + \frac{1}{g}\exp\left(\frac{E_\mathrm{D} - E_\mathrm{F} - q\varphi}{k_\mathrm{B}T}\right)\right)^{-1}$$

is the fraction of ionized donors with the donor level E_D and the ground state degeneracy of the donor impurity level g [59].

The donor density N_D evolves through diffusion,

$$\frac{\partial N_\mathrm{D}}{\partial t} - \nabla \cdot \left(\beta N_\mathrm{D} \nabla \left(\frac{\partial W_\mathrm{D}}{\partial N_\mathrm{D}} + qzf(\varphi)\varphi\right)\right) = 0, \tag{5.26}$$

where β is the donor mobility [56], and W_D is the contribution to the free energy due to defects which is assumed to be the usual free energy of mixing at small concentrations [60].

Numerical solution of (5.7), (5.9)–(5.11) and (5.25) in the one dimensional case gives a stable solution shown in Fig. 5.11 where charge carrier and defect densities

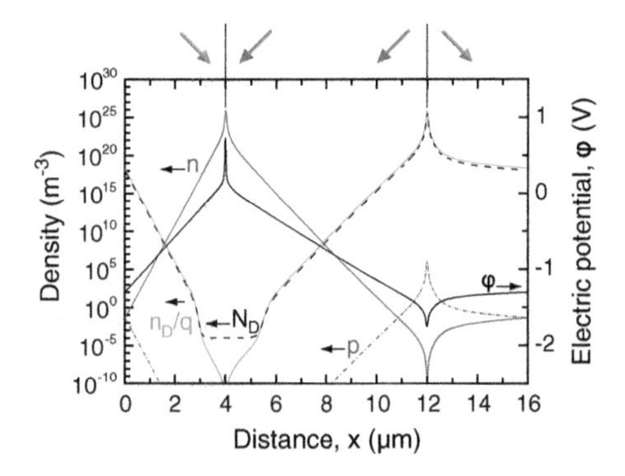

Fig. 5.11 One dimensional phase field simulation of charge carrier and defect densities (*left* axis) and electric potential φ (*right* axis) across the Head-to-Head and Tail-to-Tail domain walls. Compensation of polarization charge at Head-to-Head wall requires accumulation of electrons, n, and depletion of oxygen vacancies N_D. The negligible remaining density of vacancies is not ionized which lowers the charge carrier density, n_D/q. The Head-to-Head sCDW accumulates holes p and almost fully ionized oxygen vacancies N_D. The electric potential φ forms a zig-zag profile across the domain walls. The oxygen vacancies almost fully replace screening holes at the Tail-to-Tail domain wall after 10^1 hours with initial defect concentration $N_\mathrm{D}|_{t=0} = 10^{18}\,\mathrm{m}^3$. It makes the Tail-to-Tail walls significantly less conductive (details in Supplementary materials of [20])

and electric potential φ (right axis) across the Head-to-Head and Tail-to-Tail sCDWs are plotted. One can see that the polarization charge at the Head-to-Head wall is compensated by accumulated electrons, n, and depletion of oxygen vacancies N_D. The Head-to-Head sCDW accumulates holes p and almost fully ionized oxygen vacancies N_D.

5.4 Charged Domain Wall Formation: Factors Controlling the Formation Energy

The occurrence of sCDWs is controlled to a great extent by the formation energy of the walls which, in turn, is sensitive to the details of the screening mechanism and the availability of charge exchange between the ferroelectric and its exterior (electrode or/and ambient atmosphere). Both, electrons (holes) and charged ionic species can participate in the bound charge screening at the sCDW.

5.4.1 Electrically Isolated Ferroelectric: Bipolar Electron-Hole Screening

The screening of Head-to-Head and Tail-to-Tail sCDWs by electrons and holes, respectively, was addressed in [1, 30]. In the case of electron-hole screening in typical perovskite ferroelectrics, the main source of compensating free carriers is electron transfer from the valence to the conduction band over the bandgap of the ferroelectric. For a 180° wall with 2 P_S bound charge per unit area, the screening requires $2P_S/q$ (q is the elementary electron charge) free carriers per unit area. Thus, the formation energy of sCDW in non-linear regime can be evaluated as [30]:

$$W_{\mathrm{CDW}} \approx \frac{2P_S E_g}{q},$$

where E_g is the bandgap. For a bandgap of about 3 eV, which is typical for perovskites, the energies of sCDWs are some two-orders of magnitude larger than the energy W_{NDW} of neutral DWs, i.e. [30]:

$$W_{\mathrm{CDW}} \cong 10^2 W_{\mathrm{NDW}}, \tag{5.27}$$

while the energy of NDWs is, on the lines of Landau theory [61]:

$$W_{\text{NDW}} \cong 2t_{\text{NDW}} U_{\text{fer}},$$

where U_{fer} is the gain of the energy density of the ferroelectric phase with respect to the paraelectric one and t_{NDW} is the half-width of the NDW.

The carrier concentration in the wall can be evaluated as

$$\rho_{\text{CDW}} = \frac{P_S}{q t_{\text{CDW}}},$$

where t_{CDW} is the half-width of the sCDW. Taking $P_S = 0.3$ C/m^2 and $t_{\text{CDW}} = 10$ nm, one finds $\rho_{\text{CDW}} = 2 \times 10^{26}$ m^{-3}. Such a concentration vastly exceeds any realistic intrinsic equilibrium concentration of free carriers in nominally insulating ferroelectrics. Based on this fact, it was concluded in [30] that the participation of equilibrium carriers in the screening can be neglected. An important feature of this screening scenario is that sCDW formation is virtually insensitive to the equilibrium free carrier concentration (and conductivity) in the single-domain material, be it a pure or a moderately doped material.

5.4.2 Electrically Isolated Ferroelectric: Unipolar Screening

One may conceive an alternative screening scenario; [2] discussing formation of a sCDW in a $PbTiO_3$ crystal argued that the free charge needed for screening of a single sCDW in a crystal can be collected from its volume. According to this scenario, electron transfer across the bandgap is not needed. The realization of this scenario is limited even when the amount of free carries in the crystal is sufficient for screening of the bound charge of the wall. The point is that, for not too small crystals, the needed amount of free carriers can be available only due to the presence of doping impurities in the material. Once the free carriers concentrate at the walls, the space charge of the ionized impurities in the bulk implies a very strong increase of the energy of the system.

5.4.3 Electrically Isolated Ferroelectric: Screening with Photon Generated Carriers

The energy of sCDW formation by electron transfer across the band-gap may be large, but this energy can be naturally provided by externally supplied super-band-gap photons which generate directly electron-hole pairs. In this scenario, sCDWs form once the required number of free carriers has been provided. In practice electron-hole pairs can be provided by illumination with super-band-gap light.

5.4.4 Electrically Isolated Ferroelectric: Mixed Electron/ion Screening

As pointed out above, screening of sCDW through the collection of free carriers in equilibrium from adjacent domains leads to a very high sCDW formation energy due to the space charge of the ionised impurities in the bulk of the crystal. However, this energy is substantially reduced if the ionised impurities are mobile, i.e. if they can be redistributed during the formation of the sCDW. In this case, the mobile carriers have both polarities, and the situation resembles the electron-hole screening discussed above. For example, in the case of an n-type material, the equilibrium electrons in the conduction band can screen the positive bound charge while the ionised donors can screen the negative bound charge. If the total amount of charge in the crystal is large enough to neutralize the bound charge at the walls, such screening mechanism is energetically more favourable than the pure electron-hole screening since no energy penalty is paid for the electron transfer across the band-gap. In this case, the sCDW formation energy, W_{mix} , can be evaluated as the energy associated with the deviation of the polarization in the wall region from its spontaneous value (c.f. the discussion from [30]), i.e.

$$W_{mix} \cong 2t_{CDW} U_{fer} \cong \frac{t_{CDW}}{t_{NDW}} W_{NDW}. \tag{5.28}$$

For typical perovskite ferroelectrics, t_{CDW} is expected to be one order of magnitude larger than t_{NDW} [30]. Thus, comparing the estimate of (5.28) with (5.27), we see that in perovskite ferroelectrics, sCDW formation with mixed electron-ion screening mechanism is more favourable energetically than bipolar electron-hole screening in which a sCDW requires an additional treatment. Note that (5.28) is not valid for ferroelectrics near rotational phase transition in which the formation energy of anomalously thick sCDWs [36] (see Sect. 5.3.1.2) is even smaller.

An essential feature of this screening scenario is that, for given charge densities of mobile carriers of both signs, ρ_+ and ρ_-, and the component of spontaneous polarization normal to the walls, P_N, the average domain width should exceed a certain minimum value L_{min}. Relation between these parameters can be found from the condition that all mobile carriers of the material are used for the screening of sCDWs:

$$P_N = L_{min} min[|\rho_+|, \rho_-],$$

yielding

$$L_{\min} = \frac{P_{\mathrm{N}}}{min[|\rho_+|, \rho_-]}. \tag{5.29}$$

Two more characteristic features of the mixed screening mechanism are evident. First, it requires substantial redistribution of the ionized impurities, implying a relatively slow sCWD formation. Second, a strong difference in the conduction of Head-to-Head and Tail-to-Tail walls is expected. In view of their much higher electronic mobility, the walls screened by electronic carriers will exhibit substantially higher conductivity than those screened by the significantly less mobile charged defects. Experimental data on sCDWs in bulk crystals of $BaTiO_3$ [20], strongly suggest that it is the mixed screening mechanism which takes place in this system. Here, Head-to-Head 90° sCDWs artificially created in the materials were found highly conductive (10^9 times more than the domains) while Tail-to-Tail sCDWs did not exhibit increased conductivity. This situation matches perfectly the typical n-type conduction in a material in which oxygen vacancies ($V\ddot{o}$) play the role of donors. Since the mobility of the latter is much smaller than that of electrons, the Tail-to-Tail walls screened with $V\ddot{o}$ do not exhibit considerable conductivity contrast with respect to the domains themselves.

5.4.5 Screening of sCDW with Charge Provided from External Source

Charge injection from the exterior of the ferroelectric material may reduce strongly the formation energy of sCDW [1, 30, 62]. Depending on the band gap of the ferroelectric and on the ferroelectric/metal work function difference, the energy penalty associated with the carrier generation is necessarily reduced compared to the electron transfer over the band gap either for electrons or holes [1, 30, 62]. For certain combination of these parameters, the energy penalty can be even negative, implying negative formation energy of sCDW [47, 48]. Estimates show that in $BaTiO_3$ with Pt electrodes, Tail-to-Tail sCDWs should have negative formation energy, thus forming spontaneously. Such a phenomenon has never been documented experimentally, which can be explained by surface effects associated with surface states and surface termination [30] not taken into account in the model. Mobile charged defects like oxygen vacancies$V\ddot{o}$ due to oxygen loss from the sample may give similar effects to those resulting from the electronic screening discussed above. The formation of Head-to-Head and Tail-to-Tail configurations supported by purely ionic compensation has been documented for $LiNbO_3$ and $LiTaO_3$ crystals (see [63] and references therein). In thin films, compensating free carriers can be injected from a sharp tip of an AFM probe due to the tip-geometry related enhanced tunneling [18, 64].

5.5 Charged Domain Wall Engineering

Naturally occurring CDWs are extremely rare (except those in improper and hybrid improper ferroelectrics) and randomly distributed objects which makes them inconvenient for experimental investigation and any future development. The positioning of inclined charged domain walls such as those appearing during domain nucleation (Fig. 5.6) can be controlled into certain extent, but, on the other hand, these sCDWs are mostly unstable. Therefore domain engineering method which would allow precise control of stable sCDW (their density or nanoscale manipulation) were sought after.

The artificially engineered sCDW were created by so-called frustrative poling [19–21]. An electric field, E, was applied to a $(110)_c$ (or $(111)_c$) plate of tetragonal $BaTiO_3$ in a $[110]_c$ (or $[111]_c$)-like direction. This leads to two (or three) equally preferred ferroelectric-ferroelastic domain states (Fig. 5.12a) which can be separated either by $(110)_c$ neutral domain walls or $(1\bar{1}0)_c$ sCDW. The theoretical energy density of sCDW is significantly larger than that of the neutral domain walls [30], but their total energy in a thin plate sample can be smaller due to their smaller total surface. The sCDW are stabilized by elastic compatibility of adjacent ferroelastic states, therefore their density can be controlled by favorable poling history and boundary conditions [19, 21].

The frustrated poling of $BaTiO_3$ samples was applied during slow cooling over the paraelectric-ferroelectric phase transition (Fig. 5.12a, b) which leads to formation of planar sCDW and zig-zag partly charged domain walls, Fig. 5.12c, d.

The formation of sCDW with controlled periodicity was achieved by means of frustrative poling in samples with predetermined Vö concentration (obtained with high temperature annealing in vacuum) and by modification of electric field magnitude and poling history [21] (Fig. 5.13). The data-points in Fig. 5.13b, within their error margins, are distributed along the line which is determined by the minimal domain size as predicted by the electron/ion screening scenario, (5.29).

Deterministic nanoscale manipulation with sCDWs was first achieved in 45 nm La doped $BiFeO_3$ epitaxial thin film deposited on 5 nm $SrRuO_3$ electrode on $DyScO_3$ single-crystal substrate [18]. Here, the formation of sCDWs employs two mechanisms (i) a tri-axial writing of domain states with an AFM tip which exploits the out-of-plane poling field between the tip and the bottom electrode together with the effective trailing field produced by the line-by-line scanning over the film surface

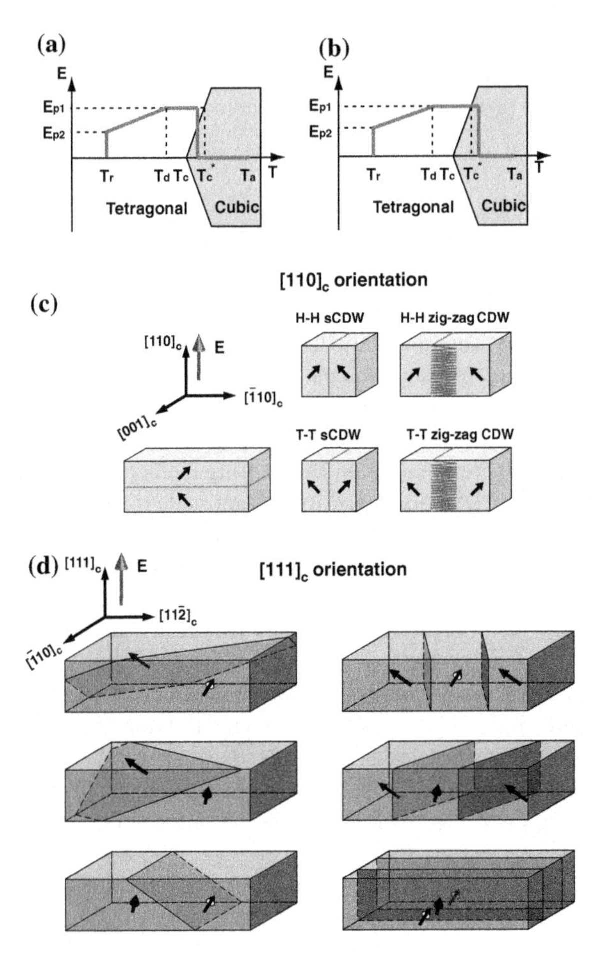

Fig. 5.12 Illustration of the frustrative poling method. An electric field is applied parallel to the $[110]_c$ or $[111]_c$ directions just above the ferroelectric phase transition. Two scenarios are then used: **a** the phase transition is crossed by cooling or **b** induced by the electric field. The field/temperature path is indicated by blue line (E_{p1} poling electric fields, E_{p2} zero field phase transition temperature, $T_c^{\star}$ E-field induced phase transition temperature, T_r room temperature, T_a temperature of annealing, T_d temperature at which the slow decrease of the electric field begun. Domain states and 90° domain walls in tetragonal $BaTiO_3$ obtained through frustrative poling. **c** Schematic views of domain states and domain walls in a $[110]_c$ poled tetragonal $BaTiO_3$ crystal observed from the $[001]_c$ direction: neutral domain wall, planar Head-to-Head sCDW, zig-zag configuration of Head-to-Head CDW, planar Tail-to-Tail sCDW, zig-zag configuration of Tail-to-Tail CDW. **d** Schematic views of domain states and domain walls in a $[111]_c$ poled tetragonal $BaTiO_3$ crystal: three orientations of NDWs and three orientations of sCDWs with Head-to-Head and Tail-to-Tail configurations are possible. *Green* planes depict inclined NDWs while *red* and *blue planes* represent sCDWs which are perpendicular to the surface. *Black arrows* denote the directions of spontaneous polarization [21]

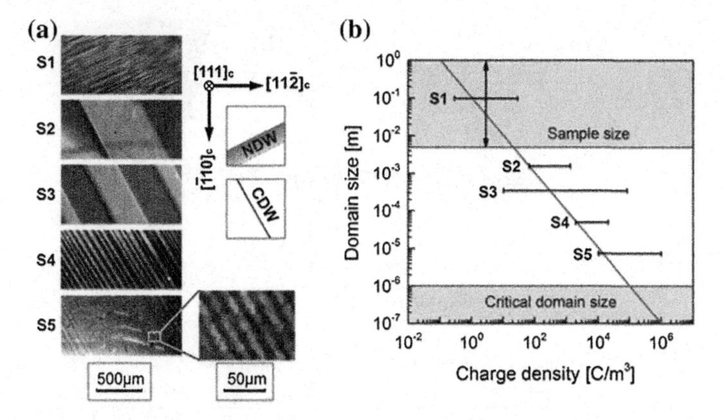

Fig. 5.13 Controlled sCDW period in $(111)_c$ $BaTiO_3$ crystal plates with different Vö density. **a** Optical micrographs of domain patterns with drawings of allowed orientations of domain walls: CDW charged domain walls, NDW neutral domain walls; **b** Dependences of the domain size versus charge density in single-domain material, red line determines the minimal domain size as a function of the charge density in the single-domain material as predicted by the electron/ion screening scenario according; the *upper horizontal line* is the limitation by the sample size; *lower horizontal line* is given by the minimal domain size consistent with the stability of ferroelectricity in the domains, evaluated with the results from [21, 34]

with an applied voltage (illustrated in Fig. 5.14) [18, 65] and (ii) the tip-geometry-enhanced electron tunneling into the ferroelectric film that provides screening of the polarization charge [18, 64] (illustrated in Fig. 5.15).

5.6 Charged Domain Wall Conductivity

As predicted by theory and explained above, electronically compensated sCDWs should display a metallic-type conductivity due to the presence of a quasi-two-dimensional electron or hole gas. The latter is less likely in oxides due to the presence of positively charged mobile Vö which can entirely replace screening holes. The first evidence of non-thermally activated conductivity at CDWs was seen at transient domain walls of nucleating nanodomains in $Pb(Zr_{0.2}Ti_{0.8})O_3$ thin film [7]. Figure 5.16a shows qualitatively different non-thermally activated temperature dependence at nanodomains in comparison with domain interior (macrodomain) and neutral domain walls. Figure 5.16b illustrates the charging of domain walls at a nanodomain. In this experiment, the magnitude of conductivity at nanodomains temporarily exceeded the macrodomain conduction by three orders of magnitude [7].

As introduced above, the sCDWs were later stabilized by ferroelastic clamping in $BaTiO_3$ single crystals [20]. While the conduction between electrodes connected

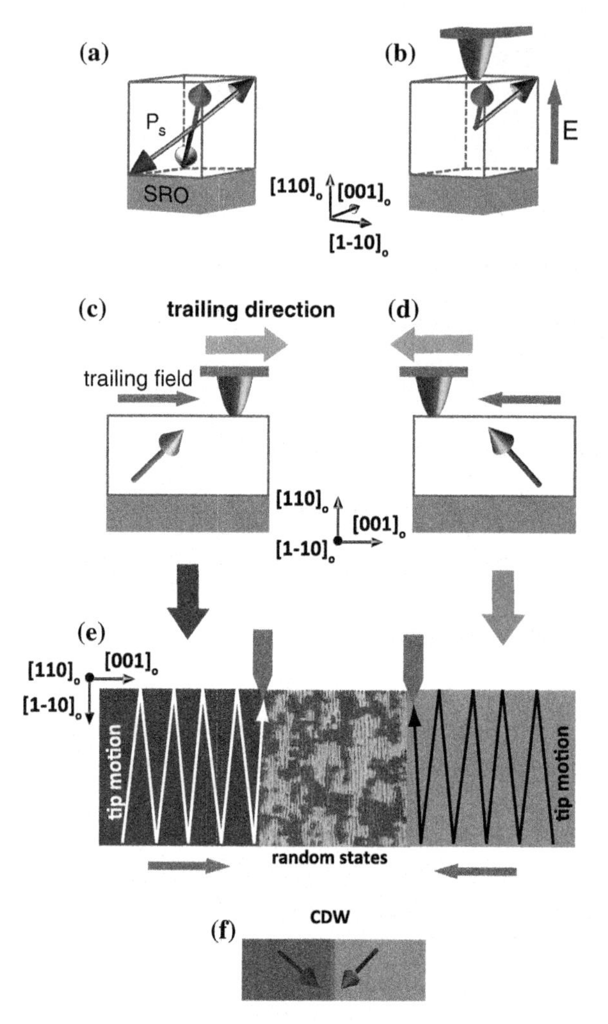

Fig. 5.14 Principle of controlled charged domain wall creation in a ferroelectric thin film. **a** Representation of the four as-grown polarization states (indicated by *arrows*) in a La doped $BiFeO_3$ (45 nm)/$SrRuO_3$ (5 nm)//$DyScO_3$ heterostructure. **b** The application of a tip-$SrRuO_3$ electric field E selects the two states pointing toward the tip. **c**, **d** The selection of only one of these two polarization states is achieved with trailing field (indicated by *red arrows*) created by the tip trailing motion (indicated by *grey arrows*). **e** Controlled switching of the original random domain structure (*center*) to the selected monodomain states with opposite in-plane directions indicated by violet and yellow colour. **f** Joining of the two mono-domain states creates Head-to-Head charged domain wall (CDW) [18]

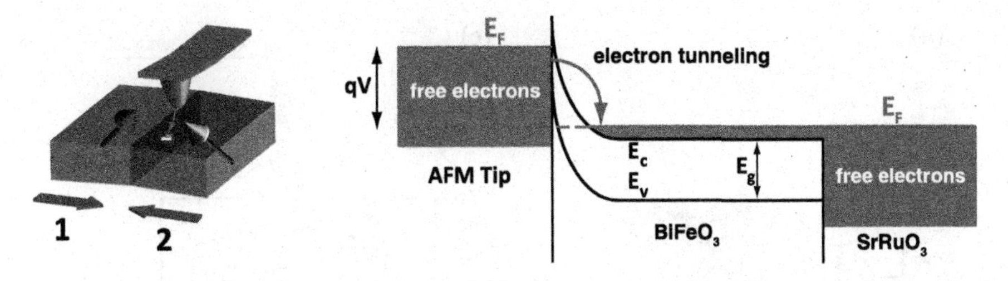

Fig. 5.15 Schematic illustration of a $BiFeO_3$ band diagram during the formation of a Head-to-Head sCDW. The screening of the polarization charge is achieved by the tip-enhanced electron injection from the tip to the $BiFeO_3$ conduction band [18]

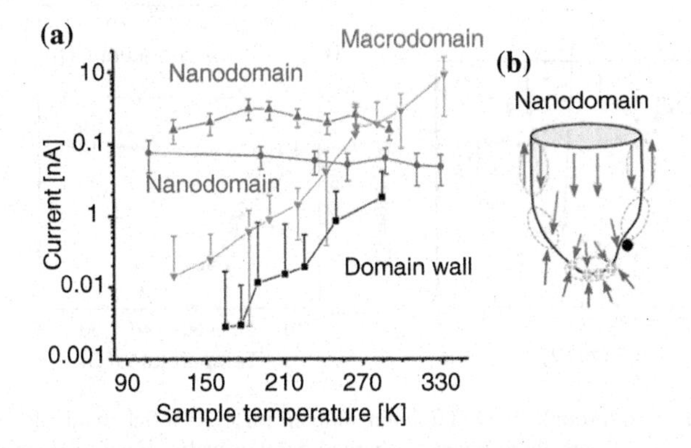

Fig. 5.16 **a** Temperature-dependence of local current obtained from averages of 100 hysteretic IV curves on $Pb(Zr_{0.2}Ti_{0.8})O_3$ nanodomains (*blue*, measured at 4.5 V), 50 resistive curves on nanodomains (*red*, measured at 3.5 V), 100 non-hysteretic IV *curves* on macrodomains (*green*, measured at 7 V) and one current image from the domain walls (measured at 2.6 V) at each temperature. The non-thermally activated current at nananodomains indicate the presence of free carriers due to polarization charge at the domain nucleus as illustrated in (**b**) [7]

by a Tail-to-Tail sCDW was almost identical to the conduction through the bulk, the conduction between electrodes touching Head-to-Head sCDW was reproducibly and steadily (for >120 hours) $10^4 - 10^6$ times higher, Fig. 5.17. Figure 5.17b shows room-temperature I-V curves of the bulk and the cases when a single Head-to-Head or Tail-to-Tail domain wall is present between 200 μm diameter top Pt electrode and a full bottom Pt electrode. The steady difference between conductance measured with and without the Head-to-Head sCDW is more than six orders of magnitude (at $V = 100$ V across 200 μm thick sample after >660 min) [20]. Assuming the thickness of sCDW is ~10–100 nm, its intrinsic conductivity is $10^8 - 10^{10}$ times higher than the conductivity of the bulk.

Conductivity of sCDWs was measured in $BiFeO_3$ thin films [18] where current flowing through stable Head-to-Head sCDWs was more than three orders of magnitude higher than current through the domain or neutral domain walls. Figure 5.18 shows PFM images of the $BiFeO_3$ domain structure and correspond-

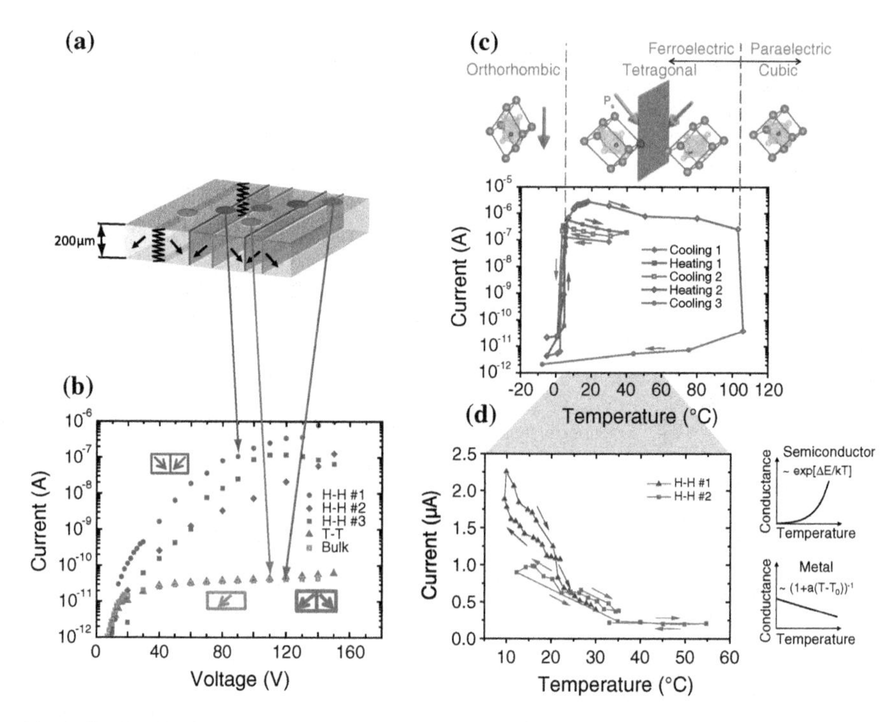

Fig. 5.17 Conduction through charged domain walls in $BaTiO_3$. **a** Schematic diagram of $(110)_c$ plate of $BaTiO_3$ crystal with three types of charged domain walls: Head-to-Head (*red*), Tail-to-Tail (*blue*) and zig-zag (*black*). The colored circles represent Pt cathodes at which the current through the sample thickness was measured. **b** Semilogarithmic-scale room-temperature I-V characteristics showing up to 10^5 times higher conduction through electrodes touching Head-to-Head (H-H) charged domain walls compared to the bulk and Tail-to-Tail (T-T) domain walls. **c** Current-temperature dependence shows switching of the metallic-type conductivity when a 90° Head-to-Head (H-H) charged domain wall is created and annihilated at phase transitions. The domain walls formed in tetragonal $BaTiO_3$ cannot exist in the orthorhombic and paraelectric phases as illustrated in the cartoon. The charged domain wall is annihilated at both transitions from the tetragonal phase. The annihilation at the transition to the paraelectric phase is permanent in this case. The conductivity characteristic shows change from metallic-type temperature dependence and magnitude to thermally activated conduction typical for wide bandgap semiconductor bulk $BaTiO_3$. **d** Linear-scale current-temperature dependence showing pronounced positive temperature coefficient indicating thermally non-activated (i.e. metallic-type) conduction at domain walls as illustrated on the *right* [20]

ing current maps collected at a tip-bottom electrode bias of 2.5 V. Conducting lines can be clearly seen at Head-to-Head sCDWs.

Figure 5.19 shows that, in this case, sCDWs exhibit non-thermally activate type of conduction in current-temperature measurement, which clearly contrasts with the thermally activated conductivity of 71° neutral domain walls.

The elevated conductivity was observed also at topologically protected CDWs in single crystal of improper ferroelectrics $ErMnO_3$ [27]. The Tail-to-Tail CDWs

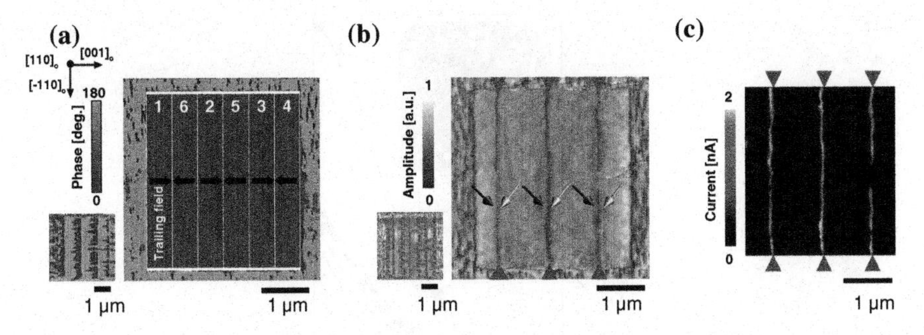

Fig. 5.18 Charged domain walls created in $BiFeO_3$ thin film. **a** Vertical (large image) and lateral (small image) PFM phase micrographs of six regions successively poled up with opposite trailing field directions as indicated by the *red arrows*. **b** The corresponding PFM amplitude images. The *violet* and *yellow arrows* denote the direction of the polarization in the six regions. The *blue arrows* point at the position of the Head-to-Head sCDWs. **c** The current map collected at a tip-bottom electrode bias of 2.5 V on the CDWs array shows pronounced conduction at the Head-to-Head sCDWs [18]

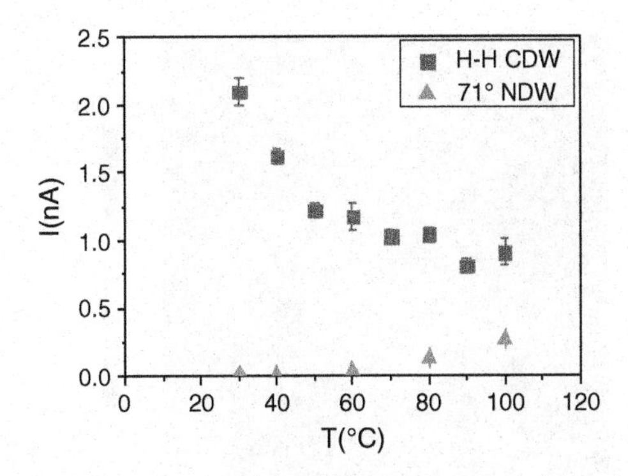

Fig. 5.19 Metallic-like conductivity of charged domain walls. Current versus temperature at $V_{DC} =$ 3 V on a Head-to-Head sCDW (*red*) compared with a 71° neutral domain wall (*orange*). The metallic-type trend of the sCDW is opposite to the thermally activated trend at the neutral domain wall pointing out the different mechanisms at the origin of their conduction—polarisation charge as a quasi-dopant at the former and defect-assisted at the latter [18]

transported ~ three times higher current than domains and Head-to-Head CDWs, Fig. 5.20.

Similar behavior of CDWs was observed in a single crystal of oxygen-deficient hybrid improper ferroelectric $Ca_{2.46}Sr_{0.54}Ti_2O_7$ [28]. One can see in Fig. 5.21 that Head-to-Head CDWs conduct about one order of magnitude more than domains.

Interestingly, in all cases, except in improper ferroelectrics $ErMnO_3$ where the trend was exactly opposite [27], elevated conductivity was seen at Head-to-Head CDWs while Tail-to-Tail CDWs displayed similar behavior as bulk or even smaller

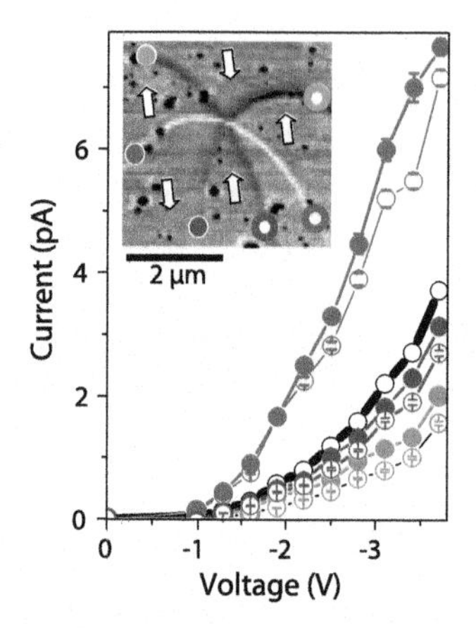

Fig. 5.20 Anisotropic electrical conductance of CDWs in improper ferroelectric $ErMnO_3$. Local conductance of the domain walls at the positions indicated in the conductivity map in the inset. The *arrows* indicate orientation of spontaneous polarization [27]

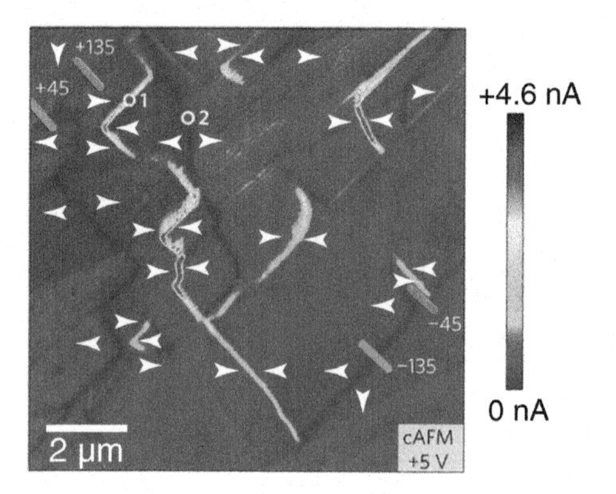

Fig. 5.21 Conductivity of CDWs in oxygen-deficient hybrid improper ferroelectric $Ca_{2.46}Sr_{0.54}Ti_2O_7$ crystal at room temperature. Conductive atomic force microscope (cAFM) image of (001) surface with a voltage of $V_{tip} = +5$ V applied to the tip. Colour scale depicts the magnitude of the measured cAFM current. The *white arrows* denote in-plane polarization conponent [28]

conductivity. As explained earlier, this trend might be attributed to the electronic compensation of polarization charge at Head-to-Head CDWs and defect compensation at Tail-to-Tail CDWs or to the contact barriers between probing electrode and the ferroelectric discussed in e.g. in [20], but otherwise little studied.

References

1. B.M. Vul, G.M. Guro, I. Ivanchik, Ferroelectrics **6**(1–2), 29 (1973)
2. Z. Surowiak, J. Dec, R. Skulski, E.G. Fesenko, V.G. Gavrilyatchenko, A.F. Semenchev, Ferroelectrics **20**(3–4), 277 (1978)
3. C.A. Randall, D.J. Barber, R.W. Whatmore, J. Mater. Sci. **22**(3), 925 (1987)
4. V.Y. Shur, E.L. Rumyantsev, A.L. Subbotin, Ferroelectrics **140**(1), 305 (1993)
5. S.B. Lang, V.D. Kugel, G. Rosenman, Ferroelectrics **157**(1), 69 (1994)
6. A.A. Grekov, A.A. Adonin, N.P. Protsenko, Ferroelectrics **13**(1–4), 483 (1976)
7. P. Maksymovych, A.N. Morozovska, P. Yu, E.A. Eliseev, Y.H. Chu, R. Ramesh, A.P. Baddorf, S.V. Kalinin, Nano Lett. **12**(1) (2012)
8. H.Y. Hwang, Y. Iwasa, M. Kawasaki, B. Keimer, N. Nagaosa, Y. Tokura, Nat. Mater. **11**(2), 103 (2012)
9. G. Catalan, J. Seidel, R. Ramesh, J.F. Scott, Rev. Mod. Phys. **84**(1), 119 (2012)
10. C.L. Jia, S.B. Mi, K. Urban, I. Vrejoiu, M. Alexe, D. Hesse, Nat. Mater. **7**(1), 57 (2008)
11. Y. Qi, Z. Chen, C. Huang, L. Wang, X. Han, J. Wang, P. Yang, T. Sritharan, L. Chen, J. Appl. Phys. **111**(10) (2012)
12. L. Li, P. Gao, C.T. Nelson, J.R. Jokisaari, Y. Zhang, S.J. Kim, A. Melville, C. Adamo, D.G. Schlom, X. Pan, Nano Lett. **13**(11), 5218 (2013)
13. P. Gao, C.T. Nelson, J.R. Jokisaari, S.H. Baek, C.W. Bark, Y. Zhang, E. Wang, D.G. Schlom, C.B. Eom, X. Pan, Nat. Commun. **2** (2011)
14. P. Gao, J. Britson, J.R. Jokisaari, C.T. Nelson, S.H. Baek, Y. Wang, C.B. Eom, L.Q. Chen, X. Pan, Nat. Commun. **4** (2013)
15. Y.L. Tang, Y.L. Zhu, Y.J. Wang, W.Y. Wang, Y.B. Xu, W.J. Ren, Z.D. Zhang, X.L. Ma, Sci. Rep. **4** (2014)
16. N. Balke, M. Gajek, A.K. Tagantsev, L.W. Martin, Y.H. Chu, R. Ramesh, S.V. Kalinin, Adv. Funct. Mater. **20**(20), 3466 (2010)
17. R.K. Vasudevan, A.N. Morozovska, E.A. Eliseev, J. Britson, J.C. Yang, Y.H. Chu, P. Maksymovych, L.Q. Chen, V. Nagarajan, S.V. Kalinin, Nano Lett. **12**(11), 5524 (2012)
18. A. Crassous, T. Sluka, A. Tagantsev, N. Setter, Nat. Nanotechnol. **10**, 614 (2015)
19. S. Wada, K. Yako, K. Yokoo, H. Kakemoto, T. Tsurumi, Ferroelectrics **334**, 293 (2006)
20. T. Sluka, A.K. Tagantsev, P. Bednyakov, N. Setter, Nat. Commun. **4** (2013)
21. P. Bednyakov, T. Sluka, A. Tagantsev, D. Damjanovic, N. Setter, Sci. Rep. **5**, 15819 (2015)
22. M. Schroeder, A. Haussmann, A. Thiessen, E. Soergel, T. Woike, L.M. Eng, Adv. Funct. Mater. **22**(18), 3936 (2012)
23. V.Y. Shur, I.S. Baturin, A.R. Akhmatkhanov, D.S. Chezganov, A.A. Esin, Appl. Phys. Lett. **103**(10) (2013)
24. T. Kaempfe, P. Reichenbach, M. Schroeder, A. Haussmann, L.M. Eng, T. Woike, E. Soergel, Phys. Rev. B **89**(3) (2014)
25. T. Choi, Y. Horibe, H.T. Yi, Y.J. Choi, W. Wu, S.W. Cheong, Nat. Mater. **9**(3) (2010)
26. W. Wu, Y. Horibe, N. Lee, S.W. Cheong, J.R. Guest, Phys. Rev. Lett. **108**, 077203 (2012). doi:10.1103/PhysRevLett.108.077203, http://link.aps.org/doi/10.1103/PhysRevLett.108.077203
27. D. Meier, J. Seidel, A. Cano, K. Delaney, Y. Kumagai, M. Mostovoy, N.A. Spaldin, R. Ramesh, M. Fiebig, Nat. Mater. **11**(4) (2012)
28. Y.S. Oh, X. Luo, F.T. Huang, Y. Wang, S.W. Cheong, Nat. Mater. **14**, 407 (2015)
29. F. Kagawa, S. Horiuchi, N. Minami, S. Ishibashi, K. Kobayashi, R. Kumai, Y. Murakami, Y. Tokura, Nano Lett. **14**(1), 239 (2014)
30. M.Y. Gureev, A.K. Tagantsev, N. Setter, Phys. Rev. B **83**(18) (2011)
31. E.A. Eliseev, A.N. Morozovska, G.S. Svechnikov, V. Gopalan, V.Y. Shur, Phys. Rev. B **83**(23) (2011)
32. E.A. Eliseev, A.N. Morozovska, Y. Gu, A.Y. Borisevich, L.Q. Chen, V. Gopalan, S.V. Kalinin, Phys. Rev. B **86**(8) (2012)
33. Y. Zuo, Y.A. Genenko, B.X. Xu, J. Appl. Phys. **116**(4) (2014)

34. T. Sluka, A.K. Tagantsev, D. Damjanovic, M. Gureev, N. Setter, Nat. Commun. **3**, 748 (2012)
35. P. Mokry, A.K. Tagantsev, J. Fousek, Phys. Rev. B **75**(9) (2007)
36. P.V. Yudin, M.Y. Gureev, T. Sluka, A.K. Tagantsev, N. Setter, Phys. Rev. B **91**, 060102 (2015)
37. I. Stolichnov, L. Feigl, L. McGilly, T. Sluka, X.K. Wei, E. Colla, A. Crassous, K. Shapovalov, P. Yudin, A. Tagantsev, N. Setter, Nano Lett. **15**(12), 8049 (2015)
38. S. Liu, F. Zheng, N.Z. Koocher, H. Takenaka, F. Wang, A.M. Rappe, J. Phys. Chem. Lett. **0**(ja), (2015)
39. A. Tagantsev, L. Cross, J. Fousek, *Domains in Ferroic Crystals and Thin Films* (Springer, New York, 2010)
40. J. Fousek, V. Janovec, J. Appl. Phys. **40**(1) (1969)
41. B.B. van Aken, T.T.M. Palstra, A. Filippetti, N.A. Spaldin, Nat. Mater. **3**(3) (2004)
42. Y. Geng, N. Lee, Y.J. Choi, S.W. Cheong, W. Wu, Nano Lett. **12**(12), 6055 (2012)
43. B. Mettout, P. Tolédano, M. Lilienblum, M. Fiebig, Phys. Rev. B **89**, 024103 (2014)
44. A.P. Levanyuk, D.G. Sannikov, Uspekhi Fizicheskikh Nauk **112**(4), 561 (1974)
45. A.K. Tagantsev, Ferroelectrics **79**, 351 (1988)
46. F. Kadlec, J. Petzelt, V. Zelezny, A.A. Volkov, Solid State Commun. **94**(9), 725 (1995)
47. G.M. Guro, I. Ivanchik, N.F. Kovtonyu, Sov. Phys. Solid State, USSR **11**(7), 1574 (1970)
48. V.F. Krapivin, E.V. Chenskii, Sov. Phys. Solid State, USSR **12**(2), 454 (1970)
49. Y.L. Li, L.Q. Chen, Appl. Phys. Lett. **88**(7) (2006)
50. J. Hlinka, P. Ondrejkovic, P. Marton, Nanotechnology **20**(10) (2009)
51. S. Semenovskaya, A.G. Khachaturyan, J. Appl. Phys. **83**(10), 5125 (1998)
52. M.J. Haun, E. Furman, S. Jang, L. Cross, Ferroelectrics **99**, 45 (1989)
53. N.A. Pertsev, V.G. Kukhar, H. Kohlstedt, R. Waser, Phys. Rev. B **67**, 054107 (2003)
54. P. Marton, I. Rychetsky, J. Hlinka, Phys. Rev. B **81**(14) (2010)
55. M. Zgonik, P. Bernasconi, M. Duelli, R. Schlesser, P. Gunter, M.H. Garrett, D. Rytz, Y. Zhu, X. Wu, Phys. Rev. B **50**(9), 5941 (1994)
56. H.I. Yoo, M.W. Chang, T.S. Oh, C.E. Lee, K.D. Becker, J. Appl. Phys. **102**(9) (2007)
57. M. Choi, F. Oba, I. Tanaka, Appl. Phys. Lett. **98**(17) (2011)
58. D. Smyth, Classic dielectric science book series, *The Defect Chemistry of Metal Oxides* (Oxford University Press, New York, 2000)
59. Y. Xiao, V.B. Shenoy, K. Bhattacharya, Phys. Rev. Lett. **95**(24) (2005)
60. D. Porter, K. Easterling, *Phase Transformations in Metals and Alloys*, 2nd edn. (Taylor & Francis, 1992)
61. B. Strukov, A. Levaniyuk, *Ferroelectric Phenomena in Crystals: Physical Foundations* (Springer, Berlin, 1998)
62. I.I. Ivanchik, Ferroelectrics **145**(1), 149 (1993)
63. G. Rosenman, V.D. Kugel, D. Shur, Ferroelectrics **172**(1), 7 (1995)
64. P. Maksymovych, S. Jesse, P. Yu, R. Ramesh, A.P. Baddorf, S.V. Kalinin, Science **324**(5933), 1421 (2009)
65. N. Balke, S. Choudhury, S. Jesse, M. Huijben, Y.H. Chu, A.P. Baddorf, L.Q. Chen, R. Ramesh, S.V. Kalinin, Nat. Nanotechnol. **4**(12), 868 (2009)

Chapter 6
Extended Defects in Nano-Ferroelectrics: Vertex and Vortex Domains, Faceting, and Cylinder Stress

James F. Scott

Abstract The rapid success of density functional theory (DFT) has created the impression in the scientific community that most problems of interest involving ferroelectric and multiferroic structures can be solved via DFT. Unfortunately, this is not the case: DFT invariably requires the assumption of periodic boundary conditions, which can often be tantamount to throwing out the baby with the bath water. Problems of vertex and vortex structures, of faceting, of preferential nucleation and switching around the boundaries of nano-crystals (Bessel-function-like propagation) and many other problems are intrinsically aperiodic, as well as being inherently time dependent (nucleation and growth); some are both nonlinear and non-equilibrium (kinetic rather than thermodynamic, for which the Landau free-energy approach is also not optimum). In this chapter we examine some of those problems.

6.1 Introduction

The two most important things about ferroelectric oxides are that they generally are not insulators. Often (as in the old USSR) the syllabus for physics students offered options of semiconductor physics or ferroelectrics but not both. However, the popular perovskite oxides have bandgaps below those in wide-gap III-Vs such as GaN or II-Vis such as ZnO (e.g., $PbTiO_3$ is at ca. 2.87 eV [1]). When ferroelectric devices were bulk ceramics mm thick, this was satisfactory, but devices made from 100-nm thin films conduct, and it becomes necessary to understand their majority carriers, effective masses, mobilities, and whether their conduction is space-charge-limited, Schottky-limited, Poole-Frenkel, Fowler-Nordheim, etc [2]. The ferroelectrics literature is replete with serious errors that have delayed engineering progress. One group

J.F. Scott (✉)
Cavendish Laboratory, Department of Physics,
University of Cambridge, Cambridge CB3 0HE, UK
e-mail: jfs4@st-andrews.ac.uk

J.F. Scott
Departments of Chemistry and Physics,
St. Andrews University, St. Andrews, Fife KY16 9ST, UK

J. Seidel (ed.), *Topological Structures in Ferroic Materials*, Springer Series in Materials Science 228, DOI 10.1007/978-3-319-25301-5_6

in Switzerland has published errors for the most popular PZT (lead zirconate-titanate) material, asserting [3] that it is usually p-type and then later giving an electron effective mass of $m^* = 1.0$ m(e) [4]. In fact PZT is usually n-type with $m^* =$ ca. 6 m(e); $BaTiO_3$ and $SrTiO_3$ have $m^* = 5.1$–6.5 m(e) [5]; this cannot be due to differences between band masses and transport mobility masses, since Mahan has shown that the band mass m^* and transport mass m^{**} are equal in these systems [6]. A second huge error from both Belgium [7] and Yokohama [8, 9] was the claimed value of the Schottky barriers for BST/Pt or PZT/Pt or strontium bismuth tantalate SBT/Pt; these groups were in error by approximately 1.0 eV, giving 5 eV as the (erroneous) bandgap in SBT; this error was corrected by the present author based upon correct electron affinity measurements by Dixit et al. [10]. It is difficult to understand how one can have $1 million in kit with 10 meV accuracy and get such numbers wrong by 1 eV. Finally the field has been plagued by the erroneous belief that ferroelectric films much thinner than a micron would be unstable and useless because surface depolarization fields would destabilize the polarization. This originated in IBM in the 1960s [11], led to the decision by Rolf Landauer to terminate ferroelectric memory work there, [12] was exacerbated in Japan [13] and in North Carolina, [14] and delayed progress for three decades. In fact, ferroelectric thin film devices have progressed despite these researchers, not because of them.

Six is bigger than three. The actual limit in film thickness for stable ferroelectricity in most oxides is 2–3 nm, [15, 16] not a micron. This was first shown by Marty Gregg's group in Belfast [17]. That number is very important because ferroelectric tunnel junctions can function up to ca. 6 nm thickness, even with direct tunnelling, as shown beautifully by the work of Barthelemy, Bibes, Garcia et al. at Thales [16–18]. The fact that 6 is bigger than 3 implies that one can fabricate very good voltage-driven tunnel junctions.

6.2 Definitions: Vertex, Vortex, and Kosterlitz-Thouless Melting

One should be careful with definitions. In ferroelectrics some authors use the terms "Vertex domains" and "Vortex domains" interchangeably. In my opinion this is very unwise. Vortex domains require a non-zero polarization curl; whereas vertex domains can be simple intersections with a resulting pure divergence of P nearby, as illustrated in Mermin's famous review article [18]. (Mermin also shows that the distinction between vertex and vortex is NOT closely related to winding numbers, and that +1 winding numbers can involve a rotation-like curl or a pure divergence of P and do not discriminate between the two in any way.) This view is different from that published by Cheong [19].

Another point of interest is the possible occurrence of Kosterlitz-Thouless melting and hexatic [2D] phases in ferroelectrics [20]. In my judgment, KT-melting has not been shown, and there are theoretical reasons to believe that it requires a Potts model with n = 5 or greater, [21] whereas the systems studied thus far are describable with either Potts models or clock models (vector Potts models) of n = 3 or 4 (recall that a Potts model of n = 2 is just the standard Ising model). One should not be misled by the lattice symmetry, especially if this exhibits a threefold or sixfold axis; as shown below, faceting and domain shapes can exhibit symmetries higher than the lattice symmetry. Although it is true that ferroelectric thin films often exhibit hexagonal faceting and approximate two-dimensional system if very thin, hexagonal is not hexatic, and these systems do not appear to exhibit KT-melting.

6.3 Basic Theory: Landau-Lifshitz-Kittel as Extended by (a) Lukyanchuk; (b) Catalan, Schilling, Scott et al.

The basic theory of stripe domain structures and closure domain (vertex) structures was given by Landau and Lifshitz in 1935 [22] (two years before their more famous paper on order parameters and free energies), and independently re-derived and extended by Kittel in 1946 [23]. They derived from the balance of axial stress and depolarization fields the simple relationship that the domain stripe width w is proportional to the square root of the film thickness. This has been satisfied from nm to mm thicknesses in a wide range of ferroelectrics and ferromagnets (Fig. 6.1). However these authors did not calculate the proportionality constant. Much more recently this was done independently by Lukyanchuk's group at Amiens [24] and by Catalan and Scott [25]. The resulting express in (6.1) gives an exact formula with a dimensionless constant of order unity, by calculating the ratio of stripe width to domain wall thickness. The latter is not easy to measure, because thermal motion gives in most X-ray studies a misleading time-average that is spatially widened by thermal excursions (similar to a Debye-Waller factor for ions). Note that (6.1) has no adjustable parameters but does involve the anisotropy of the dielectric constant. A graph of w versus d for several common ferroelectrics has been presented in the review by Catalan et al. Note that these values of domain width w are the thermal equilibrium values. If one tries to alter widths w, for example by periodic poling of $LiNbO_3$, there will always be a tendency to relax back to these equilibrium values over time and use; this gives rise to degradation of periodically poled devices for second harmonic generation or other nonlinear optics applications [26].

Catalan-Scott-Lukyanchuk expression for stripe domain widths was functions of film thickness d and domain wall thickness δ:

$$w^2/d\,\delta = (2\,\pi^3/21\,\zeta(3))\,[\,\chi(y)/\,\chi(z)]^{1/2} \tag{6.1}$$

where ζ is the Riemann zeta function of argument 3 and χ is the susceptibility normal or parallel to the polarization.

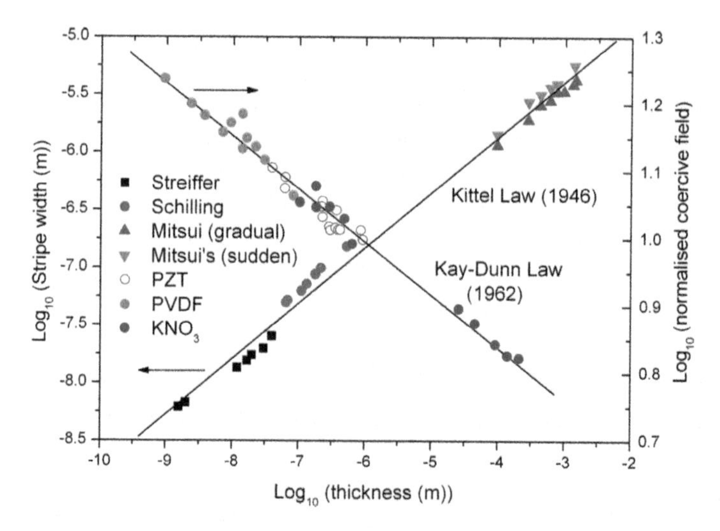

Fig. 6.1 Plot of stripe domain width versus film thickness for ferroelectrics of large lateral area, showing agreement with the Landau-Lifshitz-Kittel Law from nm to mm. The other *line* is coercive field versus *thickness*, showing agreement with the Kay-Dunn theory. Figure from Catalan et al.

6.4 Statics

(a) clock models compared with Potts models (Srolovitz); (b) comparison with hexatic arrays and Kosterlitz-Thouless models; (c) Landau-Lifshitz-Kittel model; (d) Roytburd model; (e) Arlt model.

Figure 6.2 shows schematically two possible closure or vertex structures for ferroelectric domains (Srolovitz and Scott [27]). In one case (Potts model), the adjacent pairs of threefold vertices are stable, and any fourfold coalescence will be unstable and short-lived; this is the experimental situation found by McQuaid et al. very recently [28]. Parenthetically we note that the vertex-vertex trajectories observed by

Fig. 6.2 Closure domains and vertex structure in ferroelectrics of rectangular geometry. From Schilling, Prosandeev et al.

McQuaid closely follow the equipotential calculated by Francu for this geometry due to stress. The threefold vertex-antivertex pairs were also studied (and created by AFM tip voltages) by Ivry, Durkan et al. [29, 30]. In the second case (clock model—vector Potts model)), the fourfold vertex is stable and the adjacent threefold vertex-antivertex pairs will coalesce. This situation has not been observed in ferroelectrics yet. Parenthetically we emphasized in the sections above that Kosterlitz-Thouless hexatic structures and KT-melting requires Potts models higher than $n = 4$, so that these vertex structures in ferroelectric thin films do NOT lead to Berezinsky-Kosterlitz-Thouless ("BKT") melting.

These considerations for ferroelectric domains all apply to ferroelastic twinning as well, as shown in detail by Roytburd [31]. The most precise numerical application of these models has been given by Arlt [32].

6.5 More statics

(a) Ising-like compared with Bloch- or Neel-like; (b) Off-centering (Schilling, Prosandeev et al.); (c) Hoop stress (Scott); (d) Kolmogorov-Avrami-Ishibashi theory; (e) Fitting of KAI theory by Scott and by Shur; (e) Topological-defect and nucleation-limited (X. Du and I-Wei Chen; Tagantsev).

For several decades it was assumed that ferroelectric domain walls were Ising-like and almost atomically thin (a few unit cells) with no curvature. However, rather early Lajzerowicz [33] showed that ferroelectric domain walls were wide enough to exhibit phase transitions within the walls at temperatures different from bulk. This theoretical work was extended recently by Daraktchiev et al. [34, 35] and this year by Iniguez et al. [36] and by Salje and Scott [37]. The best experimental studies are reviewed by Catalan et al. and form the basis of what is now termed domain-wall nano-electronics. This was stimulated by the photovoltaic studies of Seidel et al. [38].

One of the surprising studies of nanodomain structures is the off-centering of vertex patters in rectangular nanocrystals, [39] reported by Schilling, Prosandeev et al. (Fig. 6.2) The modelling of Prosandeev and Bellaiche shows that this can be described as a geometrically driven phase transition.

The off-center vertex site loci have not been calculated but presumably are determined by surface polarization energies on the two inequivalent sides of the rectangles. Their positions depend upon the aspect ratio of the rectangle, and the vertex loci resemble those for field minima in the rectangular model calculation of Francu [40].

6.5.1 *Hoop Stress*

As discussed above, the Landau-Lifshitz-kittel Law for stripe domain width is based upon the balance of axial stress and depolarization fields. If there were only axial stress, materials would usually exhibit single-domain states. However, breaking up into narrow domains saves depolarization energy at the surfaces. But such domains increase wall energy. In their calculations Landau and Kittel ignored "hoop stress" (or cylinder stress), because they were modelling parallel-plate capacitors of infinite lateral area. Of course, modern physics and device engineering emphasize nano-crystals of small lateral size (the crystals studied by Scott and Kumar are only 8 nm in diameter). This finite size implies that "hoop stress" (or cylinder stress) is not negligible. This is well known to architects and mechanical engineers. This is an azimuthal or tangential stress the increases the circumference of a ring or disk. The important thing is that it varies not as reciprocal area of the base, but as the circumference. Putting this extra term in 1/r into the Landau free energy, together with the original 1/r-squared stress term gives the result shown in Fig. 6.3 [41] and (6.2) below

$$Uw^2 = Bd - cw/[(d/w) - 1], \tag{6.2}$$

For small r (nano-disks) the dependence of w upon d is linear (Fig. 6.3). This seems to be confirmed in very new data from Lichtensteiger et al. [42] (Fig. 6.4).

6.5.2 *KAI Theory*

For some years the standard theory for nucleation and growth was the KAI theory (Kolmogorov, Avrami, Ishibashi) [43, 44]. Originally designed for grain growth, this model fit domain growth and switching extremely well. Particularly convenient is the fact that there is a self-consistency check on the model: For a given dimensionality the product of the peak displacement current i(t) = dD/dt (approx. = dP/dt) multiplied by the time t(peak) at which this occurs, gives a precise fraction of the spontaneous polarization P(s) [45]. See (6.3a and 6.3b) below:

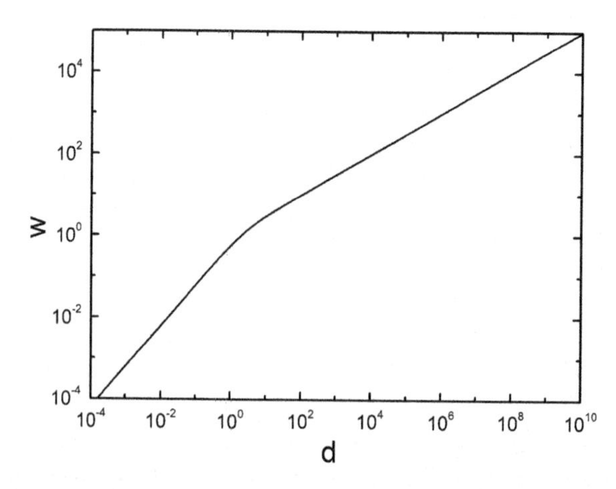

Fig. 6.3 Calculated stripe width versus film thickness including hoop stress, from (6.2), with B, c, d set equal to unity. Figure courtesy of P. Zubko

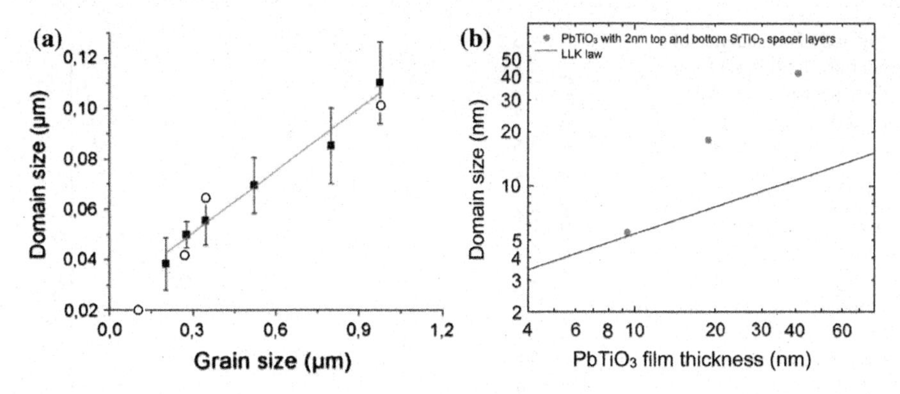

Fig. 6.4 **a** Measured dependence of stripe width versus grain diameter for $K_{1/2}Na_{1/2}NbO_3$ [M. Eriksson, Ph.D. thesis, Stockholm]. **b** Lead titanate domain widths versus film thickness (Note that 3a is versus grain size, not film thickness) showing apparent linear relationship in contrast with Landau-Lifshitz-Kittel square root prediction Lichtensteiger et al. [42]

$$i(max)\ t(max)/Ps = constant = B \qquad (6.3a)$$

where the dimensionless constant is not freely adjustable but can be calculated for a given dimensionality from the model and depends strongly on the dimensionality of the domain growth (e.g., needle-like).

This equation is fitted simultaneously with

$$i(t) = (2P_sAn/t_o)(t/t_o)^{n-1}Pexp[-(t/t_o)^{n]} \qquad (6.3b)$$

in Ishibashi's theory. The latter gives the dimensionality of the domain growth n (typically between 1.0 and 3.0) and scaled dimensionless time t/to = u. A is capacitor area.

This works very well in KNO_3, PZT, and some other ferroelectrics, [46] and the dimension of the domains is found to be nearly 3.0 in some cases and as low as 1.0–2.0 in others. Fits of (6.3b) for KNO_3 give u = 1.79 and n = 1.63; for such values the model of Ishibashi and Takagi predict the constant B = 0.68 in (6.3a); the measured value is 0.67 + / − 0.08 for 300-nm thick films. The theory predicts needle-like domain growth with dimension 1.6 + / − 0.2. But for very thin potassium nitrate films, the dimensionality is found to be 3.1 + / − 0.2 (three-dimensional spherical-growth is the rate-limiting step). The results are completely self-consistent. The model has two cases: One with pre-existing nano-domains ("nuclei"), and the other, without. There are a few unphysical assumptions made to make the model analytically tractable: notably, the domain wall speed v is assumed independent of domain radius r; this is generally untrue, and v is typically 1/r, decreasing with growth.

A different model, based upon nucleation as the rate-limiting step, was developed by Du and Chen [47, 48]. Jung et al. (Fig. 6.5) show that it gives a better fit to most ferroelectrics than does the KAI theory, [49] although KAI is still used widely

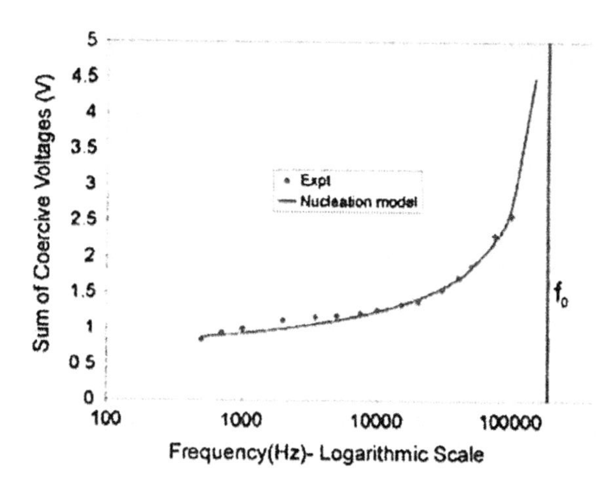

Fig. 6.5 Dependence of coercive field $E_c(f)$ for a ferroelectric perovskite oxide upon frequency of applied electric field (Jung et al.). The data exhibit a divergence at the reciprocal of the domain nucleation time, in accord with the theory of Du and Chen. The fit to the KAI theory does not satisfy that and gives no frequency divergence at all

by Shur's group, especially for $LiNbO_3$ and $LiTaO_3$ [50]. Note that the Du-Chen model predicts a strong divergence of coercive field with the frequency f of the ac applied voltage V. For BST Jung found that $E_c(f)$ increases from 0.8 to 2.5 V as f goes from 50 Hz to 100 kHz. Even greater increases are known for the $BaMF_4$ family of ferroelectrics, as shown in Japan [51]. This means that newly reported ferroelectrics such as $GaFeO_3$ or $LaTaO_4$ should be tested for switching at 50 Hz or lower, not 100 kHz; by 100 kHz the coercive fields may surpass the breakdown fields. The physical reason for this frequency divergence in coercive field is long nucleation times: Jung showed in PMN-PT a nucleation time of ca. 500 μs; hence as f approaches 200 kHz there is insufficient time in each cycle to reverse polarization domains, and E_c therefore becomes very large.

The work of Du and Chen was "rediscovered" by Tagantsev with very similar conclusions and published without citations to the original work [52, 53].

6.5.3 *Faceting*

(a) Lukyanchuk, Gruverman et al.; (b) faceting oscillations (Scott and Kumar; Ahluwalia, Ng et al.).

Faceting in nature is a very deep problem that transcends ferroelectric films. However in the case of such films it is especially interesting: For very thin films the aspect ratio is such that we might hope to approximate them as two-dimensional systems. However, for such [2D] systems the perimeter is one-dimensional, and long-range order is not thermodynamically stable in one dimension [54]. Thus faceting is likely to be kinetically limited and non-equilibrium. It may also be nonlinear. So at the outset we are faced with a nonlinear, non-equilibrium problem [55, 56].

Textbooks sometimes wax poetic on why snow-flakes have hexagonal symmetry and state that it is obvious that this arises from the fundamental bond angles in molecular water. This is not only not obvious, it is not true. Sometimes facets transcend the lattice atomic symmetry and produce square or pentagonal facets in materials with rhombohedral or hexagonal lattice symmetry. We shall see why.

The easiest approach is to begin with Thiele's 1971 theory of faceting in magnetic bubble domains [57]. Under normal conditions magnetic bubble domains are circular disks (or cylinders). However, with applied fields or stress they switch to elliptical with a twofold axis. This is reversible and can be used to encode a "1" (elliptical") or a "zero" (circular) as the binary Boolean algebra in a magnetic bubble memory. Although these were touted as "the memory of the future" when I was at Bell Labs in 1970, they failed because they are very slow (sequential access rather than random access).

Recently Lukyanchuk et al. have extended Thiele's kinetic theory of magnetic bubble domain faceting to ferroelectric films [58]. The important point is that the larger susceptibility permits facets of 4, 5, 6, … to be accessed kinetically in a metastable state. There are macroscopic analogies known in plastics. Illustrations are shown in Fig. 6.6 from Gruverman. Even more surprising is the faceting oscillation from disks to hexagons observed by Scott and Kumar [59] under HRTEM e-beam illumination (Fig. 6.7). The hexagonal faceting has been very precisely simulated by Ahluwalia,

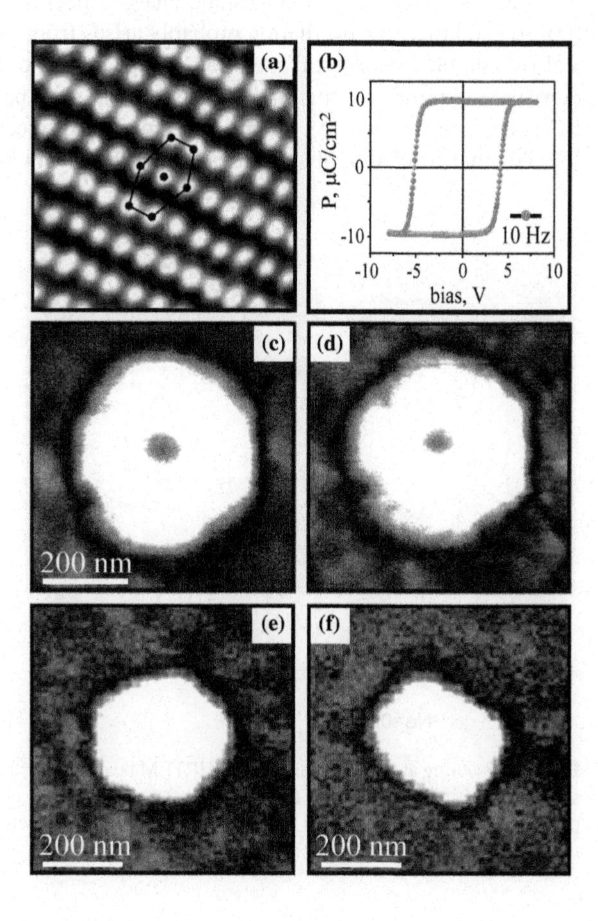

Fig. 6.6 Polygon faceting in PVDF-TFE thin films. Figure from Lukyanchuk et al.

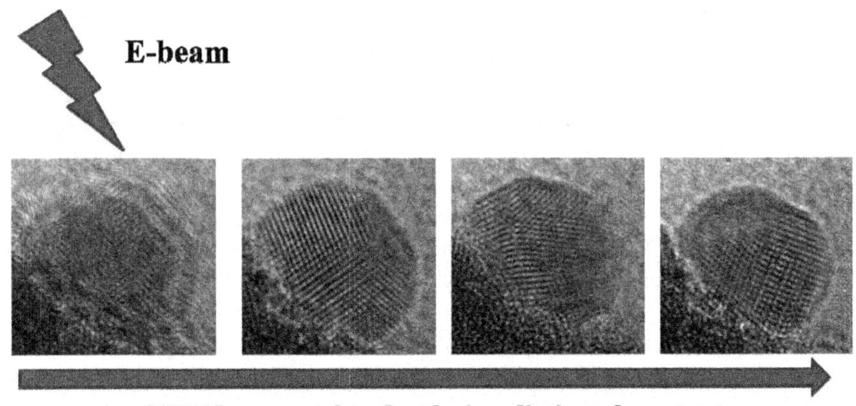

Fig. 6.7 Faceting at edges of 8-nm nanodisk: Experiment (Scott and Kumar)

Ng et al. as arising from electrostatic charge injection from the e-beam (Fig. 6.8), [60, 61] and the slow oscillation probably arises from piezoelectricity and mechanical overshoot. Malozemoff has reported analagous overshoot effects in magnetic domains and points out that in that case the very slow period is not a linear response LC resonance (inductance L coming from charge injection, as shown by Jonscher [62, 63]) but from a highly nonlinear non-equilibrium kinetic response.

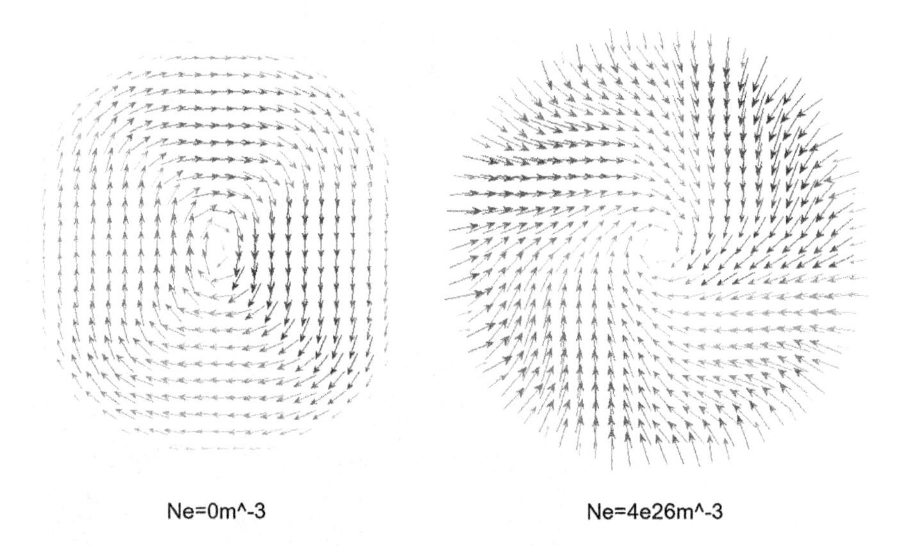

Fig. 6.8 Faceting at edges of nanodisk in HRTEM beam due to charge injection: Model simulation showing faceting with parallel domain polarization and more circular disk with normal domain polarizations (Ahluwalia, Ng, et al.) The numbers Ne are the electron concentration per cubic meter injected by the HRTEM beam

It is believed that hexagonal facets are special in some way, and those have been known since the days of the Schwartz-Hora Effect [64, 65]. They may relate to Plateau's Laws for soap bubble raft formation, which favor hexagonal geometries [66]. It is most important to note in Fig. 6.7 that the stripe domains are primarily aligned normal to the perimeter for the circular geometries, whereas they flop 90° to realign parallel to the facets in the hexagonal geometries. This shows that the exterior faceting is driven by domain wall rotation to minimize depolarization fields at the edges.

6.5.4 Toroidal Domains (Ginzburg, Kopaev, et al.; Fiebig)

Back in the 1960s Soviet scientists including Zeldovich, [67] and especially Kopaev, Gorbatsevich and Ginzburg, [68, 69] considered a special kind of previously undiscovered domain—toroidal domains in which both time and space reversal were violated. Athough Zeldovich was not a condensed matter theorist, he was greatly interest in symmetry-breaking, stimulated by the discovery of parity non-conservation in that era. These toroidal domains were first measured by Fiebig, and good reviews exist on the topic [70]. This completes the possible symmetries of domains in multiferroics: Ferroelectric domains symmetric with regard to time inversion but antisymmetric with regard to spatial inversion; ferromagnetic domains with space inversion symmetry but temporal asymmetry; and toroidal domains (Fig. 6.9 from Gorbatsevich) with antisymmetric temporal and spatial symmetry.

(a)

(b)

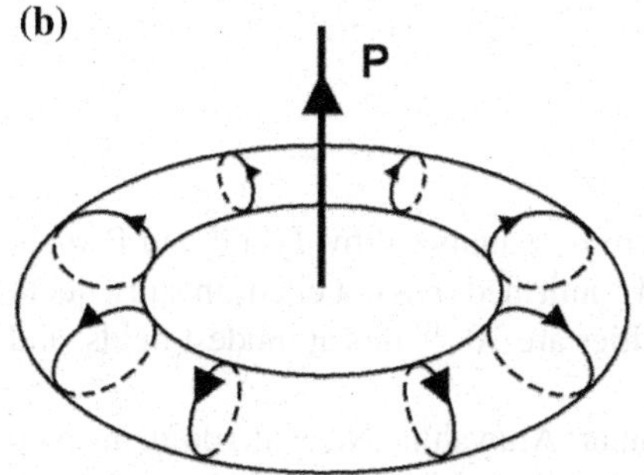

Fig. 6.9 Schematic diagram of toroidal domain polarization. **a** From X. Fu; **b** From Gorbatsevich et al.

6.6 Dynamics:

6.6.1 Comparison with Magnetic Domains

Ferroelectric domains traditionally were considered to be nearly the same as magnertic domains. Although they exhibit some static similarities, their dynamics are radically different: Magnetic domain wall motion satisfies the Landau-Lifshhitz-Gilbert equations [71], which are first-order in time, so magnetic walls have no momentum or inertia; when the magnetic field H is turned off, the spin precession stops, since there is no "coasting" to a stop with first-order time-dependent differential equations. In contrast to this, ferroelectric walls follow Newton's Second Law, which is second-order in time, so that walls "coast" as much tens of microns after the applied electric field E terminates. A second point of qualitative difference is that magnetic walls can be supersonic (Democritov, Kreines et al., 1984) [72] whereas ferroelectric walls cannot be; when the magnon spin wave velocity exceeds the acoustic phonon speed, coherent emission of phonons at a Cerenkov-like "bow wave" angle results. In comparison, supersonic ferroelectric domain walls would cause fracture of the crystal from their shock waves, since real ionic motion is required. Early textbooks sometimes inferred that ferroelectric walls were supersonic, simply because the peak displacement current dD/dt was faster than the transit time d/v of a sound wave from cathode to anode in a sample of thickness d; but this naively assumed that all domains nucleated at an electrode surface, which is simply not true (as Ishibashi showed, nucleation of reversed domains occurs throughout the interior of the dielectric).

A general theory of domain wall dynamics in multiferroics has been given only very recently by Lukyanchuk, Sidorkin et al. [73]. It combines the limiting cases of Newton's Laws for ferroelectric walls and the Landau-Lifshitz-Gilbert Equation for magnetic walls. However, the resulting equation is not trivial to fit experimental data to.

One of the characteristics of domain wall motion that has been elucidated relatively recently is that wall motion can often be described with ballistic models and viscous drag. This was initially demonstrated by Dawber and Scott [74] and subsequently confirmed by Gruverman et al. [75] and by Baudry et al. [76].

6.6.2 Domain Wall Creep

Domain wall creep and creep exponents were first analyzed by Tybell and Paruch; [77] much of their work was reproduced and confirmed (but not cited) in subsequent work in Korea [78]. The typical creep velocities are 10^{-10} m/s at modest fields, and the creep exponent is very nearly unity.

Under e-beam illumination (Scott and Kumar; Aluwahlia, Ng et al.) domain creep controls faceting. Figure 6.7. Simulation studies of Aghluwalia and Ng model this in Fig. 6.8; and earlier experiments by Ganpule et al. [79] and by Gruverman et al.

[80] illustrate some examples (however, the origin in the early work was not obvious because the films were [111]-oriented with pseudo-sixfold symmetry). The earlier work lacked atomic resolution provided by the recent HRTEM studies. The latter show that the faceting is driven by HRTEM injected charge and not thermal heating, and more importantly, that the internal ferroelectric domain wall alignment is predominantly normal to the perimeter in disks but parallel to the facets in hexagonally faceted nano-crystals. Hence the faceting is driven by 90° realignment or flopping of stripe domains. This arises from P.n depolarization fields at the perimeter, where n is the unit vector orthogonal to the perimeter. One should keep in mind that HRTEM are two-dimensional images of three-dimensional polarization patterns, so little information exists concerning the out-of-plane components. It is quite possible that the stripe domain reorientation occurs via out-of-plane rotation. The speed of the domain wall reorientation in the experiments of Scott and Kumar is ca. 1 nm/s, comparable to the creep velocities inferred by Paruch, Tybell et al.

Other very recent measurements include vertex-vertex collisions (McQuaid, Gregg et al.) in real time [81]. In this case (Fig. 6.10) one vertex domain is driven (probably by strain incurred from depolarization fields along the edges of the sample. The trajectory of the moving domain follows closely an equipotential surface (Fancu). The true coalescence of adjacent threefold vertex structures to form one X-shaped fourfold closure domain vertex does NOT occur, as explained by the model of Srolovitz and Scott. Notable is the velocity v(x) of one vertex as a function of distance x to the adjacent veretex:

$$V(x) = ax \tag{6.4}$$

which gives rise to an exponential dependence of position with respect to time. This is the exact result expected from the Standard Linear Model in mechanics, which arises from an equivalent circuit model consisting of a Maxwell element and a Kelvin-Voigt element (i.e., springs and dashpots in both series and parallel), which is expected for linear strain.

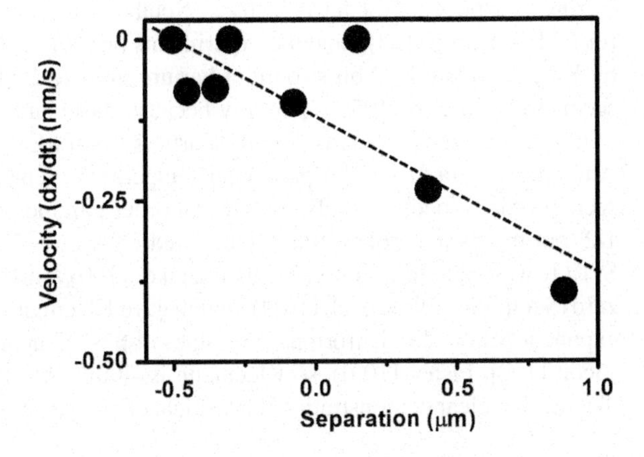

Fig. 6.10 Domain wall vertex motion in real time near an adjacent vertex, satisfying (6.4). The trajectory follows an equipotential due to stress (Francu), and the stress arises at the edges, probably due to depolarization fields

Domain walls can also be injected (L. W. Chang, McQuaid, Gregg et al. [82]; Ivry, Durkan et al. [83]); as shown by Ivry et al., these are usually injected as nearby vortex-antivortex pairs. These can be used to create multi-level memory elements by offering specific quantized conduction trajectories in resistive RAMs (McQuaid et al.). A related but more complex process involving domain wall injection is the domain wall "ping-pong" process developed by Jiang, Hwang et al. in which domains injected from a cathode have their travel to the anode synchronized with the applied ac electric field; this permits the domains to be "volleyed" back and forth without entering the dead layer of high loss near each electrode and thereby enhances the dielectric constant by as much as x100 [84]. Such an enhancement in electric susceptibility would permit memory elements in FRAMs with much smaller size or "footprints".

6.6.3 Pinning

The pinning of domains is generally not by point defects, which are traditionally treated well by theoretical models, but via extended defects—arrays of oxygen vacancies, screw dislocations, etc. (Ivry, Salje, Durkan et al.) [85]. Such defect dynamics have been theoretically modeled and their exponents predicted (Levanyuk and Sigov; Sigov et al. [86, 87]). Typically the exponents are numerically much larger than for intrinsic models (specific heat exponent ca. 1.0; ultrasonic attenuation exponent ca. 1.5–3.0 or even larger,; and these are in very good agreement with those measured experimentally in $BaMnF_4$, oxygen-18 isotopic $SrTiO_3$, PMN-PT, TSCC, and other ferroelectrics (Fritz [88]; Bobnar et al. [89], Scott [90]; Scott, Pirc et al. [90]). Most recently the exponent n describing thermal expansion at Tc was measured for a second-order ferroelectric phase transition (Lashley et al. [91]; Scott [92]) and found (Fig. 6.11) to be $-1.51 +/- 0.02$, in exact agreement with the Sigov prediction of 3/2 (Table 6.1).

There are still experiments and theories that claim non-mean field (and non-extrinsic) exponents for ferroelectrics. Notable are the theoretical work of Kornev for PZT (I. Kornev [93]), and the experiments on PZT (Z-G Ye et al. [94]). The work by Kleemann and Dec on strontium barium niobate ("SBN"—$Sr_{0.61}Ba_{0.39}Nb_2O_6$) seems to be in error, [95–97] simply because these authors assumed the transition to be second-order, whereas [98] it clearly is first-order (as admitted by Dec [99]). Attempts to fit a first-order parameter temperature dependence curve near Tc to a second-order model must always give an order parameter exponent much less than 1/2, but this is a simple artifact. Unrelated errors by Kleemann for 0–18 isotopic $SrTiO_3$ were due to a more serious mistake, [100] tantamount to dividing zero by zero (Scott, Pirc, Blinc et al. [101]), which gave Kleemann a value of order parameter exponent beta = 2.0. Unfortunately values of $\beta > 2/3$ appear to violate hyperscaling (Scott [102], Fisher [103]), so Kleemann is simply wrong. It is worth pointing out that the first clear demonstration that critical exponents near Tc in ferroelectrics were

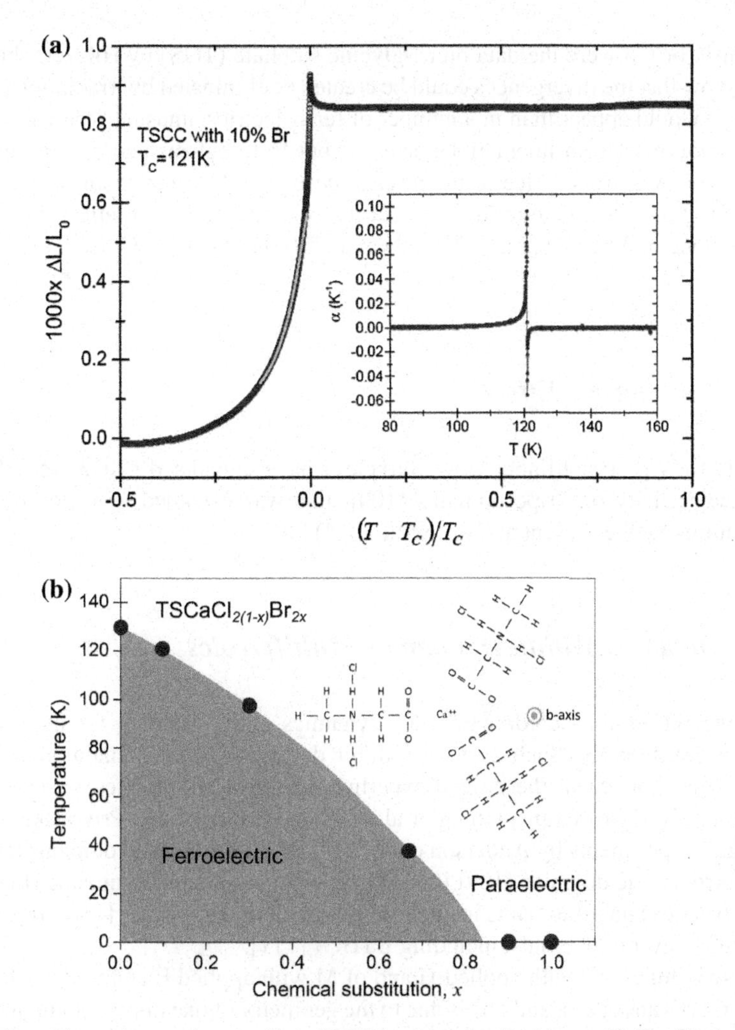

Fig. 6.11 **a** Thermal expansion near Tc in TSCC ferroelectrics (Lashley et al.). The fitted exponent is $-1.51 +/- 0.02$, agreeing exactly with the defect model of $-3/2$ from Sigov et al. This material is of special interest because under bromination to replace Cl-ions, its Curie temperature goes to $T = 0$, a Quantum Critical point, **b** Phase diagram for TSCC:Br

Table 6.1 Topological defect-dominated exponents near T_c [t = reduced temperature (T-Tc)/Tc]

Exponent	Topological defect model	Material
Isothermal susceptibility γ	5/2	PMN-PT, $BaMnF_4$
Specific heat α	1.0–1.5	$KMnF_3$[a], CsH_2PO_4[b], $BaMnF_4$
Ultrasonic attenuation η	5/2	$BaMnF_4$
Thermal expansion n	−3/2	TSCC

[a] S. A. Kishaev, G. A. Smolensky, A. K. Tagantsev, Pis'ma Zh. Eksp. Teor. Fiz. 43, 445 (1986)
[b] E. D. Yakushkin, A. I. Baronov, and L. A. Shuvalov, Pis'ma Zh. Eksp. Teor. Fiz. 33, 27 (1981)

of extrinsic origin were the data on tri-glycine sulphate (TGS) by Hilczer; she was able to show that the divergences could be created or eliminated by irradiation [104].

Thus it would appear than in a number of ferroelectrics, unusual "critical" exponents do not arise from fluctuations near Tc, but from defects and pinning effects that are wholly extrinsic, due to topological defects. PZT remains an unexplained puzzle, but that may relate to the fact that its local structure is complex, involving not only none monoclinic phase (Noheda et al. [105]a), but two (Glazer et al., 2014 [105]b).

6.6.4 *Pyroelectric Effects*

Lane Martin's group (Illinois, now Berkeley) have elucidated (2013) the effects on piezoelectricity of stripe domains [106]; this was extended to vertex domain contributions by the present author (Scott 2014) [107].

6.6.5 *Domains Within Domains—Multiferroics*

All of the discussions of domain wall dynamics above become far more complex for multiferroics. Such systems exhibit domains within domains and walls within walls: For recent theory and experiment readers are directed to: (a) Scott, Salje et al. [108]; (b) Evans, Gregg et al. [109]; (c) Janovec and Privratska [110]; (d) Prague experiments by Anderson et al. [111]. Both ferroelastic domains (twins) inside ferroelectric domains [109] in lead iron-tantalate-zirconate-titanate (PFTZT, and ferroelectric nano-domains inside ferroelastic domains (twins) [110] have been observed in several materials, including $BaTiO_3$ [111].

The switching of P with applied H and of M with applied E is not equivalent or symmetric (Evans et al.), and this is due to the geometry of one kind of domain being nestled within another. Figure 6.12. Gregg's group [82] had previously illustrated this nesting and self-similarity of domains.

6.7 Transport

Domain walls can provide very high electrical conduction in rather insulating materials. Initially shown by Seidel et al. (Fig. 6.13), [112] this has been measured in different materials by Kalinin et al. [113], by Noheda et al. [114], and by Paruch et al. [115], with very recent studies from Lausanne (McGilly, Stolichnov et al. 2014) [116]. A review of work up to 2013 is given by Catalan et al. It has not yet been determined whether the electrical conductivity in such domain walls is

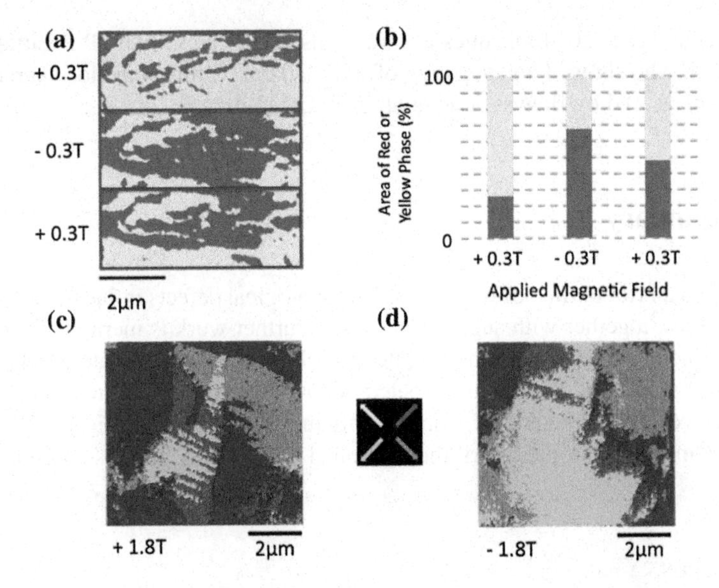

Fig. 6.12 Switching of ferroelectric polarization domains with an applied magnetic field (Evans et al.)

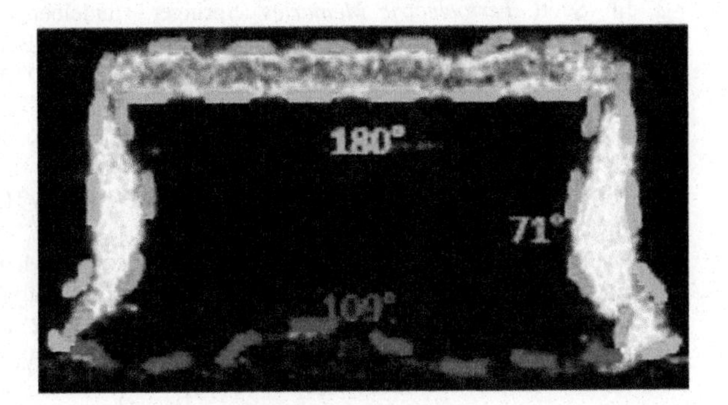

Fig. 6.13 Micrograph showing electrical conduction along domain walls, from Seidel et al.

Shlovskii-type variable hopping with a $V^{1/4}$ dependence (unlikely at $T = 300\,K$), Mott hopping mechanisms with $V^{1/2}$ dependence, or not hopping at all, but experiments are underway at Warwick (Alexe et al.) to answer this important question.

6.8 Domain Wall Oscillation

A final question to touch on is the oscillation of domain walls. This was first analysed by Sidorkin [117]; work in 2013 by Pakhmanov, Lukyanchuk et al. [118] extended Sidorkin's early work with considerable detail. It appears that oscillations occur far from Tc at frequencies of order hundreds of GHz (some cm^{-1}), but decrease to ca.

30 GHz near Tc. In 2014 this question was considered carefully by P B Littlewood et al. [119], who showed a gap energy of order 30 cm^{-1}; this is in close agreement with that measured by Banys, Scott et al. in TSCC [120].

6.9 Summary

I have given a brief outline of ways in which topological defects influence ferroelectric dynamics, together with suggestions where further work is merited. Topics not discussed include Mott transitions in systems such as $NdNiO_3$ nickelates [121], gated injection of superconductivity in ferroelectric O-18 $SrTiO_3$ [122], and domain wall behaviour very near Quantum Critical points just above zero Kelvin [123]. These and other unexplored topics merit the continued effort of the ferroics community.

References

1. S. Piskunov, E. Heifets, R.I. Eglitis, G. Borstel, Computational. Mater. Sci. **29**, 165 (2004)
2. J.F. Scott, *Ferroelectric Memories* (Springer, Heidelberg, 2000). Translations: Japanese (Springer-Japan, Tokyo, 2003), Chinese (Tsinghua Press, Beijing, 2004)
3. I. Stolichnov, A.K. Tagantsev, J. Appl. Phys. **84**, 3216 (1998); see the correct situation in J. Robertson, C.W. Chen, Appl. Phys. Lett. **74**, 1168 (1999)
4. m* = 1.4 m(e) is erroneously given by I. Stolichnov, A.K. Tagantsev, N. Setter, J.S. Cross, M. Tsukuda, Appl. Phys. Lett. **175**, 1790 (1999)
5. Correct m* values for $BaTiO_3$ (m* = 6.00 m) are given by H.P.R. Frederikse, G.A. Candela, Phys. Rev. **147**, 583 (1966)
6. Mahan shows m* band mass = m* transport mass: G.D. Mahan, Phys. Rev. **153**, 983 (1967)
7. D.J. Wouters, G.J. Willems, H.E. Maes, Microelectron. Eng. **29**, 249 (1995)
8. K. Uchino, E. Sadanaga, T. Hirose, J. Am. Ceram. Soc. **72**, 1555 (1989)
9. J. Robertson, C.W. Chen, W.L. Warren, C.D. Gutleben, Appl. Phys. Lett. **69**, 1704 (1996); C.D. Gutleben, MRS Proc. **433**, 109 (1996)
10. A.V. Dixit, N.R. Rajopadhye, S.V. Bhoraskar, J. Mater. Sci. **21**, 2798 (1986)
11. I.P. Batra, B.D. Silverman, Solid State Commun. **11**, 291 (1972); R. Mehta, B. Silverman, J.T. Jacobs. J. Appl. Phys. **44**, 3379 (1973)
12. R. Landauer (private communication)
13. C. Basceri, S.K. Streiffer, A.I. Kingon, R. Waser, J. Appl. Phys. **82**, 2497 (1997)
14. M. Dawber, P. Chandra, P.B. Littlewood, J.F. Scott, J. Phys. Condens. Matter **15**, L393 (2003)
15. J. Junquera, P. Ghosez, Nature **422**, 506 (2003)
16. A. Lookman, R.M. Bowman, J.M. Gregg et al., J. Appl. Phys. **96**, 555 (2004)
17. F.A. Cuellar, Y.H. Liu, J. Salafranca et al., Nat. Commun. **5**, 4215 (2014)
18. A. Chanthbouala, A. Crassous, V. Garcia et al., Nat. Nano **7**, 101 (2012)
19. See for example S.-W. Cheong, Bull. Am. Phys. Soc. 59, paper W39.00004 (2014); APS Conference Proceedings, February 27–March 2, abstract #Y-19.004 (2012)
20. S.C. Chae, Y. Horibe, D.Y. Jeong, S. Rodan, N. Lee, S.W. Cheong, Proc. US Nat. Acad. Sci. **107**, 21366 (2010). Phys. Rev. Lett. **110**, 167601 (2013)
21. Y. Zhao, W. Li, B. Xi, Z. Zhang, X. Tan, S.-J. Ran, Y. Liu, G. Su, Phys. Rev. E **87**, 032151 (2013)
22. L.D. Landau, E. Lifshitz, Phys. Z. Sowjetunion **8**, 153 (1935)

23. C. Kittel, Phys. Rev. **70**, 965 (1946)
24. F. De Guerville, M. El Marssi, I. Lukyanchuk, L. Lahoche, Ferroelectrics **359**, 14 (2007); F. De Guerville, I. A. Luk'yanchuk, L. Lahoche. Mater. Sci. Eng. **B120**, 16 (2003)
25. G. Catalan, J.F. Scott, A. Schilling, J.M. Gregg, J. Phys. Condens. Matter **19**, 022201 (2007)
26. G. Catalan, J. Seidel, R. Ramesh, J.F. Scott, Rev. Mod. Phys. **84**, 119 (2012)
27. D.J. Srolovitz, J.F. Scott, Phys. Rev. B **34**, 1815 (1986)
28. Y. Ivry, D.P. Chu, C. Durkan, Nanotechnology **21**, 065702 (2010)
29. Y. Ivry, C. Durkan, D.P. Chu, J.F. Scott, Adv. Funct. Mater. **24**, 5567 (2014)
30. J.R. Whyte, R.G.P. McQuaid, P. Sharma, C. Canalias, J.F. Scott, A. Gruverman, J.M. Gregg, Adv. Mater. **25**, 03567 (2013)
31. A. Roitburd, Phys. Status Solid **A37**, 329 (1976)
32. G. Arlt, J. Mater. Sci. **25**, 2655 (1990)
33. J. Lajzerowicz, J.J. Niez, J. Physique Letteres **40**, L165 (1979)
34. M. Daraktchiev, G. Catalan, J.F. Scott, Ferroelectrics **375**, 122 (2008)
35. M. Daraktchiev, G. Catalan, J.F. Scott, Phys. Rev. B **81**, 025022 (2010)
36. J. Wojdel, J. Iniguez, Phys. Rev. Lett. **112**, 247603 (2014)
37. E.K.H. Salje, J.F. Scott, Appl. Phys. Lett. (in press)
38. J. Seidel, L.W. Marrtin, Q. He et al., Nat. Mater. **8**, 229 (2009)
39. A. Schilling, S. Prosandeev, R. McQuaid et al., Phys. Rev. B **84**, 064110 (2011)
40. A. Francu, P. Novakov, P. Janicek, Eng. Mech. **19**, 45 (2012)
41. J.F. Scott, J. Phys. Condens. Matter **26**, 212202 (2014)
42. C. Lichtensteiger, S. Fernandez-Pena, C. Weymann et al., Nano Lett. **14**, 4205 (2014)
43. Y. Ishibashi, H.A. Orihara, Integr. Ferroelectr. **9**, 57 (1995)
44. Y. Ishibashi, Y. Takagi, J. Phys. Soc. Jpn. **31**, 506 (1971)
45. K. Dimmler, M. Parris, D. Butler et al., J. Appl. Phys. **61**, 5467 (1987)
46. J.F. Scott, B. Pouligny, K. Dimmler et al., J. Appl. Phys. **62**, 4510 (1987)
47. X. Du, I.-W. Chen, J. Appl. Phys. **83**, 7789 (1998); Ferroelectruics **208**, 237 (1998)
48. X. Du, I.-W. Chen, Appl. Phys. Lett. **72**, 1923 (1998)
49. D.J. Jung, M. Dawber, J.F. Scott et al., Integr. Ferroelectr. **48**, 59 (2002)
50. V.Y. Shur, J. Mater. Sci. **41**, 199 (2006)
51. K. Shimamura, E.C. Villora, H. Zeng, M. Nakamura, S. Takekawa, K. Kitamura, Appl. Phys Lett. **89**, 232911 (2006)
52. S. Hong, E.L. Colla, E. Kim, D.V. Taylor, A.K. Tagantsev et al., J. Appl. Phys. **86**, 607 (1999)
53. S. Hong, E.L. Colla, E. Kim, A.K. Tagantsev et al., Ferroelectrics **223**, 143 (1999)
54. B. Berge, L. Facheux, K. Schwab, A. Libchaber, Nature **350**, 322 (1991)
55. A.P. Malozemoff, Appl. Phys. Lett. **21**, 149 (1972)
56. A.P. Malozemoff, J.C. Deluca, Appl. Phys. Lett. **26**, 12 (1975)
57. A.A. Thiele, Bell Syst. Tech. J. **48**, 3287 (1969)
58. I. Lukyanchuk, P. Sharma, T. Nakajima et al., Condmat (2013). arXiv:1309.0291; Nano Lett. (in press 2014)
59. J.F. Scott, A. Kumar, Appl. Phys. Lett. **105**, 052902 (2014)
60. N. Ng, R. Ahluwalia, D.J. Srolovitz, Acta Mater. **60**, 3632 (2012)
61. R. Ahluwalia, N. Ng, A. Schilling et al., Phys. Rev. Lett. **111**, 165702 (2013)
62. A.K. Jonscher, *Dielectric Relaxation in Solids* (Chelsea Press, London, 1983)
63. A.K. Jonscher, Nature **250**, 191 (1974)
64. H. Schwarz, H. Hora, Appl. Phys. Lett. **15**, 349 (1969)
65. J.F. Scott, R.A. O'Sullivan, Nature **382**, 305 (1996)
66. J.A.F. Plateau, Statique Experimentale et Theorique des Liquides Soumis aux Forces Moleculaires (Gauthier-Villars, Paris, 1873); note that Plateau was a blind experimentalist
67. B. Zel'dovich, Sov. Phys. JETP **6**, 1184 (1958); see also V.M. Dubovik, V.V. Tugushev, Phys. Rep. **187**, 145 (1990)
68. A.A. Gorbatsevich, Y.V. Kopaev, Ferroelectrics **161**, 321 (1994); JETP Lett. **39**, 684 (1984); JETP Lett. **40**, 1076 (1984); Fiz. Tverd. Tela **28**, 1155 (1986); J. Magn. Magn. Mat. **54**, 632 (1986)

69. V.L. Ginzburg, A.A. Gorbatsevich, Y.V. Kopaev, B.A. Volkov, Solid State Commun. **50**, 339 (1984)
70. M. Fiebig, J. Phys. D **38**, R123 (2005); I. Naumov, L. Bellaiche, X. Fu, Nature **432**, 737 (2004)
71. T.L. Gilbert, Phys. Rev. **100**, 1243 (1955); S. Prosandeev, I. Ponomareva, I. Kornev, I. Naumov, L. Bellaiche. Phys. Rev. Lett. **96**, 237601 (2006)
72. S.O. Demokritov, A.I. Kirilyuk, N.M. Kreines, A.S. Borovik-Romanov, JETP Lett. **48**, 294 (1988)
73. I. Lukyanchuk, A. Pakhmanov, A. Sidorkin (2014). arXiv:1410.3124
74. M. Dawber, D.J. Jung, J.F. Scott, Appl. Phys. Lett. **82**, 436 (2003)
75. A. Gruverman, D. Wu, J.F. Scott, Phys. Rev. Lett. **100**, 097601 (2008)
76. L. Baudry, I. Luk'yanchuk, J.F. Scott, Phys. Rev. B **90**, 024102 (2014)
77. T. Tybell, P. Paruch, T. Giamarchi, J.-M. Triscone, Phys. Rev. Lett. **89**, 097601 (2002)
78. J.Y. Jo, S.M. Yang, T.H. Kim, H.N. Lee, J.-G. Yoon, S. Park, Y. Jo, M.H. Jung, T.W. Noh, Phys. Rev. Lett. **102**, 045701 (2009)
79. C.S. Ganpule, A.L. Roytburd, V. Nagarajan et al., Phys. Rev. B **65**, 014101 (2002)
80. J.F. Scott, A. Gruverman, D. Wu, I. Vrejoiu, M. Alexe, J. Phys. Condens. Matter **20**, 452222 (2008)
81. R.G.P. McQuaid, A. Gruverman, J.F. Scott, J.M. Gregg, Nano Lett. **14**, 4230 (2014)
82. A. Q. Jiang, C. S. Hwang, J. F. Scott et al., Sci. Rpts. (in press 2015)
83. L.W. Chang, V. Nagarajan, J.F. Scott, J.M. Gregg, Nano Lett. **13**, 2553 (2013)
84. Y. Ivry, D.P. Chu, J.F. Scott, C. Durkan, Nano Lett. **11**, 46719 (2011)
85. Y. Ivry, J.F. Scott, E.K.H. Salje, C. Durkan, Phys. Rev. B **86**, 20542 (2012)
86. A.P. Levanyuk, A.I. Morosov, A.S. Sigov, Ferroelectrics **79**, 337 (1988); **22**, 725 (1978); **37**, 533 (1981); Izvest. Akad. Sci. SSSR Ser. Fiz. **45**, 1640 (1981)
87. A.I. Morozov, A.S. Sigov, Fiz. Tverd. Tela **25**, 1352 (1983); Comments condens. Mater. Phys. **18**, 279 (1998); Y.M. Kishinetz, A.P. Levanyuk, A.I. Morosov, A.S. Sigov. Ferroelectrics **79**, 321 (1988)
88. I.J. Fritz, Phys. Rev. Lett. **35**, 1511 (1975)
89. V. Bobnar et al., J. Appl. Phys. **107**, 084104 (2010)
90. J.F. Scott, J. Appl. Phys. **108**, 086107 (2010)
91. J.C. Lashley et al., Adv. Mater. **26**, 3860 (2014)
92. J.F. Scott, EPL **104**, 36004 (2014)
93. E. Almahmoud, I. kornev, and L. Bellaiche, Phys. Rev. Lett. **102**, 105701 (2009)
94. Z.-G. Ye, A.A. Bokov et al. (in press)
95. W. Kleemann, J. Dec, V.V. Shvartsman, Z. Kutnjak, T. Braun, Phys. Rev. Lett. **97**, 065702 (2006); see also these authors' early work on domain precursors within the paraelectric phase (this also suggests a first-order transition): J. Dec, V.V. Shvartsman, W. Kleemann. Appl. Phys. Lett. **89**, 212901 (2006)
96. T. Chen, S.-J. Sheih, J.F. Scott et al., Ferroelectrics **120**, 115 (1991)
97. J.F. Scott, S.-J. Sheih, T. Chen, Ferroelectrics **117**, 1 (1991)
98. K. Samanta, A.K. Arora, T.R. Ravindran, S. Ganesamoorthy, K. Kitamura, S. Takekawa, Vib. Spectrosc. **62**, 273 (2012)
99. J. Dec, Invited talk (EMF, Lake Bled, Slovenia, 1996); see also very recently P. Ondrejkovic, M. Kempa, J. Kulda et al., Phys. Rev. Lett. **113**, 167601 (2014)
100. J. Dec, W. Kleemann, M. Itoh, Phys. Rev. B **71**, 144113 (2005)
101. J.F. Scott, R. Pirc, A. Levstik, R. Blinc, J. Phys. Condens. Matter **18**, L205 (2006)
102. J.F. Scott, J. Phys. Condens. Matter **18**, 7123 (2006)
103. D. Fisher (private communication cited in [99])
104. B. Hilczer, M. Michalczyk, Ferroelectrics **22**, 721 (1978)
105. (a) B. Noheda; (b) K. Z. Baba-Kishi, A.M. Glazer, J. Appl. Cryst. **47**, 1688 (2014)
106. M.T. Kesim, J. Zhang, S.P. Alpay et al., Appl. Phys. Lett. **105**, 052901 (2014)
107. J.F. Scott, Ferroelectrics (special issue dedicated to S. Lang, 2014)
108. E.K.H. Salje, O. Aktas, M.A. Carpenter, J.F. Scott, Phys. Rev. Lett. **111**, 247603 (2013)

109. D.M. Evans, J.M. Gregg, A. Kumar, J.F. Scott, Nat. Commun. **4**, 1534 (2013)
110. V. Janovec, J. Privatska, Multiple domain structures: domains in domains, walls in walls, EMF (Krakow, 2013); B. Houchmandzadeh, J. Lajzerowicz, E. Salje, J. Phys. Cond. Mat. **3**, 5163 (1991)
111. J. Bornarel, J. Lajzerowicz, J.F. Legrand, Ferroelectrics **7**, 313 (1974); see also B.J. Rodriguez, L.M. Eng, A. Gruverman, Appl. Phys. Lett. **97**, 042902 (2010); D. Maurya, M. Murayama, A. Pramanick, W.T. Jr Reynolds, K. An, S. Priya. J. Appl. Phys. **113**, 114101 (2013)
112. J. Seidel, L.W. Martin, Q. He et al., Nat. Mater. **8**, 229 (2009)
113. N. Balke, B. Winchester, W. Ren et al., Nat.Phys. **8**, 81 (2012)
114. S. Farokhipoor, B. Noheda, Phys. Rev. Lett. **107**, 127601 (2011)
115. J. Guyonnet, I. Gaponenko, S. Gariglio et al., Adv. Mater. **23**, 5377 (2011); I. Stolichnov, M. Iwanowska, E. Colla et al., Appl. Phys. Lett. **104**, 132902 (2014)
116. L. Feigl, L.J. McGilly, N. Setter, Ferroelectrics **465**, 36 (2014)
117. V.N. Fedosov, A.S. Sidorkin, Fiz. Tverd. Tela **18**, 1661 (1976)
118. A. Pakhmanov, I. Lukyanchuk, A. Sidorkin, Ferroelectrics **444**, 177 (2013)
119. R.T. Brierly, P.B. Littlewood, Phys. Rev. B **89**, 184104 (2014)
120. R. Mackeviciute, M. Ivanov, J. Banys, N. Novak et al., J. Phys. Condens. Matter **25**, 212202 (2013)
121. R. Scherwitzl, P. Zubko, L. Gutierrez et al., Adv. Mater. **22**, 5517 (2010)
122. H. Tamura, A. Yoshida, S. Hasuo, Appl. Phys. Lett. **59**, 298 (1991)
123. S.E. Rowley, J.C. Lashley, J.F. Scott et al. (in press 2015)

Chapter 7
Ferroelectric Domain Walls and their Intersections in Phase-Field Simulations

J. Hlinka, V. Stepkova, P. Marton and P. Ondrejkovic

Abstract This chapter deals with phenomenological theoretical description of ferroelectric domain walls in perovskite ferroelectrics in the framework of Ginzburg-Landau-Devonshire theory. Its first part focuses mostly on the Bloch-like ferroelectric walls in comparison with the more standard Ising-like domain wall profiles as well as hypothetical Néel-like domain walls. Its second part is devoted to line defects that occur at the intersection of ferroelectric domain walls. The overview of the various concepts and recent studies is illustrated by polarization profiles obtained using phase-field simulations for the model ferroelectric perovskite crystal $BaTiO_3$.

7.1 Introduction

Ferromagnetic and ferroelectric domain walls are often quoted as well known examples of topological defects. For example, recent papers suggested to exploit analogies between the ferroelectric domain walls or their intersections and the topological defects such as cosmic strings considered in modern cosmological theories [1]. The ferroelectric and ferromagnetic substances are indeed accessible for various laboratory experiments and they are rather common materials in technology and materials science since several decades. Consequently, the accumulated understanding the problem of domain walls is enormous. In particular, systematic approach to the classification of their symmetry is available for example in the Volume D of the International Tables of Crystallography [2], and a broad coverage of the ferroelectric domain wall research can be found for example within the Tagantsev, Cross and Fousek's monograph devoted to experimental and theoretical studies of domains in ferroelectric crystals and thin films [3].

Here we shall focus on phenomenological modeling of ferroelectric domain wall structures in the simplest ferroelectric perovskite crystals. Among others, perovskite ferroelectrics are the key materials in the current ferroelectric epitaxial thin film

---

J. Hlinka (✉) · V. Stepkova · P. Marton · P. Ondrejkovic
Institute of Physics, Czech Academy of Science, Na Slovance 2,
18221 Praha 8, Czech Republic
e-mail: hlinka@fzu.cz

J. Seidel (ed.), *Topological Structures in Ferroic Materials*, Springer Series in Materials Science 228, DOI 10.1007/978-3-319-25301-5_7

research and considerable effort has been already put in the investigations of their structure and transport properties by the forefront microscopy techniques such as high-resolution electron microscopy or the atomic force microscopy techniques.

From the structural point of view, one of the most intriguing results of the past decade is the prediction of so-called Bloch-like domain walls in $BaTiO_3$ and $PbTiO_3$ ferroelectrics. These predictions were based on Ginzburg-Landau-Devonshire (GLD) theory [4, 5], but also ab-initio calculations [6, 7] as well as ab-initio based atomistic modeling [7]. These results are now well understood and widely accepted, but as far as we know, the experimental evidence is still missing. In this situation, it seems worth reviewing the progress in theoretical studies of this problem. Perhaps even more attention has been recently paid to the charged domain walls and electronic properties of ferroelectric domain walls in general. These latter subjects have been covered for example in the review paper by Catalan et al. [8] and so these subjects are not included here.

It is worth stressing that the extremely narrow thickness of ferroelectric walls is an advantage for microscopic modeling and it definitely promises to achieve smaller domain sizes needed for miniaturization of future devices. On the other hand, this narrow thickness is also the most essential bottleneck issue for probing their structure experimentally. In this situation, phenomenological theories allowing to predict domain walls properties are of a great importance. The modeling of domain wall properties within the framework of the phenomenological GLD theory appears nowadays as one of the most efficient tools for guiding the research in this area.

The aim of this chapter is to review the current state of the understanding domain boundaries in $BaTiO_3$ and similar defects of polarization fields in perovskite ferroelectrics from the perspective of the GLD theory. Since the phase-field modeling and the GLD theory are nowadays standard tools, we decided to avoid the technical details of the phase-field methods and refer the reader towards the information provided e.g. in [9] and references therein and/or to the details given directly in the original research publications.

The chapter is organized as follows. In Sect. 7.2 we introduce the concept of ferroelectric Bloch-like and Néel-like domain walls and possible definitions of these domain walls are briefly discussed. Section 7.3 is devoted to the phase transition between Bloch-like and Ising-like domain walls and manifestation of this transition in phase-field simulations. General comments on Ginzburg-Landau model used for domain wall studies are given in Sect. 7.4. The importance of crystallographic orientation of the domain walls is addressed in Sect. 7.5. Section 7.6 contains a review of possible issues related to ferroelastic domain walls in ferroelectric perovskites. Line defects, occurring at the intersection of domain walls, are considered in Sect. 7.7. Last section before general conclusions deals with polarization switching of closed-circuit domain states in ferroelectric nanorods and with processes of field-induced domain wall transformations.

7.2 Bloch Versus Ising Versus Néel-like Domain Wall

Distinction between Bloch-like, Ising-like and Néel-like domain walls is usually based on the comparisons with idealized profiles of boundaries separating domain states with opposite polarization, i.e. 180-degree domain walls. Such idealized examples assume a 180-degree domain wall parallel to the spontaneous polarization of the adjacent domains. This is the most natural domain wall orientation, because the spatially averaged charge density at the domain wall vanishes (i.e. such a domain wall is electrically "neutral") and a huge extra electrostatic energy penalty for uncompensated charge density is avoided.

The idealized polarization profiles across the Bloch, Ising and Néel ferroelectric domain walls are sketched in Fig. 7.1. The wall in which the polarization reversal is achieved by polarization rotation within the plane of the domain wall is designated as the Bloch wall. This domain wall is chiral and it exists in two enantiomorphic variants depicted in panels (a) and (b). In contrast, within the idealized Ising wall shown in panel (c) the polarization remains strictly parallel or antiparallel to the spontaneous polarization of the adjacent domains. Finally, the panel (d) shows the idealized Néel wall in which the polarization reversal is achieved by polarization rotation within the plane containing both the domain wall normal and the spontaneous polarization. The ideal Néel wall also exists in two variants, because the rotation can be accomplished by a clockwise and anticlockwise manner, but this wall is not chiral, as it has a mirror symmetry.

The structure of neutral ferroelectric domain walls can be conveniently described with the help of three auxiliary orthogonal unit vectors depicted in the bottom part of Fig. 7.1b. The domain wall *side vector* **s** is normal to the domain wall and its sense allows to distinguish opposite sides of the wall (by definition, the vector points from the first to the second domain). The polarization *reversal vector* **r** defines the direction of the polarization switched on by the translation of the wall along its normal **s**. In case of neutral ferroelectric domain walls, **r** and **s** are always perpendicular and, in combination with their cross-product $\mathbf{t} = \mathbf{r} \times \mathbf{s}$, they form a right-handed orthogonal set $\{\mathbf{r},\mathbf{s},\mathbf{t}\}$.

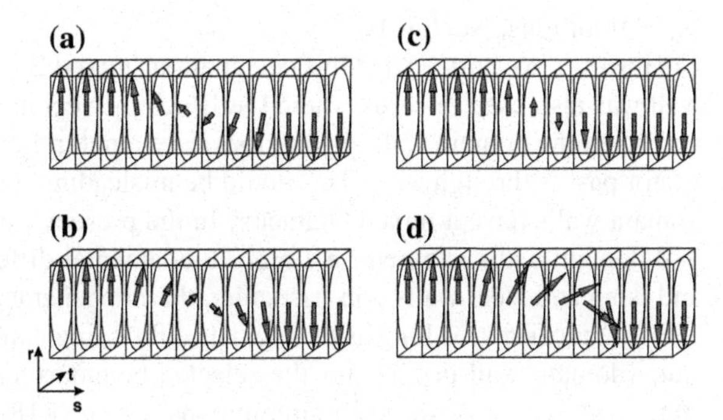

Fig. 7.1 Polarization profiles of idealized Bloch (**a**, **b**), Ising (**c**) and Néel-like (**d**) domain walls. The inset in the *bottom* part of (**b**) indicates the auxiliary orthogonal set of unit vectors $\{\mathbf{r},\mathbf{s},\mathbf{t}\}$, allowing to define convenient symmetry-adapted coordinates described and utilized in the text

Within the phenomenological framework of Landau theory, domain walls are identified with specific trajectories in the order parameter space. It is practical to describe these trajectories in terms of polarization components parallel to the {**r**,**s**,**t**} directions. The idealized Bloch wall can be described as polarization rotation in the $P_r - P_t$ plane, Néel wall is polarization rotation in the $P_r - P_s$ plane, and idealized Ising wall corresponds to a path restricted to the one-dimensional P_r subspace only.

Strictly speaking, none of these idealized forms of electrically neutral 180-degree domain walls can be realized in real ferroelectric perovskites. Typically, P_t and P_s components are drastically suppressed by electrostriction and electrostatic interaction driven mechanisms, respectively. Thus, P_r components are prevailing and the 180-degree ferroelectric domain walls appear to be closest to the Ising profile of Fig. 7.1c. We shall discuss here the exceptional cases in which domain wall profiles do resemble Bloch-like solution, but it is worth stressing that the ideal circular trajectories can never be realized in perovskites since there will be always some perturbation by the crystalline anisotropy. In some cases, the domain wall structure has a mirror symmetry plane perpendicular to **t** and then P_t will vanish identically. On the other hand, strictly speaking, the neutral 180-degree domain wall cannot have a mirror symmetry plane perpendicular to **s** and so one can always expect some nonzero P_s close to the neutral 180-degree domain wall. This actually implies that the ideal Ising wall restricted to the one-dimensional P_r subspace is always just an approximation to the real profile. As far as we know, all recent realistic microscopic calculations of the 180-domain walls in perovskites indeed reveal some small P_s component in agreement with this basic symmetry argument [10, 11].

In order to emphasize the difference between the ideal, single-component Ising wall considered above, and the more realistic domain wall determined for example from microscopic calculations, some authors described the latter one as a "mixed Bloch-Ising-Néel" 180-domain wall [10]. In many cases, it is instructive to analyze the departures from the ideal Ising wall by exploring the tilt of the polarization vector out of the **r** direction and the variation of this tilt across the domain wall profile. The maximal amplitudes of these tilt angles towards **t** and **s** directions, denoted as *Bloch angle* ϕ_B and *Néel angle* ϕ_N, respectively, are then useful quantitative parameters for detailed description of the non-ideal domain walls. Clearly, both angles vanish for the ideal Ising wall while $\phi_B = \pi/2$, $\phi_N = 0$ for ideal Bloch wall and $\phi_N = \pi/2$, $\phi_B = 0$ for ideal Néel wall.

However, we noticed that this designation "mixed Bloch-Ising-Néel" 180-domain wall was also used in cases where the Bloch or Néel angles are few degrees or less and when the domain wall visibly does have a natural center where the polarization vector passes through zero. This could be misleading, because in this sense, all real domain walls have a mixed character. In the present work, our classification of the 180-domain walls is based on a more fundamental difference between ideal Bloch and Ising and Néel walls—in particular, their degeneracy and chirality.

More precisely, as long as there exists only a single, non-degenerate domain-wall state (domain-wall profile) for the selected boundary conditions (adjacent domain states and vector **s**), we shall simply consider such 180-domain wall as *Ising-like* domain wall, irrespectively on Bloch or Néel angles and in agreement with the nomenclature of our earlier works [4, 5, 12, 13]. In this way, the non-degenerate domain walls are sharply distinguished from another class of domain walls discussed e.g. in [4, 12, 14], where a pair of degenerate, symmetry-related solutions exists for fixed boundary conditions. If the two symmetry-related profiles for the fixed boundary conditions form an enantiomorphic pair analogical to those of Fig. 7.1a and b, it is natural to denote these walls as *Bloch-like* walls. In these cases, the modulus of local polarization remains finite within the domain wall, and although is far from being constant, it is still possible to describe the situation in terms of polarization rotation. Obviously, one can analogically anticipate existence of achiral degenerate *Néel-like* ferroelectric 180-domain walls, but the existence of these walls in perovskite ferroelectrics was not analyzed yet and we shall thus not develop this idea here either.

7.3 Bloch-Ising Phase Transition

The above stated fundamental difference between Bloch-like and Ising-like domain walls implies possibility of a phase transition between these two cases. The conditions for the existence of this transition are obviously more restrictive than the conditions for the existence of ferroelecric Bloch-like domain walls themselves, nevertheless, there have been already few hints about when such a phase transition might be expected to occur. Probably the first material-specific prediction of Bloch-Ising domain wall transition has been reported for 180-degree domain wall of rhombohedral $BaTiO_3$ with $\mathbf{s} \parallel [\bar{2}11]$ and $\mathbf{r} \parallel [111]$ [13]. The GLD modeling predicts for this domain wall rather robust Bloch character, which is maintained in the entire temperature stability range of the rhombohedral ferroelectric phase. Therefore, in order to induce the Bloch-Ising domain wall transition, an external pressure of uniaxial symmetry was assumed. The obtained profiles of the P_r and P_t components across the domain wall at $T = 50\,K$ at different in-plane compressive pressure levels are shown in Fig. 7.2. One can see sizable value of the P_t polarization component at the ambient pressure, as well as its pronounced suppression upon approaching the

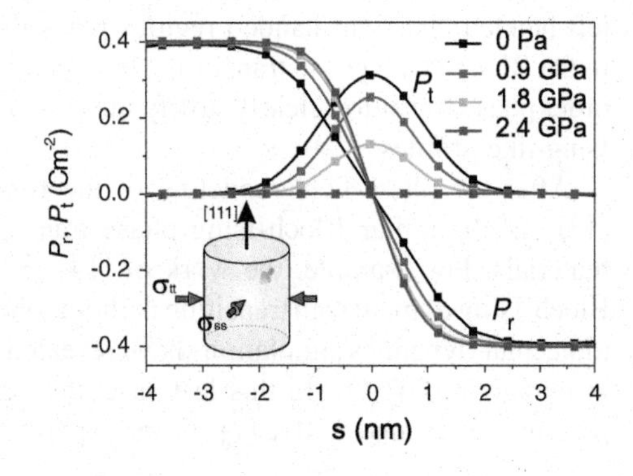

Fig. 7.2 Bloch-Ising domain wall transition in rhombohedral $BaTiO_3$ from simulations of [13]. The GLD model parameters are taken for 50 K. The inset indicates the orientation of the applied pressure of uniaxial symmetry

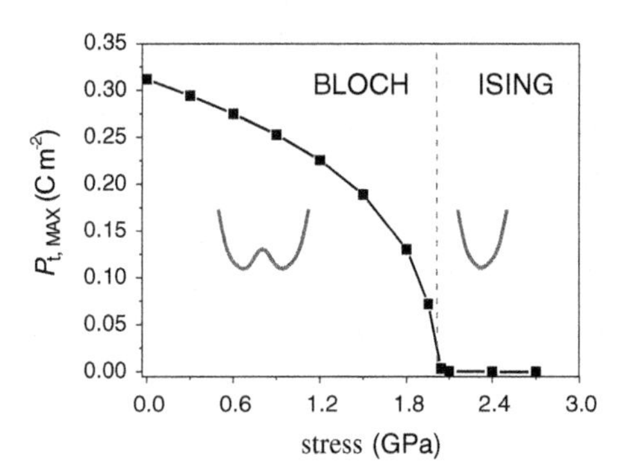

Fig. 7.3 Order parameter of the Bloch-Ising domain wall transition in rhombohedral $BaTiO_3$. Order parameter is the maximum of the P_t component from the domain wall profiles simulated in [13]. Some of them are also shown in Fig. 7.2

critical pressure of about 1.9 GPa. Although this level of stress is considerably high, it almost does not change the magnitude of the spontaneous polarization within the domain and it is still well comparable with stresses readily achievable in epitaxial ferroelectric films.

Within the phase-field simulations, this stress-induced transition appears as a second-order phase transition. As an order parameter, one can take the maximum value of the P_t component. Indeed, the transition is associated with appearance of the nonzero P_t component which is strictly absent in the Ising phase. Its pressure dependence is shown in Fig. 7.3. One can clearly see that this component vanishes continuously at the critical pressure value.

This transition can be also considered as a ferroelectric phase transition, with new component of polarization emerging mostly within the domain wall. The transition is therefore associated with divergence of the corresponding component of the dielectric constant. This divergence has been also studied in phase-field simulations [15].

The new component of polarization breaks the m_t mirror symmetry present in the Ising domain wall. In the center of the Bloch wall, the polarization vector is parallel or antiparallel to the vector **t**. If $\{\mathbf{r},\mathbf{s},\mathbf{t}\}$ is chosen consistently to the above introduced convention ($\partial P_r/\partial s < 0$ and $\mathbf{t} = \mathbf{r} \times \mathbf{s}$), then positive value of the P_t component of the polarization within the center of the wall correspond to the right-handed variant of the domain wall (the sense of the winding resemble to the usual right-handed screw as in the ideal case shown in Fig. 7.1a) and *vice versa*. The line separating the left-handed and right-handed regions plays the role of a ferroelectric domain wall in the Bloch-Ising phase transition [16]. Summarizing, the signatures of the second-order phase transition clearly justify the essential difference between Bloch-like and Ising-like profiles.

After the phase-field modeling of the Bloch-Ising phase transition in $BaTiO_3$ [13, 15], a similar Bloch-Ising phase transition has been also searched in other materials. For example, the work of [14] explored possibilities of {211} oriented Bloch-Ising domain wall transition in rhombohedral PZT. Most remarkably, atomistic molecular dynamics simulations have revealed a Bloch-like 180-degree domain wall in tetragonal $PbTiO_3$. In this last case, the transition could be reached at ambient pressure, at around 330 K [7]. From experimental point of view, the temperature-driven transition is obviously much more accessible. Hopefully, these findings will encourage experimental efforts devoted to the detection of the Bloch-Ising phase transition.

7.4 Predictive Value of Phase-Field Simulations

Let us stress that here and other places in this chapter, we often give quite precise numbers for simulation conditions, in order to facilitate follow-up studies or comparisons among various results. But the precision of these numbers are only relevant with respect to the selected model. When comparing to the actual *experimental* conditions, the numerical values should be considered only as indicative ones, since the model potential and Landau theory itself rely on important approximations and there is obviously also a large uncertainty in the model parameters. For example, this remark concerns the above predicted 1.9 GPa value of the critical pressure needed for achieving the Bloch-Ising phase transition in $BaTiO_3$.

For the same reason (allowing comparisons) we tried to avoid updating our basic GLD model parameters unless it was useful for demonstrating some dependence. In 2009, however, we have used phase-field simulations for analysis of piezoelectric properties of multidomain $BaTiO_3$, and we have realized [17] that there was an overlooked confusion about a factor of 2 in the papers we used as a source of electrostriction tensor component q_{1212}. Therefore, we have updated the value of q_{1212} but otherwise we have kept the same elastic constants as well as the same stress-free Landau potential as before. Here in this chapter we use the results obtained either with the former set of parameters model A (the basic set of parameters with the former q_{1212}) or the model B (the basic set of parameters with the updated q_{1212}). As this change does have a definite impact on some domain wall properties, the list of references to our earlier phase-field results obtained with the model A and B are listed in Table 7.1. In this chapter, we of course use the better tuned model B, unless stated differently.

Table 7.1 Usage of alternative sets of GLD model parameters in original papers devoted to domain wall properties of $BaTiO_3$

Model	q_{1212}	References
A	0.79×10^9 J m C^{-2}	[4, 6, 9, 12, 18–20]
B	1.57×10^9 J m C^{-2}	[13, 15, 17, 21–24]

Parameter sets denoted as model A and B have same stress-free Landau expansion and elastic tensors, the only difference is the value of the q_{1212} electrostriction tensor component, updated in the model B

7.5 180-Degree Domain Walls with Different Crystallographic Orientation

In general, properties of domain walls depend not only on the adjacent domain states, but also on the crystallographic orientation of the domain wall. This dependence could be quite pronounced even in case of neutral 180-degree domain walls. For example, we can keep the same rhombohedral domain states with polarization parallel or antiparallel to the $\mathbf{r} \parallel \langle 111 \rangle$ direction as in Fig. 7.2 and let varying the orientation of the domain wall normal $\mathbf{s}$ in the plane perpendicular to $\mathbf{r}$.

Lets consider two cases, with $\mathbf{s} \parallel \langle 1\bar{1}0 \rangle$ and $\mathbf{s} \parallel \langle 2\bar{1}\bar{1} \rangle$, depicted in Fig. 7.4, both having $\mathbf{s} \cdot \mathbf{r} = 0$. Typical domain-wall profiles obtained for such domain-wall orientations in rhombohedral $BaTiO_3$ are shown in Figs. 7.5 and 7.6. These profiles are calculated with the same conditions, in particular, same GLD potential as in [4, 6, 9, 12, 18–20], i.e. model A and the same temperature ($T = 118\,K$). At a first sight, there is a clear difference between profiles of Figs. 7.5a and 7.6a, corresponding to the domain wall orientations shown in Fig. 7.4a and b, respectively. Note that the Bloch domain wall profiles are highly asymmetric in Fig. 7.5a and b, while those of in Fig. 7.6a and b are symmetric. Moreover, one can see that the P_r component of the Bloch-like profiles shown in Fig. 7.5a actually seems to drop down in two steps; in other words, the domain wall appears to be composed of two nearby domain walls in a small distance of the order of 2–3 nm. This picture is a perfectly fair description of the situation, the Bloch-like walls we observe can be indeed understood as bound states of more elementary domain walls. For example, phase-field simulations indicate that a sufficiently large external electric field with nonzero components along $\mathbf{t}$ is capable to split such a domain wall in two well separated domain walls, which then define a few nm thick slab that could be perfectly considered as an independent intermediate ferroelectric domain [12]. On the top of it, the $\mathbf{s} \parallel \langle 1\bar{1}0 \rangle$ domain orientation allows also Ising-like singlet profile shown in Fig. 7.5c, which has somewhat higher energy, while no coexisting Ising-like domain wall solutions were found for $\mathbf{s} \parallel \langle 1\bar{1}0 \rangle$ domain orientation, in agreement with second-order nature of the Bloch-Ising phase transition discussed in the Sect. 7.3.

There is even more important difference between the $\mathbf{s} \parallel \langle 1\bar{1}0 \rangle$ and $\mathbf{s} \parallel \langle 2\bar{1}\bar{1} \rangle$ 180-degree domain walls if the phase-field simulation is made with model parameter set

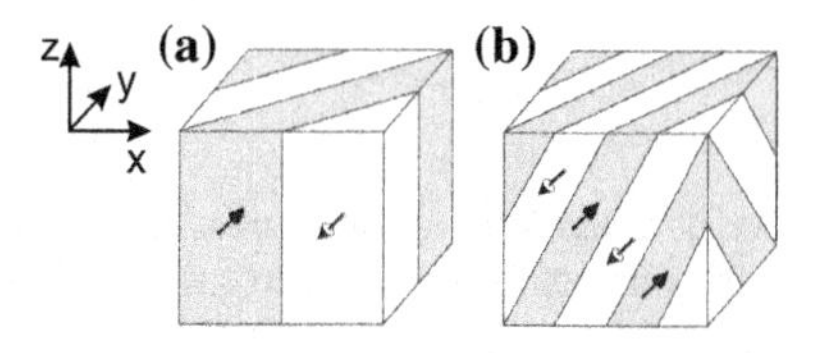

Fig. 7.4 Schematic illustration of possible periodic arrays of electrically neutral 180-degree ferroelectric domain walls in rhombohedral phase of $BaTiO_3$. White and grey areas correspond to the domains with opposite polarization, edges of the cubes are parallel to the pseudocubic axes, the illustrated cases (**a**) and (**b**) correspond to $\mathbf{s} \parallel \langle 1\bar{1}0 \rangle$ and $\mathbf{s} \parallel \langle 2\bar{1}\bar{1} \rangle$, respectively

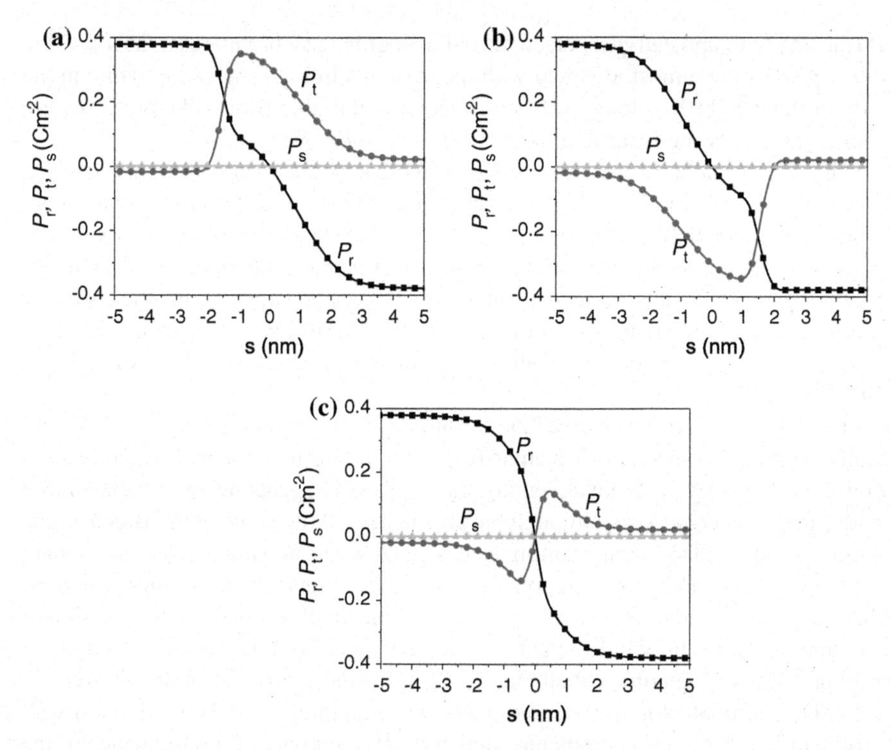

Fig. 7.5 Domain-wall profiles obtained for antiparallel ferroelectric walls with **s** ∥ $\langle 1\bar{1}0 \rangle$ orientations. Calculation was carried out for rhombohedral $BaTiO_3$ with GLD model of [18]. Panels (**a**) and (**b**) show *left-handed* and *right-handed* Bloch walls, corresponding to the lowest energy domain wall profile. Last panel (**c**) shows Ising-like domain wall with a metastable (locally stable but higher-energy) singlet profile with a pronounced lateral transverse component P_t, sometimes denoted as a bichiral domain wall

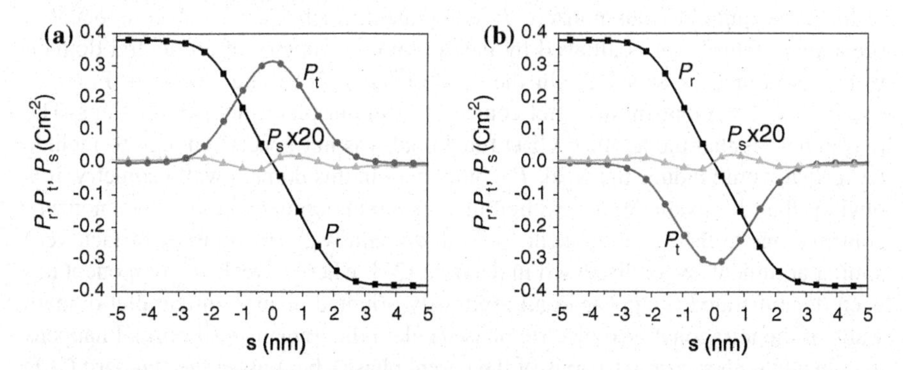

Fig. 7.6 Domain-wall profiles obtained for antiparallel ferroelectric walls with **s** ∥ $\langle 2\bar{1}\bar{1} \rangle$ orientation. Calculations were carried out for rhombohedral $BaTiO_3$ with GLD model of [18]. Panels (**a**) and (**b**) show *left-handed* and *right-handed* Bloch walls, corresponding to the lowest energy domain wall profile

B (the one with updated q_{1212} electrostriction coefficient). In this case, the **s** $\parallel \langle 2\bar{1}\bar{1}\rangle$ domain walls are almost identical with those shown in Fig. 7.6a and b, while in the case of the **s** $\parallel \langle 1\bar{1}0\rangle$ domain wall, no stable or metastable Bloch-like profiles were found, (here only the Ising-like solution, similar to Fig. 7.5c, exists).

These phase-field simulations results are in full agreement with an earlier reported independent study of [25], where the same domain wall was explored in the framework of the model B as well as in a more complex model involving charge compensation and flexoelectric coupling. In other words, the contemporary GLD theory (with the experimentally more relevant value of the q_{1212}) suggests that the character of the antiparallel domain walls of rhombohedral $BaTiO_3$ would change their character depending on the domain wall normal: the **s** $\parallel \langle 1\bar{1}0\rangle$ is Ising-like, **s** $\parallel \langle 2\bar{1}\bar{1}\rangle$ is Bloch-like.

The Ising-like profile of Fig. 7.5c is interesting also as an example of domain wall profile with a sizable Bloch angle ($\phi_B \doteq 40°$). This domain wall structure has a chiral symmetry by itself, but reversing sign of P_t would generate an unstable profile with a higher energy. Also, although the profile has a finite value of the Bloch angle, there is no net polarization rotation accumulated when passing across this domain wall. Indeed both P_t and P_r components are odd functions with a common center, so that the polarization rotates in a way that the same angle winded on the one side of the domain wall is un-winded again on the other side. For this reason, such domain wall profiles are sometime called bichiral [26]. In comparison, P_t is strictly zero for **s** $\parallel \langle 2\bar{1}\bar{1}\rangle$ rhombohedral 180-degree walls (for example, see 2.4 GPa domain wall profile in Fig. 7.2.). This completely vanishing P_t component is consequence of the fact that in this case the m_t plane coincides with a symmetry plane of the parent phase structure. For other orientations between **s** $\parallel \langle 1\bar{1}0\rangle$ and **s** $\parallel \langle 2\bar{1}\bar{1}\rangle$, the nonzero P_t component will reappear. In this sense, the bichiral profile of P_t is a rather common situation.

The simulations in Fig. 7.6a, b are also showing a weak, but after magnification of the scale quite obvious nonzero P_s component. Such nonzero P_s component is often very small but it is allowed by the common symmetry of all neutral domain walls (as argued in Sect. 7.2). In the case of the wall profile shown in Fig. 7.5a and b, this P_s component does not come out from our phase-field simulations. The reason here is that the adopted GLD functional was not general enough to include terms that would induce the weak P_s component in this domain wall geometry. It is obvious that, for example, the explicit flexoelectric coupling would induce nonzero contributions to the P_s component for both domain wall orientations. In fact, very similar arguments were discussed in detail in [27], where a weak P_s component has been demonstrated to appear in all arbitrarily oriented neutral antiparallel domain walls of the tetragonal ferroelectric phase (unless the domain-wall normal happens to coincide with a symmetry axis of the parent phase), but only if the standard GLD expansion is complemented by the flexoelectric coupling.

7.6 Electrically Neutral Ferroelastic Walls

Degenerate chiral domain-wall profiles are not restricted to 180-degree ferroelectric domain walls. In fact, our interest in phase-field simulations of ferroelectric domain walls in $BaTiO_3$ has been largely motivated by the fact that $BaTiO_3$ perovskite crystal allows a rich collection of distinct types of ferroelectric walls including those between domains with nonequal spontaneous strain tensor orientation with respect to the local parent pseudocubic lattice coordinates (ferroelastic domain walls). Here we have in mind the electrically neutral and mechanically compatible ferroelectric domain walls of the rhombohedral, orthorhombic and tetragonal phases of bulk single crystal. The temperature range of these ferroelectric phases and the temperature dependence of the Landau energy of the stress-free $BaTiO_3$ single crystal as calculated from the GLD model are shown in Fig. 7.7. It is well known that with respect to the parent paraelectric cubic phase, there are eight possible inequivalent rhombohedral ferroelectric domains, twelve orthorhombic ferroelectric domains and six tetragonal ferroelectric domains. Domain walls are intuitively understood as objects defining the interfaces between different domains. As a rule, they have rather low curvature at the atomic scale and so it is natural to study planar domain walls first. Type of the planar domain wall is most often specified by the domain-wall normal **s** and an ordered pair of domain states (e.g. assuming that **s** points from the first to the second domain state). We have previously denoted different types (classes) of such walls by a letter specifying the phase (R/O/T), the approximate value of the angle (in degrees) by which the polarization rotates when passing from one domain to the other, and by the orientation of the domain-wall normal (in pseudocubic setting)—for example, O 120{110}.

Most often, the domain walls naturally try to reach orientations, in which they become electrically neutral and mechanically compatible walls. For such domain walls, right-handed orthogonal set {**r**,**s**,**t**} can be defined in the same way as above.

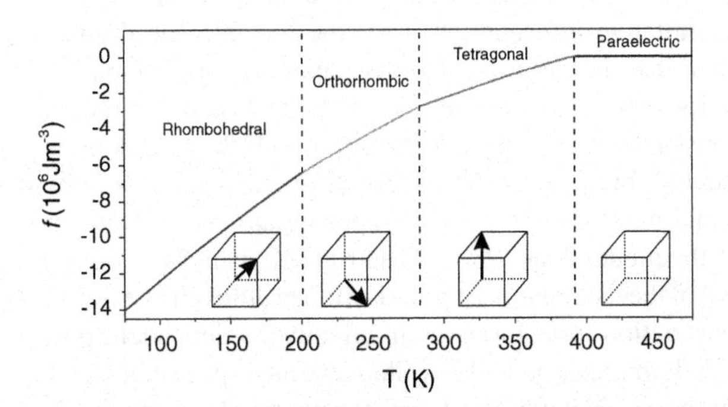

Fig. 7.7 Temperature dependence of the absolute minimum of the Landau-energy density for the stress-free $BaTiO_3$ single crystal as calculated from the GLD theory (same for parameter sets of model A and B). Insets show schematically the orientation of polarizatation vector of the representative domain state with respect to the unit cell of the cubic reference lattice

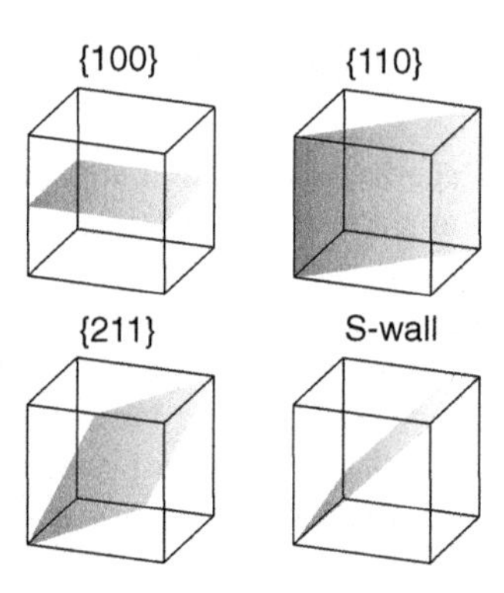

Fig. 7.8 Nonequivalent orientations of domain walls encountered in a systematic survey of orientations of electrically neutral and mechanically compatible ferroelastic domain walls of three ferroelectric phases of $BaTiO_3$

Previously, we have closely analyzed all types of electrically neutral and mechanically compatible ferroelastic domain walls of $BaTiO_3$, and, in addition, we have also analyzed electrically neutral non-ferroelastic (180-degree) ferroelectric domain walls for few selected, high-symmetry orientations. Encountered domain wall orientations in this survey are those shown in Fig. 7.8. Note that two of them, {100} and {110}, coincide with symmetry planes of parent cubic phase and {211} plays a special role as it contains both three-fold and two-fold symmetry axis of the parent cubic phase. Finally, one of the compatible ferroelastic domain wall types in orthorhombic phase requires consideration of a noncrystallographic orientation (so called S-wall) with the domain-wall vector **s** depending on numerical values of spontaneous strain tensor components at given temperature.

In particular, we have calculated numerically the profiles of these walls as variational solutions of the GLD functional by imposing the boundary condition ensuring that very far from the domain wall the polarization and strain coincide with the selected single domain spontaneous values and that there are no charges ($div\mathbf{P} = 0$).

Resulting domain-wall profiles can be used to evaluate various basic domain wall properties, such as their domain wall thickness or planar energy densities. Most condensed quantitative information about the ferroelectric domain wall structure is the associated path in the primary order-parameter space (space of polarization). Such polarization paths are shown in Fig. 7.9. Only P_r and P_t components are shown, since P_s is constant under $div\mathbf{P} = 0$ condition. All the calculations are made with the same potential, but in order to access different phases, we obviously selected Landau potential parameters within the appropriate phase. Data shown in Fig. 7.9 correspond to the model A at 298, 208, and 118 K, as in [4].

The method of the calculation of polarization profiles can be formulated as searching for the least action trajectories in an effective, **s**-dependent potential similar to Landau potential, that can be called Euler-Lagrange potential [4, 18]. The equipotential lines of the associated Euler-Lagrange potentials within the $P_s = const$ order parameter plane are also shown.

It is natural to identify the Bloch-like domain walls as those where two equivalent trajectories exist. In Fig. 7.9, Bloch-like solutions were found as lowest energy paths in 5 out of 12 cases (Bloch-like domain walls in Fig. 7.9 are those denoted

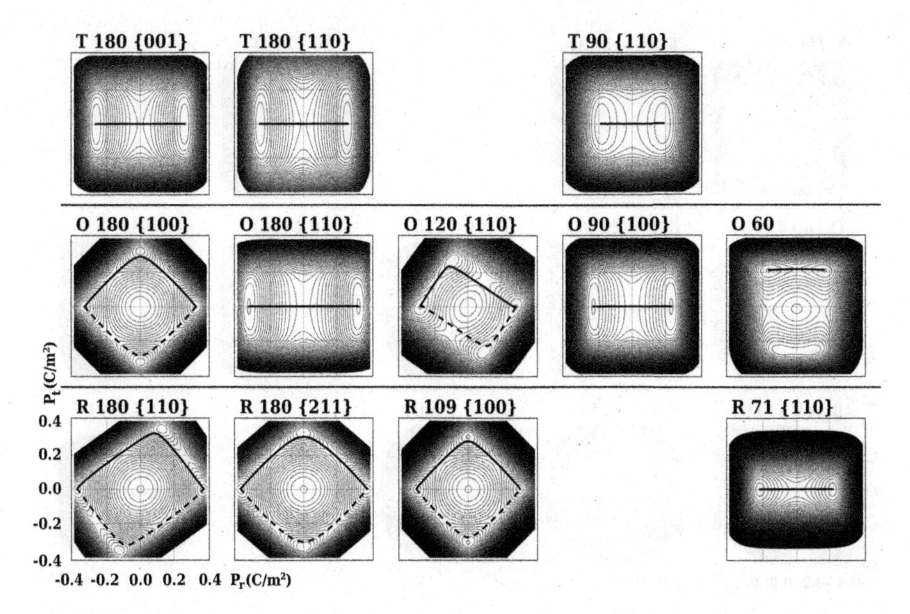

Fig. 7.9 Domain wall paths and Euler-Lagrange potentials for electrically neutral and mechanically compatible domain walls of three ferroelectric phases of $BaTiO_3$. Results are obtained for parameter sets of the model A. All panels have same axes shown in the *left bottom* corner of the figure. Equipotential contours are plotted at each 0.5 MJ/m^3. All domain wall paths are lowest energy solutions with prescribed boundary conditions perfectly compensated bound charge. In case of Bloch-like domain walls, two energetically equivalent paths exist. They are indicated by the full *line* and the *dashed line*, respectively

as O 180{100}, O 120{110}, R 180{110}, R 180{210}, R 109{100}). It is clear from Fig. 7.9, that the Bloch-like domain walls are favored when there is a symmetric Euler-Lagrange potential landscape and additional minima in the potential. Nevertheless, the influence of the model parameters is quite subtle. For example, passing from the model A to the model B is not changing the stress-free Landau potential, nor the data shown in Fig. 7.7, but the Euler-Lagrange potentials do show a small change and this small change can easily tip between Bloch and Ising-like profiles, as we have seen it above. This is also apparent from the comparison of Figs. 7.9 and 7.10, which show equivalent calculations but with the potential A and B, respectively.

7.7 Domain Wall Intersections

Properties of isolated domains and permissible domain walls are certainly the most basic ingredients for understanding the domain-structure related phenomena. One of the next possible steps in domain wall studies is to study line defects formed by domain wall intersections or by domain wall junctions.

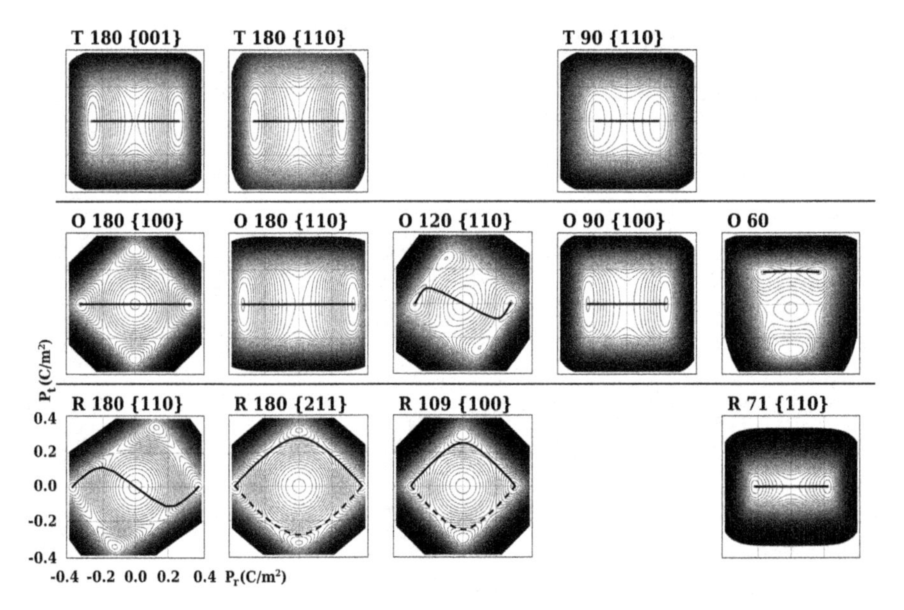

Fig. 7.10 Same as in Fig. 7.9, but calculated for parameter sets of model B (updated value of the q_{1212} parameter)

They are frequently encountered in experiments as well as in phase-field simulations, but so far, relatively little attention has been payed to them. Typically, domain wall intersections or domain wall junctions are formed in phase-field simulations started from random noise polarization configuration, as in the examples shown in Fig. 7.11.

Perhaps the most common line defects of this kind appear at the intersections of ferroelastic and non-ferroelastic domain walls. In the tetragonal phase, such defect lines are located at the junction of two 180-degree domain walls and two 90-degree domain walls. Many of such junctions are present for example in the domain structure shown in the left panel of Fig. 7.11. In this case, two adjacent 180-degree domain walls reach the junction at the (nominally) right angle, while the two ferroelastic domain walls are located at a common plane that bisects this right angle. This arrangement forms the principal motif of the characteristic "herringbone" domain pattern structures of tetragonal ferroelectrics like $PbTiO_3$ or $BaTiO_3$. Let us stress that such configuration allows to merge four different ferroelectric domain states, that nevertheless belong to a pair of ferroelastic domain states only. This results in a fully mechanically and electrostatically compatible state.

On the other hand, intersections of ferroelastic domain walls force a whole quadruplet of ferroelastic domain states to coexist at a single line. Detailed analysis of [28] shows that in general, within the rhombohedral, orthorhombic and tetragonal domain species of $BaTiO_3$-like ferroelectrics, formation of such line defects actually always induces additional stresses. Nevertheless, the violation of mechanically compatibility is energetically less prohibitive in some particular cases within the orthorhombic

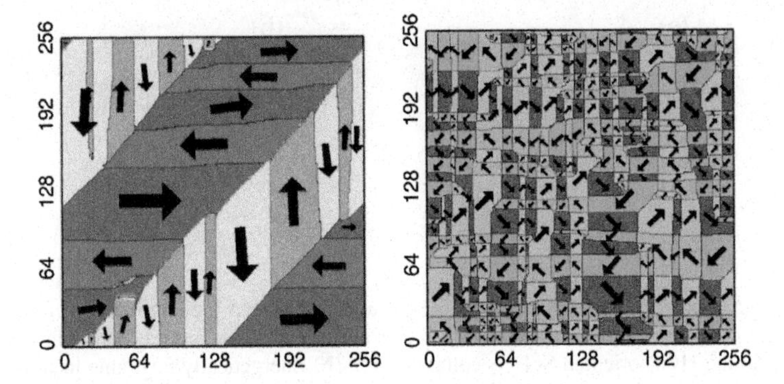

Fig. 7.11 Phase-field simulations of ferroelectric domain structure using the GLD model A. Different *colors* indicate individual domains with polarization direction indicated by *thick arrows*. *Left* panel shows simulation within tetragonal phase at 298 K, *Right* panel shows simulation in orthorhombic phase at 208 K. Data displayed in the pictures show domain structure as naturally evolved from the in-plane but random initial polarization distribution and under periodic mechanically clamped boundary conditions. Edges of the 256 × 256 discrete simulation field are parallel to the cubic directions of the parent phase, spatial step is 1 nm. For simplicity, polarization was restricted to in-plane components only. Further details related to these simulations are described in [19]

phases [28] and also in phases with small spontaneous stresses. Indeed, the intersections of orthorhombic ferroelastic domain walls have been sometimes observed in experiments (see references quoted in [28]) as well as in phase-field simulations [19] (as, for example, in the right part of Fig. 7.11). At the same time, we are not aware of convincing examples of ferroelastic domain wall intersections in tetragonal ferroelectric perovskites—there the elastic incompatibility is probably too limiting condition, even though the domain structures with such incompatible ferroelastic domain wall intersections could perhaps be one of the key elements responsible for enhanced piezoelectric activity of domain-engineered $BaTiO_3$ single crystals [17, 21, 29].

Having these conditions in mind, we were looking for situations under which the mechanically incompatible intersections of ferroelastic domain walls could occur in $BaTiO_3$ at ambient temperature. An interesting conceptual possibility exploiting the impact of ferroelectric-paraelectric interfaces in a $BaTiO_3$-$SrTiO_3$ nanocomposite has been reported in [23, 24]. A suitable nanocomposite geometry naturally favoring an intersection of two ferroelastic domain walls within $BaTiO_3$ is depicted in Fig. 7.12a. Ferroelectric $BaTiO_3$ is assumed to have a shape of a nanosized rod, which is surrounded in a suitable dielectric material which provides insulating boundary conditions and mechanical clamping. For example, this could be achieved if $BaTiO_3$ nanorods are epitaxially embedded in a thin film of $SrTiO_3$. Mechanical clamping in this geometry is favoring a low-strain rhombohedral domain state of over the tetragonal one, and if the axis of the nanorod is along the fourfold axis of the cubic parent phase, the spontaneous polarization tends to form flux-closure arrangements with polarization being tangential to the $BaTiO_3$-$SrTiO_3$ interfaces. For a sufficiently small

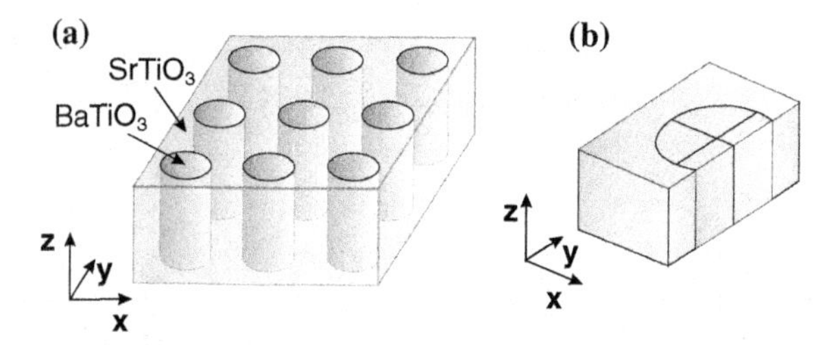

Fig. 7.12 Schema of the $BaTiO_3$-$SrTiO_3$ heterostructure designed for stabilization of vortex-like ferroelectric nanodomain pattern: (**a**) Brush-like arrangement of the $BaTiO_3$ nanorods embedded epitaxially in the ⟨100⟩ oriented $SrTiO_3$ epitaxial film; (**b**) Energetically favorable location of two intersecting ferroelastic domain walls in the natural domain state of 40 nm nanorod, as obtained in phase-field simulations [23]. Pseudocubic axes are shown in the *bottom* of each panel

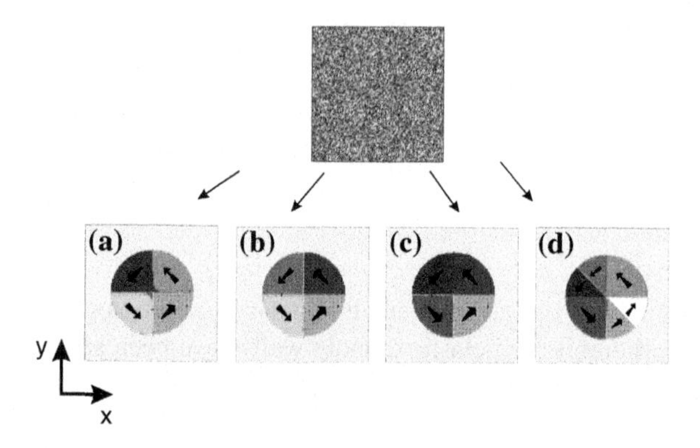

Fig. 7.13 Schema summarizing different spontaneously formed nanodomain arrangements in 40-nm diameter $BaTiO_3$ nanorods embedded in $SrTiO_3$ matrix as obtained from phase-field simulations. All images correspond to a top view of the structure displayed in Fig. 7.12. Relaxation started from a completely random initial polarization state systematically leads to one of the configurations depicted in the bottom part of the figure or equivalent ones. All final states can be described as rhombohedral domain structures. The in-plane components are always arranged in a flux-closure (vortex) manner, either clockwise (not shown) or anticlockwise, and the z-components of polarization arranged in *up-down-up-down* sequence (**a**), *up-up-down-down* sequence (**b**), or all in the same direction (**c**). In principle, a more complex higher energy state with six domains (**d**) could also occur. Images were prepared using data from [23, 24]

diameter of the nanorod (of the order of 20 nm), the optimum configuration is a state with two intersecting ferroelastic domain walls arranged as shown in Fig. 7.12b. The whole system was thoroughly investigated by phase-field simulations [23, 24]. When simulations are started from random polarization, there are several possible final states encountered. Usually, domain walls do have the geometry of Fig. 7.12b, but realized domain states would have one of the arrangements shown in the Fig. 7.13a-c or equivalent ones [23, 24].

Let us stress that all domains there have nonzero out-of-plane polarization components which is indicated in Fig. 7.13 by the enlarged or decreased size of the polarization arrow (suggesting viewing in a perspective). Consequently, these nanodomain states contain different kinds of ferroelastic domains. For example, the asymmetric domain structure of Fig. 7.13b contains 71-degree wall perpendicular to the y-axis and 109-degree wall perpendicular to the to x-axis, while there are only 109-degree walls in the $S_4(\bar{4})$ symmetry domain structure shown in Fig. 7.13c.

7.8 Polarization Switching Mediated by Bloch Wall

The net average polarization within an individual domain quadruplet state as depicted in Fig. 7.13 is either close to zero (Fig. 7.13a, b) or it is directed along the z-axis. The quadruple domain state with zero average polarization (Fig. 7.13a, b) can be poled by z-axis bias electric field into the Fig. 7.13c state. This electric field poling process can be also studied by phase-field simulations. A typical hysteresis loop as obtained from phase-field simulations of [23] is shown in Fig. 7.14.

The remanent average polarization $\langle P_z \rangle$ is comparable with bulk spontaneous values. Smooth temperature dependence of $\langle P_z \rangle$ shown in Fig. 7.15a confirms that domain quadruplet state is really robust arrangement and stable over the temperature interval of about 400 K. The magnitude of the in-plane polarization can be conveniently characterized by electric toroidal moment g_N [31, 32]. This quantity has a similar temperature dependence as $\langle P_z \rangle$, also with no additional anomaly in the entire temperature range of the ferroelectric phase (Fig. 7.15).

Interestingly, the virgin switching process (starting at the domain quadruplet state of Fig. 7.13b, with zero net polarization) proceeds through the transformation of a 109-degree Ising-like wall into a Bloch-like wall. With increasing electric field this Bloch-like wall then disintegrates into a pair of 71-degree Ising-like domain walls. This process is illustrated in Fig. 7.16. Figure 7.16a shows the initial state with zero net polarization, identical to the state shown in Fig. 7.13b. Bottom panel shows P_r, P_s and P_t polarization components traced along a trajectory crossing the 109-degree domain wall (the trajectory being indicated in the top panel). The P_t component is small and

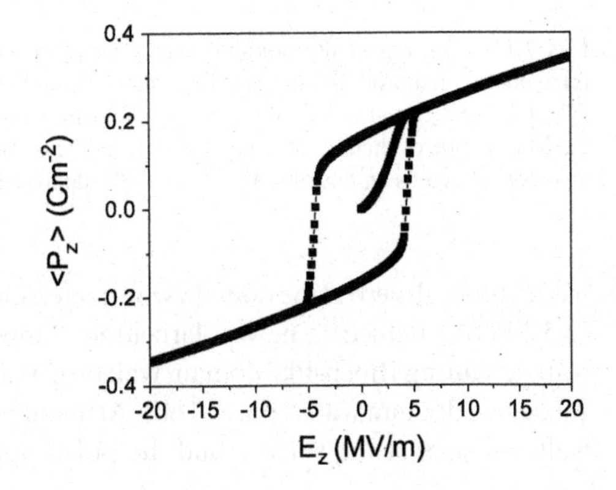

Fig. 7.14 P_z versus E_z hysteresis loop from simulations of switching cycle driven by electric field applied along the z-direction to a 20-nm $BaTiO_3$ nanorod at room temperature. Evolution of the domain structure is described in [23]

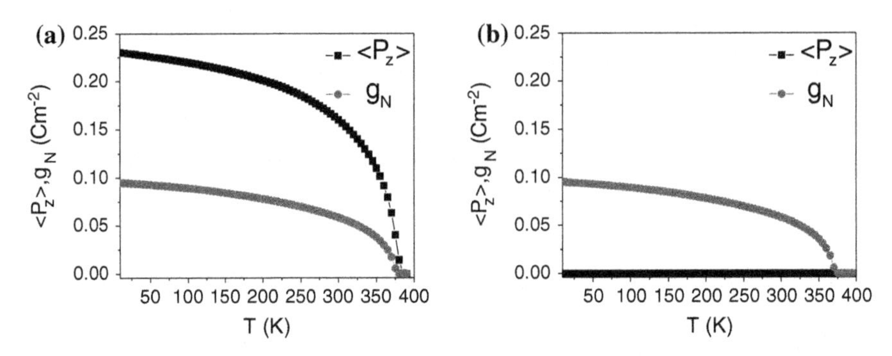

Fig. 7.15 Temperature dependence of the order parameters in the (**a**) non-polarized or anti-polar (*up-up-down-down*) and (**b**) polarized (*up-up-up-up*) states of quadruplet domain configurations within the 40 nm $BaTiO_3$ nanorod

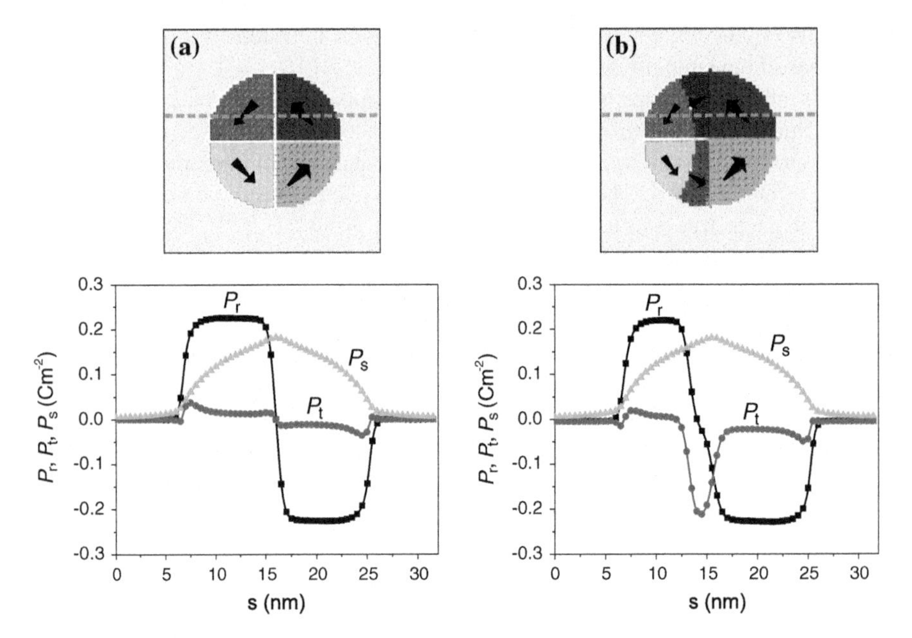

Fig. 7.16 Splitting of the domain wall in a virgin, macroscopical non-polar state of a clamped ferroelectric nanorod by the out-of-plane oriented bias electric field. The initial state (**a**) at $E = 0$ MV/m contains two intersecting Ising-like domain walls, while in the state (**b**) at $E =$ 2.5 MV/m, one of them is split into two 71-degree domain walls. This polarization profile could be also considered as an unstable Bloch-like 109-degree domain wall

passes through zero at the domain wall center. On the other hand, applied external field (2.5 MV/m) induces a new polarization component within the 109-degree domain wall. Resulting Bloch-like domain wall broadens and decomposes in two independent 71-degree domain walls (Fig. 7.16b). At about 5 MV/m, the additional curved domain wall vanishes at the surface and the poled domain structure looks like the one of

Fig. 7.13c. In the subsequent switching cycles, the switching of the P_z component starts from the border and does not influence the in-plane vortex structure [23].

7.9 Conclusions

Topological defects of the spontaneous polarization field, such as ferroelectric domain walls, their junctions, intersections etc., are interesting objects that could be manipulated by electric fields, stresses as well as by built-in interface effects at macroscopic or nanoscopic scales. Many of these topological defects are highly mobile and can be responsible even for macroscopic physical properties of multidomain materials. So far, phase-field simulations based on the GLD theory proved to be a very efficient for studying their structure and properties.

Recently, it has been argued that certain domain walls in ferroelectric crystals with a cubic parent phase may have a Bloch-like character, i.e., the polarization varies smoothly across the domain wall and remains finite within the domain wall. In real situations encountered so far, the modulus of the local polarization is far from being constant, but it is still instructive to describe the situation in terms of polarization rotation. In this chapter we have mostly focused on this subject. However, the experimental evidence of this polarization rotation, typically confined within 1–2 nm lengthscale, is currently the most challenging task in the field.

Acknowledgments This work is supported by Czech Science Foundation (Project CSF 15-04121S).

References

1. S.-Z. Lin, X. Wang, Y. Kamiya, G.-W. Chern, F. Fan, D. Fan, B. Casas, Y. Liu, V. Kiryukhin, W.H. Zurek, C.D. Batista, S.-W. Cheong, Topological defects as relics of emergent continuous symmetry and Higgs condensation of disorder in ferroelectrics. Nat. Phys. **10**, 970–977 (2014)
2. V. Janovec, J. Přívratská, *Domain structures*, International Tables for Crystallography (2006)
3. A.K. Tagantsev, L.E. Cross, J. Fousek, *Domains in Ferroic Crystals and Thin Films* (Springer, New York, 2010)
4. P. Marton, I. Rychetsky, J. Hlinka, Domain walls of ferroelectric $BaTiO_3$ within Ginzburg-Landau-Devonshire phenomenological model. Phys. Rev. B **81**, 144125 (2010)
5. P. Marton, Modelling of domain structures in ferroelectric crystals, Dissertation, Charles University (2007)
6. M. Taherinejad, D. Vanderbilt, P. Marton, V. Stepkova, J. Hlinka, Bloch-type domain walls in rhombohedral $BaTiO_3$. Phys. Rev. B **86**, 155138 (2012)
7. J.C. Wojdeł, J. Íñiguez, Ferroelectric transitions at ferroelectric domain walls found from first principles. Phys. Rev. Lett. **112**, 247603 (2014)
8. G. Catalan, J. Seidel, R. Ramesh, J.F. Scott, Domain wall nanoelectronics. Rev. Mod. Phys. **84**, 119 (2012)
9. P. Marton, J. Hlinka, Simulation of domain patterns in $BaTiO_3$. Ph. Transit. **79**, 467–483 (2006)

10. D. Lee, R.K. Behera, P. Wu, H. Xu, Y.L. Li, S.B. Sinnott, S.R. Phillpot, L.-Q. Chen, V. Gopalan, Mixed Bloch-Néel-Ising character of 180° ferroelectric domain walls. Phys. Rev. B **80**, 060102 (2009)
11. M. Li, Y. Gu, L.-Q. Chen, W. Duan, First-principles study of 180° domain walls in $BaTiO_3$: mixed Bloch-Néel-Ising character. Phys. Rev. B **90**, 054106 (2014)
12. J. Hlinka, V. Stepkova, P. Marton, I. Rychetsky, V. Janovec, P. Ondrejkovic, Phase-field modelling of 180° Bloch walls in rhombohedral $BaTiO_3$. Ph. Transit. **84**, 738 (2011)
13. V. Stepkova, P. Marton, J. Hlinka, Stress-induced phase transition in ferroelectric domain walls of $BaTiO_3$. J. Phys. Condens. Matter **24**, 212201 (2012)
14. P.V. Yudin, A.K. Tagantsev, N. Setter, Bistability of ferroelectric domain walls: morphotropic boundary and strain effects. Phys. Rev B **88**, 024102 (2013)
15. P. Marton, V. Stepkova, J. Hlinka, Divergence of dielectric permittivity near phase transition within ferroelectric domain boundaries. Ph. Transit. **86**, 103 (2013)
16. V. Stepkova, P. Marton, J. Hlinka, Ising lines: Natural topological defects within ferroelectric Bloch walls. Phys. Rev. B **92**, 094106 (2015)
17. J. Hlinka, P. Ondrejkovic, P. Marton, The piezoelectric response of nanotwinned $BaTiO_3$. Nanotechnology **20**, 105709 (2009)
18. J. Hlinka, P. Marton, Phenomenological model of a 90° domain wall in $BaTiO_3$-type ferroelectrics. Phys. Rev. B **74**, 104104 (2006)
19. J. Hlinka, P. Marton, Ferroelastic domain walls in barium titanate-quantitative phenomenological model. Integr. Ferroelectr. **101**, 50 (2008)
20. P. Marton, J. Hlinka, Computer simulations of frequency-dependent dielectric response of 90-degree domain walls in tetragonal barium titanate. Ferroelectrics **373**, 139–144 (2008)
21. T. Sluka, A.K. Tagantsev, D. Damjanovic, M. Gureev, N. Setter, Enhanced electromechanical response of ferroelectrics due to charged domain walls. Nat. Commun. **3**, 748 (2012)
22. T. Sluka, A.K. Tagantsev, P. Bednyakov, N. Setter, Free-electron gas at charged domain walls in insulating $BaTiO_3$. Nat. Commun. **4**, 1808 (2013)
23. V. Stepkova, P. Marton, N. Setter, J. Hlinka, Closed-circuit domain quadruplets in $BaTiO_3$ nanorods embedded in $SrTiO_3$ film. Phys. Rev. B **89**, 060101(R) (2014)
24. V. Stepkova, P. Marton, J. Hlinka, Peculiar domain states of cylindrical $BaTiO_3$ nanorods embedded in $SrTiO_3$ matrix. Ph. Transit. **87**, 922–928 (2014)
25. E.A. Eliseev, P.V. Yudin, S.V. Kalinin, N. Setter, A.K. Tagantsev, A.N. Morozovska, Structural phase transitions and electronic phenomena at 180-degree domain walls in rhombohedral $BaTiO_3$. Phys. Rev. B **87**, 054111 (2013)
26. B. Houchmandzadeh, J. Lajzerowicz, E. Salje, Order parameter coupling and chirality of domain walls. J. Phys. Condens. Matter **3**, 5163 (1991)
27. P.V. Yudin, A.K. Tagantsev, E.A. Eliseev, A.N. Morozovska, N. Setter, Bichiral structure of ferroelectric domain walls driven by flexoelectricity. Phys. Rev. B **86**, 134102 (2012)
28. P. Mokrý, J. Fousek, Elastic aspects of domain quadruplets in ferroics. J. Appl. Phys. **97**, 114104 (2005)
29. S. Wada, T. Tsurumi, Enhanced piezoelectricity of barium titanate single crystals with engineered domain configuration. Br. Ceram. Trans. **103**, 93–96 (2004)
30. T. Sluka, D. Damjanovic, A. Tagantsev, E. Colla, M. Mtebwa, N. Setter, The stress-assisted enhancement of piezoelectric properties due to mechanically incompatible domain structures in $BaTiO_3$, in *IEEE International Symposium on Applications of Ferroelectrics—ISAF 2010*, Edinburg, Scottland (2010)
31. I.I. Naumov, L. Bellaiche, H. Fu, Unusual phase transitions in ferroelectric nanodisks and nanorods. Nature **432**, 737 (2004)
32. J. Hlinka, Eight types of symmetrically distinct vectorlike physical quantities. Phys. Rev. Lett. **113**, 165502 (2014)

Chapter 8
Topological Defects in Ferroic Materials

Anna N. Morozovska, Eugene A. Eliseev and Sergei V. Kalinin

Abstract Using Landau–Ginzburg–Devonshire theory we explore unusual electronic, structural and polar properties of the topological defects inherent in ferroics, such as ferroelectric and ferroelastic domain walls, which can have rich and tunable internal structure. Also we underline that the existence of 2D defects in ferroelectrics is similar to the cross-tie defects in the ferromagnetic Bloch domain walls. The seeding for the modulated phase can be a topological defect, such as a structural domain wall.

8.1 Introduction

Materials with multiple coupled order parameters have emerged as one of the priority directions in condensed matter physics [1, 2] due to both intriguing physical behaviours and a broad variety of novel physical applications they enable. The unique physical properties of multiferroics originated from the complex interactions between the structural, polar and magnetic long-range order parameters [3, 4]. For instance, biquadratic coupling of the structural and polar order parameters, introduced by Haun [5, 6], Salje et al. [7, 8], Balashova and Tagantsev [9], are responsible for the unusual behaviour of the dielectric properties in ferroelastics—quantum paraelectrics. biquadratic and linear magnetoelectric couplings cause such intriguing effects as giant magnetoelectric tunability of multiferroics [10, 11].

A.N. Morozovska (✉)
Institute of Physics, National Academy of Sciences of Ukraine,
46, Pr. Nauki, Kyiv 03028, Ukraine
e-mail: anna.n.morozovska@gmail.com

E.A. Eliseev
Institute for Problems of Materials Science, National Academy of Sciences of Ukraine,
3, Krjijanovskogo, Kyiv 03142, Ukraine
e-mail: eugene.a.eliseev@gmail.com

S.V. Kalinin
Center for Nanophase Materials Science,
Oak Ridge National Laboratory, Oak Ridge, TN 37831, USA
e-mail: sergei2@ornl.gov

J. Seidel (ed.), *Topological Structures in Ferroic Materials*, Springer Series in Materials Science 228, DOI 10.1007/978-3-319-25301-5_8

The new and intriguing aspects of physics of multiferroics are related to the properties of topological defects, such as domain walls and interfaces, which are inherent to the materials [12]. Here, strong gradients in primary order parameters fields can lead to emergence of coupled behaviour. For instance, Daraktchiev et al. [13] considered the influence of biquadratic coupling between polarization and magnetization on the structure of ferroelectric domain walls in multiferroics and the reason of magnetization appearance inside the domain wall in non-ferromagnetic phases. Using *ab initio* calculations Dieguez et al. [14] have found that the behaviour of the structural order parameter at the domain walls of multiferroics determines their structure and energetics.

In this regard, intriguing phenomena emerging in the nanoscale phase-separated ferroics. Nanoscale phase separation ranging from giant magnetoresistive manganites [15–18], ferroelectric relaxors [19, 20] and morphotropic materials [21–23], martensites [24, 25] and birelaxors [26] remain one of the active topic of research in condensed matter physics. In many cases, significant uncertainties persist regarding the true ground states for these systems including various topological defects from single vortices to nano-domain textures [27]. Experiment [28] demonstrated the presence the order parameters modulation originated from the flexo-type coupling leading to the modulation instabilities at the structural domain boundaries. Super-structural dynamic order and antiferroelectric-antiferrodistortive modulation have been recently observed in antiferrodistortive incipient ferroelectrics [29].

For multiaxial ferroics with multicomponent order parameters, analysis of polarization structure at the topological defects (e.g. at domain wall (DW), interfaces, vortices) necessitates taking into account the coupling between the order parameter components [6, 30, 31] mediated by stress accommodation or gradients coupling. Situation is similar for ferromagnetic-ferroelectric, where local magnetic moment is possible at the ferroelectric DW due to either biquadratic [32] or inhomogeneous coupling [33, 34].

Despite the fact that attempts to describe polarization behavior in multicomponent ferroics were performed since the early days of ferroelectricity [35–37], the progress with understanding of its structure at the topological defects appeared very limited. However it is expected that the existence of e.g. charged topological defects, such as domain walls, in ferroelectric semiconductors should result in charge accumulation or depletion on adjacent sides of the domain wall, and should thus lead to enhanced conductivity at these sites [38]. Experimental verification of this prediction occurred only recently, by Seidel et al. who used scanning probe microscopy (SPM) methods to report room-temperature metallic conductivity of 180° and 109° domain walls in $BiFeO_3$ [39, 40]. Farokhipoor et al. have shown that nominally uncharged as-grown 71° domain walls in $BiFeO_3$ can also exhibit enhanced conductivity [41]. Nominally uncharged fabricated vortex structures in $BiFeO_3$ show an order of magnitude increase in conductivity over single domain regions [42]. Domain walls conductance in $ErMnO_3$ was reported by Meier et al. [43]. Domain walls act as conducting channels and $LiNbO_3$ [44]. Guyonnet et al. observed conductive domain walls $Pb(Zr,Ti)O_3$ [45]. Then Maksymovych et al. report about metallic conductivity of 180° domain walls in $Pb(Zr,Ti)O_3$ [46] To measure quantitatively the variability of the conductivity response at ferroelectric domain walls combined experimental studies with current-AFM (c-AFM) and Piezoresponse Force Microscopy (PFM) are successfully used [39, 42, 46–48].

Landau theory is a powerful method that has proven capable of predicting charged domain walls' static conductivity in ferroelectric-semiconductors. In 1969 [49], the conductivity mechanism was for the first time described stemming from compensation of polarization charge discontinuity by mobile carriers in the material. Analytical Landau-type theory was further developed for charged walls in uniaxial [50, 51] and multiaxial tetragonal ferroelectrics [52], nominally uncharged walls in rhombohedral ferroelectrics [53], improper ferroelectrics [54] and twin walls in incipient ferroelectrics–ferroelastics [55]. Using continuum Landau-Ginzburg-Devonshire (LGD) theory, Hlinka and Márton [56] calculated numerically the structure of twin boundaries in tetragonal perovskite crystal $BaTiO_3$. They found that the polarization component normal to the domain wall plane demonstrates a weak deviation from constant distribution. Then Marton et al. [57] considered Ising-Bloch-type 180° domain walls in rhombohedral $BaTiO_3$ for definite rotation angles. Lee et al. [58] reported about mixed Ising-Bloch-Neel type in $PbTiO_3$, $LiNbO_3$ and thin strained films of $BaTiO_3$. Notice, that Authors [56–58] did not consider the impact of the flexoelectric coupling on domain wall structure and energy.

Flexoelectric effect describes the coupling of polarization with strain gradient and polarization gradient with the strain [59–61], and was predicted by Mashkevich and Tolpygo [62]. Subsequently, a number of theoretical studies of the flexoelectric effect in conventional [63–70] and incipient [71] ferroelectrics have been performed. Experimental measurements of flexoelectric tensor components were recently carried out by Ma and Cross [72–74] and Zubko et al. [75].

Eliseev et al. [52] and Morozovska et al. [53] consider the influence on the flexoelectric effect on the 180° domain wall static conductivity in tetragonal $Pb(Zr,Ti)O_3$ and rhombohedral $BiFeO_3$. Yudin et al. [76] show that the flexoelectric coupling induces a new polarization component with a structure qualitatively different from the classical Bloch-wall structure in tetragonal $BaTiO_3$. Eliseev et al. [77] study the influence of the flexoelectric coupling on the domain wall structure, energy and possible free carrier accumulation by the wall in rhombohedral phase of $BaTiO_3$.

8.2 Thermodynamic Approach

One of the most often encountered 2D-topological defect, which polar properties are relatively well-studied, are domain walls in ferroelectrics. So, let us consider nominally uncharged 180° domain wall in the bulk of n-type ferroelectric-semiconductor. Within Landau-Ginzburg-Devonshire (LGD) theory, equations of state for polarization components and elastic stresses can be derived from the minimization of the Gibbs free energy functional, which density for the m3m parent (paraelectric) symmetry has the form [52]:

$$G = a_i P_i^2 + a_{ij} P_i^2 P_j^2 + a_{ijk} P_i^2 P_j^2 P_k^2 + \frac{g_{ijkl}}{2} \frac{\partial P_i}{\partial x_j} \frac{\partial P_k}{\partial x_l} - Q_{ijkl} \sigma_{ij} P_k P_l$$
$$+ \frac{F_{ijkl}}{2} \left(\sigma_{ij} \frac{\partial P_k}{\partial x_l} - P_k \frac{\partial \sigma_{ij}}{\partial x_l} \right) - \frac{s_{ijkl}}{2} \sigma_{ij} \sigma_{kl} - P_i E_i - \frac{\varepsilon_0 \varepsilon_b}{2} E_i^2 + e \left(N_d^+ + p - n \right) \varphi \tag{8.1}$$

Hereinafter P_i denotes electric polarization, a_i, a_{ij} and a_{ijk} are LGD-expansion coefficients of the 2nd, 4th and 6th order dielectric stiffness tensors correspondingly, gradient coefficient are g_{ijkl}, F_{ijkm} is the flexoelectric tensor; σ_{ij} are elastic stresses, Q_{ijkl} are 4th second rank electrostriction tensors coefficients, s_{ij} are elastic compliances. $E_k(\mathbf{r}) = -\partial \varphi(\mathbf{r})/\partial x_k$ are depolarization electric field components, $\varphi(\mathbf{r})$ is the electric potential, ε_b background permittivity [78] and $\varepsilon_0 = 8.85 \times 10^{-12}$ F/m the dielectric permittivity of vacuum. The space charge density is $e\left(N_d^+ + p - n\right)$, $e = 1.6 \times 10^{-19}$ C the electron charge, $n(\mathbf{r})$ is the concentration of the electrons in the conduction band; $p(\mathbf{r})$ is the concentration of holes in the valence band; $N_d^+(\mathbf{r})$ is the concentration of ionized donors. The latter term in (b) is the electrostatic energy of free charges with density $e\left(N_d^+ + p - n\right)$ in the electric field with potential φ. Electrostatic potential φ should be determined from the Poisson equation, $\varepsilon_0\, \varepsilon_b \frac{\partial^2 \varphi}{\partial x_i^2} = \frac{\partial P_i}{\partial x_i} - e\left(p - n + N_d^+\right)$. Hereinafter acceptors are regarded absent and holes concentration is regarded negligibly small in comparison with the electrons, which are improper carriers in n-type $BaTiO_3$. Note, that space charge is absent in the dielectric limit and Poisson equation becomes $\varepsilon_0\, \varepsilon_b \frac{\partial^2 \varphi}{\partial x_i^2} = \frac{\partial P_i}{\partial x_i}$.

Regarding that all physical quantities can depend only on the distance $\tilde{x}_1$ from the domain wall plane, $\tilde{x}_1 = 0$, it make sense to define them in the coordinate frame $\{\tilde{x}_1, \tilde{x}_2, \tilde{x}_3\}$ rotated with respect to the pseudo-cubic crystallographic axes $\{x_1, x_2, x_3\}$. All variables become dependent on the only variable $\tilde{x}_1$.

Euler–Lagrange equations for polarization components and equations of state elastic stresses were derived from the minimization of the free energy functional $\delta \tilde{G}/\delta \tilde{P}_i = -\tilde{E}_i$ and $\delta \tilde{G}/\delta \tilde{\sigma}_{ij} = -\tilde{u}_{ij}$, where u_{ij} are elastic strains; symbol δ stands for the variational derivative. Equations $\partial \tilde{G}/\partial \tilde{\sigma}_{ij} = -\tilde{u}_{ij}$ should be solved along with mechanical equilibrium conditions $\partial \tilde{\sigma}_{1j}/\partial \tilde{x}_1 = 0$ and compatibility relation $e_{i1l} e_{j1n} \left(\partial^2 \tilde{u}_{ln}/\partial \tilde{x}_1^2\right) = 0$. Evident form of elastic stresses and equations of state in rotated coordinate frame are listed in [79]. After the substitution of the elastic stresses in the equations of state $\delta \tilde{G}/\delta \tilde{P}_i = -\tilde{E}_i$ it becomes coupled and self-consistent with the Poisson equation $\varepsilon_0\, \varepsilon_b \frac{\partial^2 \varphi}{\partial \tilde{x}_1^2} = \frac{\partial \tilde{P}_1}{\partial \tilde{x}_1} - e\left(N_d^+ + p - n\right)$ for the determination of electrostatic potential φ. Corresponding boundary conditions are $\tilde{P}_3(\tilde{x}_1 = 0) = 0$, $\tilde{P}_3(\tilde{x}_1 \to +\infty) = +\tilde{P}_S$ and $\tilde{P}_{1,2}(\tilde{x}_1 \to \pm\infty) \to 0$, $\tilde{E}_1(\tilde{x}_1 \to \pm\infty) \to 0$.

In order to account for the electron gas degeneration at the domain wall [50–52], electron density $n(\tilde{x}_1)$ redistribution in conductive band can be estimated using the effective mass approximation:

$$n(\tilde{x}_1) \approx \int_0^{\infty} d\varepsilon \cdot g_n(\varepsilon) f(\varepsilon + E_C - E_F - e\varphi(\tilde{x}_1)), \tag{8.2}$$

where $g_n(\varepsilon) = \sqrt{2m_n^3 \varepsilon}/(2\pi^2\hbar^3)$ is density of states, m_n is the effective mass; $f(x) = (1 + \exp(x/k_B T))^{-1}$ is the Fermi-Dirac distribution function, $k_B = 1.3807 \times 10^{-23}$ J/K, T is the absolute temperature. E_F is the Fermi level, E_C is the bottom of the conductive band defined with respect to the vacuum level. Concentration of almost immobile ionized donors in the simple level approximation is

$$N_d^+(\tilde{x}_1) = N_d^0 (1 - f(E_d - E_F - e\varphi(\tilde{x}_1))). \tag{8.3}$$

Concentration of donors is N_d^0, $E_d\left(\tilde{u}_{ij}, \tilde{P}_i\right)$ is the donor level position. For mobile species Vegard strains should be included in expressions for N_d^+ [80]. Note that in general case conduction band and donor level in (8.2) and (8.3) can be polarization and strain-dependent, since ferroelectric polarization changes the band structure via polarization potential, $\sum_{i=1}^{3}\left(\frac{\zeta}{2}\tilde{P}_i^2 + \frac{\lambda}{4}\tilde{P}_i^4\right)$ [81–83]. Also deformation potential $\Xi_{ij}^C \tilde{u}_{ij}$ shifts the levels position [84, 85]. However, in the model case of the non-degenerated simple band structure and effective mass approximation validity, the shallow donor level and conductive band edge are shifted as a whole with the strain [86] or polarization, so the differences $E_F - E_C$ and $E_d - E_F$ become almost strain- and polarization independent.

Fermi level position E_F should be determined self-consistently from the electroneutrality condition $N_{d0}^+ - n_0 = 0$ valid in the single-domain region of ferroelectric, where potential vanishes ($\varphi \to 0$), strain tends to the spontaneous values $u_{ij}^S = Q_{ijkl} P_k^S P_l^S$, equilibrium density of electrons $n_0 = \int_0^{\infty} d\varepsilon \cdot g_n(\varepsilon) f(\varepsilon + E_C - E_F)$ and ionized donors $N_{d0}^+ = N_d^0 f(E_F - E_d)$. It is worth to underline that (8.2) and (8.3) reduces to $n \approx n_0 \exp(e\varphi/k_B T)$ and $N_d^+ \approx N_{d0}^+ \exp(-e\varphi/k_B T)$ in the Boltzmann approximation.

Results of numerical modeling for the domain wall energy, polarization vector structure, electric potential and charge carriers redistribution across the domain wall are discussed in the next sections using an example of classical ferroelectrics.

8.3 Domain Wall Vectorial Structure, Energy and Static Conductivity in Multiaxial Ferroelectrics

Using LGD theory the analysis of the carriers' accumulation by 180° domain wall in uniaxial [51] and multiaxial [52, 53, 77] ferroelectric-semiconductors was performed. Along with coupled LGD equations for the polarization components, the Poisson equation for the electrostatic potential was solved. Spatial distributions of the ionized shallow donors (e.g. intrinsic oxygen vacancies or protons), free electrons and holes were found self-consistently using the effective mass approximation for their energy density of states. Performed theoretical analysis shows that several scenarios of the domain wall conduction are possible depending on the wall geometry (tilt angle, domain shape and size), wall type (head-to-head or tail-to-tail), size effects (stripe and cylindrical domains), the sign and value of the flexoelectric coupling coefficient.

In particular the charge of carriers accumulated by uncharged domain wall is determined by the sign of the flexoelectric coefficient: positive coefficient leads to the accumulation of negative carriers (electrons or acceptors), negative coefficient leads to the accumulation of positive carriers (holes, donors or vacancies) [52]. The driving force of this intriguing phenomena for nominally uncharged walls is the flexoelectric coupling, which being rather high for ferroelectric perovskites [72–75], leads to the appearance of polarization components perpendicular to the wall plane and its strong gradient across the wall. The polarization component perpendicular to the wall plane leads to the appearance of the uncompensated bound charge at the wall. The charge creates strong electric field which in turn leads to the accumulation of free screening carriers across the wall. At the same time, the polarization component parallel to the wall plane is indifferent to the presence of the flexoelectric coupling and electrostriction coupling induces the narrowing of the domain wall.

The tilted wall is charged in both uniaxial and multi-axial ferroelectric-semiconductors and hence the electric field of the bound charge attracts free carriers of definite sign and repels the carriers of the opposite sign from the wall region. Carriers' accumulation is highest when the wall plane is perpendicular to the spontaneous polarization direction at the wall (perpendicular domain wall); it decreases with the bound charge decrease and reaches minimum for the parallel domain wall. In numbers, carrier accumulation in tetragonal $Pb(Zr,Ti)O_3$ leads to the strong increase of the static conductivity across the domain walls, e.g. up to 3 orders of magnitude for the perpendicular domain walls and up to 10–30 times increase for domain stripes and cylindrical nanodomains for the typical range of flexoelectric coefficients [52] (Fig. 8.1). But in contrast to thick domain stripes and thicker cylindrical domains, in which the carrier accumulation (and so the static conductivity) sharply increases at the domain walls only, nanodomains of radius less then 5–10 correlation length appeared conducting entire their cross-section. Such conductive nanosized channels may be promising for nanoelectronic concepts due to the possibility to control their spatial location by nano-manipulation with the charged probe.

According to LGD and DFT studies Ising-Bloch-type and mixed Ising-Bloch-Néel-type [58, 87] 180° walls can exist in a wide class of ferroelectric materials. The structure, energetics, and carriers accumulation by the 180 domain wall was investigated as functions of wall orientation on the example of $BaTiO_3$ [77].

Polarization vector inside a domain wall (DW) in rhombohedral phase of $BaTiO_3$ can have all three components. The component P_3, parallel to the spontaneous polarization $\pm P_S$ in the domains, is regarded as the Ising-type; the component P_2, parallel to the wall plane, but perpendicular to Ising-type component, which vanishes far from the wall, is regarded as Bloch-type component; and component P_1, normal to the

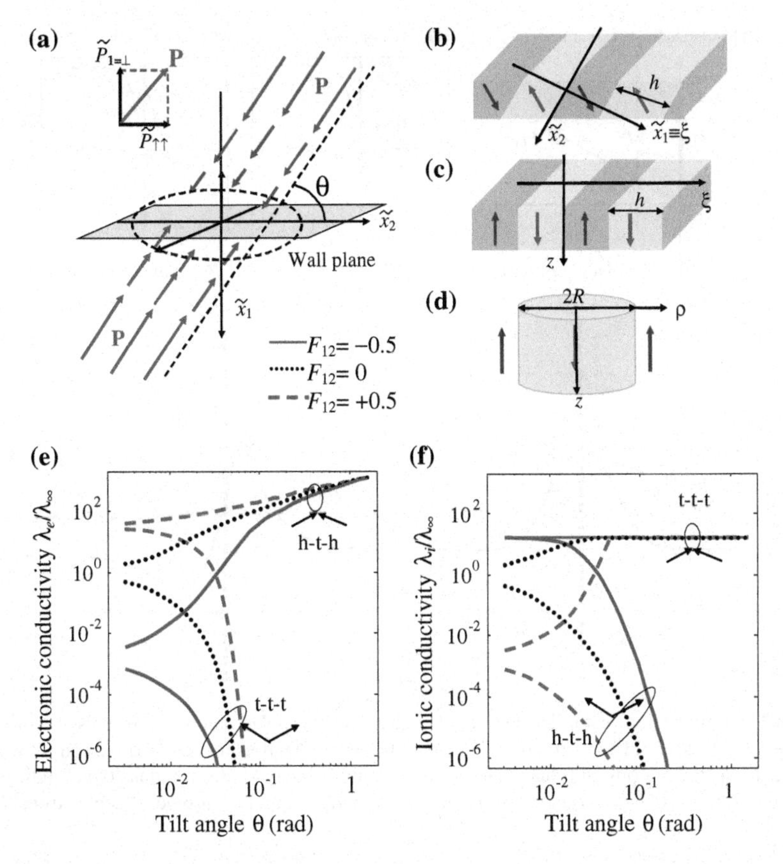

Fig. 8.1 One dimensional distribution of polarization in the vicinity of a single domain wall tilted at angle θ (**a**), tilted (**b**) and parallel (**c**) domain stripes with half-period h; (**d**) cylindrical domain of radius R. *Arrows* in plots b-d indicate the polarization direction in the center of domains. Dependence of electronic (**e**) and ionic (**f**) conductivity on the domain wall tilt angle θ between the neighboring head-to-head (h-t-h) and tail-to-tail (t-t-t) stripes and calculated for negative, zero, and positive flexoelectric coupling coefficient $F_{12} = (-0.5, 0, 0.5) \times 10^{-10}\text{m}^3/\text{C}$ (*solid*, *dotted* and *dashed curves* respectively). Other parameters corresponds to tetragonal $Pb(Zr,Ti)O_3$ and room temperature. (Adapted from [52])

wall, is regarded as Néel-type component (see Fig. 8.2a–c). Note that the Néel-type component P_1 is associated with the non-zero divergence of polarization vector and hence should be considered jointly with associated depolarization fields.

Domain walls are shown to be of mixed Ising-Bloch-Néel type for all orientations. Although the domain walls with {211} and {110} orientations are shown to have sufficiently different structures, achiral and chiral (Fig. 8.2b, c), and the phase transition from achiral to chiral state can be achieved either by varying the wall orientation at fixed temperature or by temperature change at constant orientation. The impact of

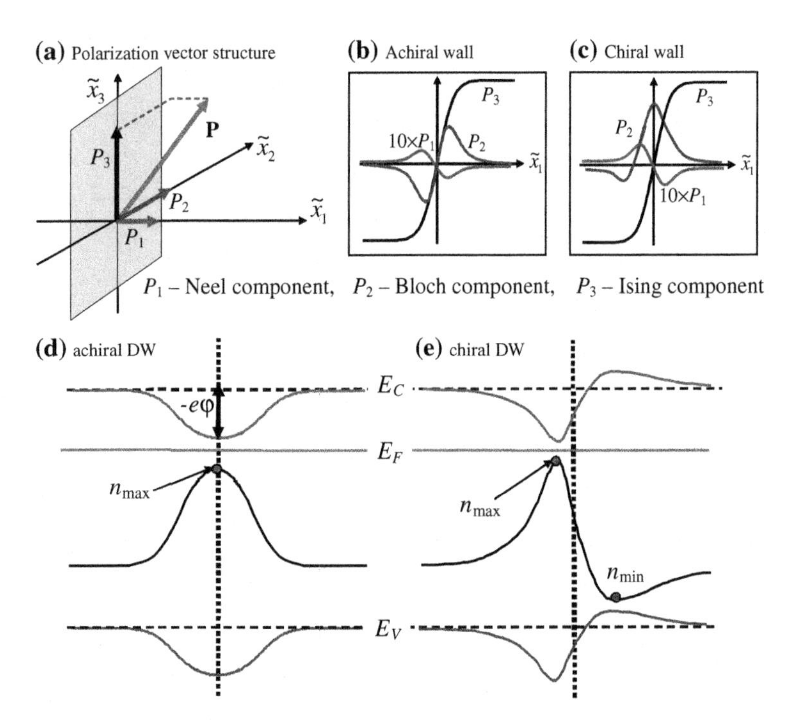

Fig. 8.2 **a** Polarization vector structure. **b**, **c** Schematics of the polarization components distribution inside achiral and chiral domain walls. Sketches of local band bending for achiral **d** and chiral **e** walls, where the spatial regions with maximal (n_{max}) and minimal (n_{min}) electron density are indicated. The concentration of holes is negligible and the conductivity is purely of n-type. (Adapted from [77])

the flexoelectric effect is that the Neel-component of polarization perpendicular to the domain wall plane is nonzero for any wall orientation.

Further analyses of such walls electronic properties suggest detecting the structural phase change inside the domain walls by c-AFM contrast due to the correlation of the domain wall structure and free charge accumulation, driven by depolarizing field. The conductivity enhancement in the domain wall is caused by the potential variation inside the wall. The potential well/hump leads to higher/lower electron concentration in the DW due to the local band bending (see sketches in Fig. 8.2d, e for chiral and achiral walls). Since $\tilde{P}_1$ profile is anti-symmetric for achiral DW (Fig. 8.2b), corresponding potential barrier $\varphi(\tilde{x}_1)$ is symmetric, while it can be asymmetric for achiral DWs. Symmetric barriers accumulate electrons only (Fig. 8.2d). Asymmetric potential barriers with double structure can attract the electrons in some spatial regions and repulse them from the other regions (Fig. 8.2e).

Figure 8.3a, b illustrate the rotation anisotropy of the relative density $n(\tilde{x}_1)/n_0$. Two sharp maxima and breaks on the figures correspond to the chiral-achiral phase transitions occurred at definite critical angles $\alpha_{cr}^{m} \approx \pi/6 \pm \pi/12 + m\pi/3$ (m is integer). Without flexoelectric coupling c-AFM contrast is equal to unity for the angles $\alpha = m\pi/3$ corresponding to the absence of the component $\tilde{P}_1$ (see Fig. 8.3a).

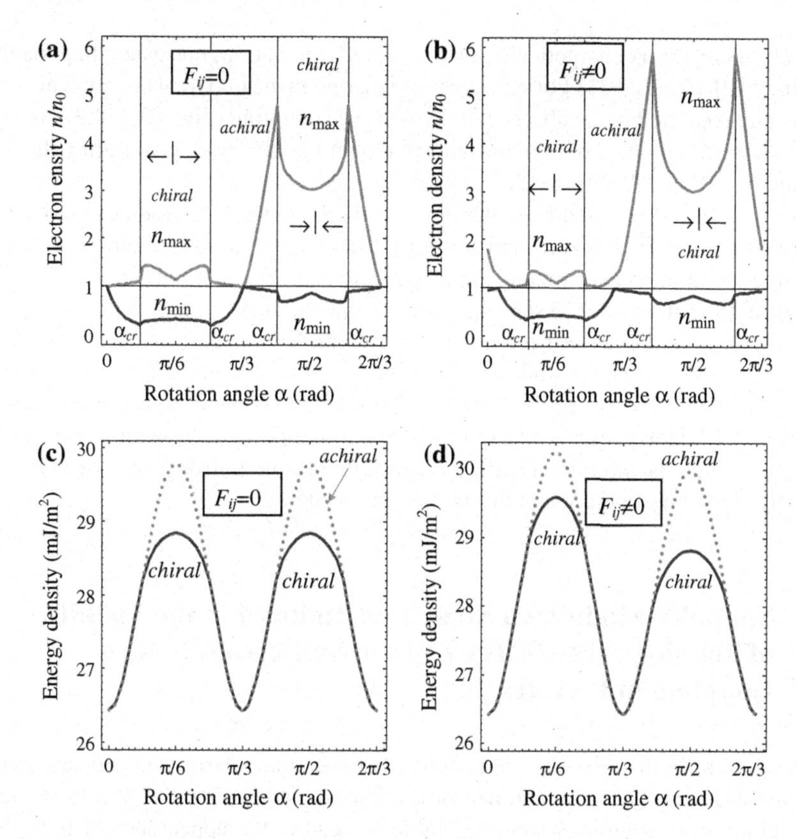

Fig. 8.3 Relative maximal $n_{\max}/n_0$ and minimal $n_{\min}/n_0$ electron density versus the DW rotation angle α calculated in rhombohedral $BaTiO_3$ at 180 K without flexoelectric coupling $F_{ij} = 0$ (**a**) and with flexoelectric coupling (**b**). α is the wall rotation angle counted from crystallographic plane $< 101 >$. tail-to-tail structure ($\leftarrow | \rightarrow$) head-to-head structure ($\rightarrow | \leftarrow$). (Adapted from [77])

Flexoelectric coupling leads to nonzero perpendicular component $\tilde{P}_1$ for all α and thus to nonzero contrast; also it slightly shifts the critical angles and create the symmetric potential structure well-barrier-well around rotation angle $\pi/3$ (see Fig. 8.3b). Strong asymmetry in the electron density distributions for angles α inside the range $\pi/6 \pm \pi/12$ and $\pi/2 \pm \pi/12$ originated from the fact, that DWs have mainly tail-to-tail structure with respect to $\tilde{P}_1$ at $\pi/6 \pm \pi/12$, and head-to-head structure at $\pi/2 \pm \pi/12$.

Let us underline that the most intriguing situation can appear in the point of the wall chiral-achiral phase transition, i.e. at rotation angles around the critical ones, α_{cr}. The chiral-achiral phase transition can be revealed by local c-AFM measurements of the cylindrical walls, since c-AFM contrast is regarded proportional to the relative electron density $n(\tilde{x}_1)/n_0$.

The numerical analysis of resultant GLD energy predicts that the flexoelectric coupling introduces additional angular anisotropy for the DW structure and energy.

Namely, there are six energetically favorable wall orientations corresponding to minima in {110}-planes, {211} orientations correspond to the maximal energy. The minima are degenerated in the absence of flexoelectric coupling (Fig. 8.3c), the coupling removes the degeneracy of maxima and split them into two triplets revealing the true symmetry of the wall (Fig. 8.3d).

These effects originated from the fact that the modulation period of polarization Neel component $\tilde{P}_1$ for zero and nonzero flexoelectric coupling, since the latter resulting in additional symmetry breaking between the maximal states [77]. The energy minima at $\alpha = m\pi/3$ stay equivalent, while the energy maxima at $\alpha = \pi/6 + m\pi/3$ for odd and even m become nonequivalent. This is seen from the different width of the area of chiral wall absolute stability and different height of the energy maximum. Note that the equivalence of the minima follows from the symmetry of the problem, which contains axis of third order along [111] and mirror plane {110}. For the maxima the situation is different since there is no mirror plane at {211} and the only symmetry operation is the axis of third order.

8.4 Spatially-Modulated Structures Induced in the Vicinity of Topological Defects by Flexo-Antiferrodistortive Coupling in Ferroics

In its initial form the flexoelectric coupling between the polarization and strain gradient is universal for macro and nanoscale objects [66, 67, 70, 88, 89]. Flexoelectric and all other couplings from the Table 8.1 lead to the appearance of improper ferroelectricity in multiferroics with the inhomogeneous spontaneous strain [60], magnetization [33, 90], aniferromagnetic [34, 91] or antiferrodistortions. Below we consider the flexo-antiferrodistortive coupling in ferroics, since antiferrodistortive modes are virtually present in all the perovskites. We will illustrate that it can cause incommensurate modulation in the vicinity of ferroelastic domain walls.

For multiferroics with the antiferrodistortive and polar long-range order parameters the conventional form of the bulk LGD Helmholtz free energy and Gibbs potential densities are:

$$F_b[\mathbf{P},\mathbf{\Phi},\mathbf{u}] = U^u_{LGD} + U^u_{Elastic} + U^u_{\xi}[\mathbf{P},\mathbf{\Phi}], \quad G_b[\mathbf{P},\mathbf{\Phi},\sigma] = U^{\sigma}_{LGD} + U^{\sigma}_{Elastic} + U^{\sigma}_{\xi}[\mathbf{P},\mathbf{\Phi}], \tag{8.4}$$

LGD-type expansions are

$$U^m_{LGD} = a^m_i P_i^2 + a^m_{ij} P_i^2 P_j^2 + \frac{g^m_{ijkl}}{2}\frac{\partial P_i}{\partial x_j}\frac{\partial P_k}{\partial x_l} + b^m_i \Phi_i^2 + b^m_{ij}\Phi_i^2\Phi_j^2 - \eta^m_{ijkl} P_i P_j \Phi_k \Phi_l + \frac{v^m_{ijkl}}{2}\frac{\partial \Phi_i}{\partial x_j}\frac{\partial \Phi_k}{\partial x_l}. \tag{8.5a}$$

Elastic energy is

$$U^{u}_{Elastic} = -q_{ijkl}u_{ij}P_kP_l + \frac{f_{ijkl}}{2}\left(\frac{\partial P_k}{\partial x_l}u_{ij} - P_k\frac{\partial u_{ij}}{\partial x_l}\right) + \frac{c_{ijkl}}{2}u_{ij}u_{kl} - r_{ijkl}u_{ij}\Phi_k\Phi_l \tag{8.5b}$$

$$U^{\sigma}_{Elastic} = -Q_{ijkl}\sigma_{ij}P_kP_l + \frac{F_{ijkl}}{2}\left(\frac{\partial P_k}{\partial x_l}\sigma_{ij} - P_k\frac{\partial \sigma_{ij}}{\partial x_l}\right) - \frac{s_{ijkl}}{2}\sigma_{ij}\sigma_{kl} - R_{ijkl}\sigma_{ij}\Phi_k\Phi_l \tag{8.5c}$$

Superscript $m = u$ for elastic strain or σ stress. Polarization components are P_i $(i = 1, 2, 3)$. Φ_i is the components of the structural antiferrodistortive order parameter, e.g. an axial tilt vector corresponding to the octahedral rotation angles [92]; u_{ij} and σ_{ij} are the strain and stress tensors correspondingly. The summation is performed over all repeated indices. Coefficients $a_i(T)$ and $b_i(T)$ temperature dependence can be fitted with Curie-Weiss law for ferroelectrics, or with Barrett law for improper ferroelectrics. Gradients coefficients g_{ij} and v_{ij} are regarded positive for commensurate multiferroics. Below we will regard $F_b\,[\mathbf{P}, \mathbf{\Phi}, \mathbf{u}]$ as **Φ-P-u** representation and $G_b\,[\mathbf{P}, \mathbf{\Phi}, \sigma]$ as **Φ-P-σ** representation.

The flexo-antiferrodistortive coupling between the polarization gradient and tilt components product should be included in the functional (8.4) in the form of Lifshitz invariant [93]:

$$U_{\xi}\,[\mathbf{P}, \mathbf{\Phi}] = \frac{\xi^{u,\sigma}_{ijkl}}{2}\left(\Phi_i\Phi_j\frac{\partial P_k}{\partial x_l} - P_k\frac{\partial\left(\Phi_i\Phi_j\right)}{\partial x_l}\right) \tag{8.6}$$

It is well-known that Gibbs and Helmholtz energies are different, but the values calculated from equations of state should be the same in both Φ-P-σ and Φ-P-u representations. Starting from the variation of the functional in any of representations, Euler-Lagrange equations for the polarization and tilt as well as equations of state for the elastic stress or strain can be derived. The equations of state give the relation between the stress and strain. After the substitution of the relation into the Euler-Lagrange equations, unambiguous relationship between the coefficients of LGD expansion for mechanically clamped $F_b\,[\mathbf{P}, \mathbf{\Phi}, \mathbf{u}]$ and free $G_b\,[\mathbf{P}, \mathbf{\Phi}, \sigma]$ systems can be established. They are listed in the Table 8.1.

Allowing for the flexoelectric coupling, one can see the gap when compare the relations between the tensorial coefficients in different presentations, summarized in the last row of the Table 8.1. Actually the flexo-antiferrodistortive coupling ξ^{σ}_{ijkl} between the tilt and polarization gradient, (8.6), is allowed by *any* symmetry. If one starts from conventional Φ-P-σ representation with zero $\xi^{\sigma}_{ijkl} \equiv 0$, then mandatory come to nonzero values $\xi^{u}_{ijkl} \equiv F_{ijmn}r_{mnkl}$ in Φ-P-σ representation, i.e. the novel coupling tensor, which strength is proportional to the convolution of the flexoelectric and rotostriction coupling tensors, appear due to the roto-flexo effect [94]. However, the direct flexo-distortive coupling tensor are unknown and cannot be determined

Table 8.1 Relations between the Φ-P-σ and Φ-P-u LGD expansion coefficients

LGD-expansion coefficient name	Relationship
Linear inverse susceptibility (stiffness)	$a_i^u \equiv a_i^\sigma \equiv a_i, \quad b_i^u \equiv b_i^\sigma \equiv b_i$
Nonlinear dielectric stiffness	$a_{ijkl}^u = a_{ijkl}^\sigma + Q_{ijmn} q_{mnkl}/2$
Nonlinear tilt expansion coefficients	$b_{ijkl}^u = b_{ijkl}^\sigma + R_{ijmn} r_{mnkl}/2$
Polarization gradient coefficient	$g_{ijkl}^u = g_{ijkl}^\sigma + F_{ijmn} f_{mnkl}$
Tilt gradient coefficient	$v_{ijkl}^u \equiv v_{ijkl}^\sigma \equiv v_{ijkl}$
Electrostriction tensor	$q_{ijmn} = Q_{ijkl} c_{klmn}$
Rotostriction tensor	$r_{ijmn} = R_{ijkl} c_{klmn}$
Flexoelectric coupling tensor	$f_{ijmn} = F_{ijkl} c_{klmn}$
Elastic constants	$s_{ijkl} c_{klmn} = \left(\delta_{im}\delta_{jn} + \delta_{in}\delta_{jm}\right)/2$
Biquadratic coupling between the tilt and polarization	$\eta_{ijkl}^u = \eta_{ijkl}^\sigma - Q_{ijmn} r_{mnkl}$
Flexo-antiferrodistortive coupling between the tilt and polarization gradient	$\xi_{ijkl}^u = \xi_{ijkl}^\sigma + F_{ijmn} r_{mnkl}$

solely from the LGD-phenomenology. Using the classical example of ferroelastic twin walls in antiferrodistortive quantum paraelectric $SrTiO_3$ the coupling strength can be estimated as the convolution of the flexoelectric and rotostriction coupling tensors, it appears about $(1-5)\times 10^{20}$V/m^2. Rigorously speaking, the coupling tensor components should be either calculated from the first principles or determined experimentally. In general case the direct flexo-antiferrodistortive coupling between the polarization gradient and tilt components product should be included in the LGD-type functionals in the form of Lifshitz invariant. The relationship $\xi_{ijkl}^u = \xi_{ijkl}^\sigma + F_{ijmn} r_{mnkl}$ (see the last raw in the Table 8.1). Similarly the polarization gradient terms $\chi_{ijkl}^u P_i P_j \frac{\partial P_k}{\partial \tilde{x}_l}$ and $\chi_{ijkl}^\sigma P_i P_j \frac{\partial P_k}{\partial \tilde{x}_l}$ are allowed by the symmetry. Corresponding relationship is $\chi_{ijkl}^u = \chi_{ijkl}^\sigma + F_{ijmn} q_{mnkl}$.

The direct coupling strongly influences on the properties of the phase diagrams of the atiferrodistortive solid solutions, primary leading to the appearance of the spatially modulated phases [93]. Incommensurate modulation appears spontaneously when the coupling strength exceeds the critical value proportional to the sum of inverse polar and structural correlation lengths (commensurate-incommensurate phase transition). Note, that modulated phase appears and strongly enlarges its stability region with the coupling strength increase in multiferroic solid solution $Bi_ySm_{1-y}FeO_3$ [28]. Further increase of the coupling strength can lead to the antiferroelectric-antiferrodistortive phase indeed observed in multiferroics $Bi_ySm_{1-y}FeO_3$ [28] and $EuTiO_3$ [29].

Emergence of spatially modulated polarization and tilt at the structural domain walls of $EuTiO_3$ are shown in the Fig. 8.4. The modulation is absent without the flexo-antiferrodistortive coupling as well as when the coupling strength is smaller that the critical value (see solid curves). It originates for the coupling strength higher

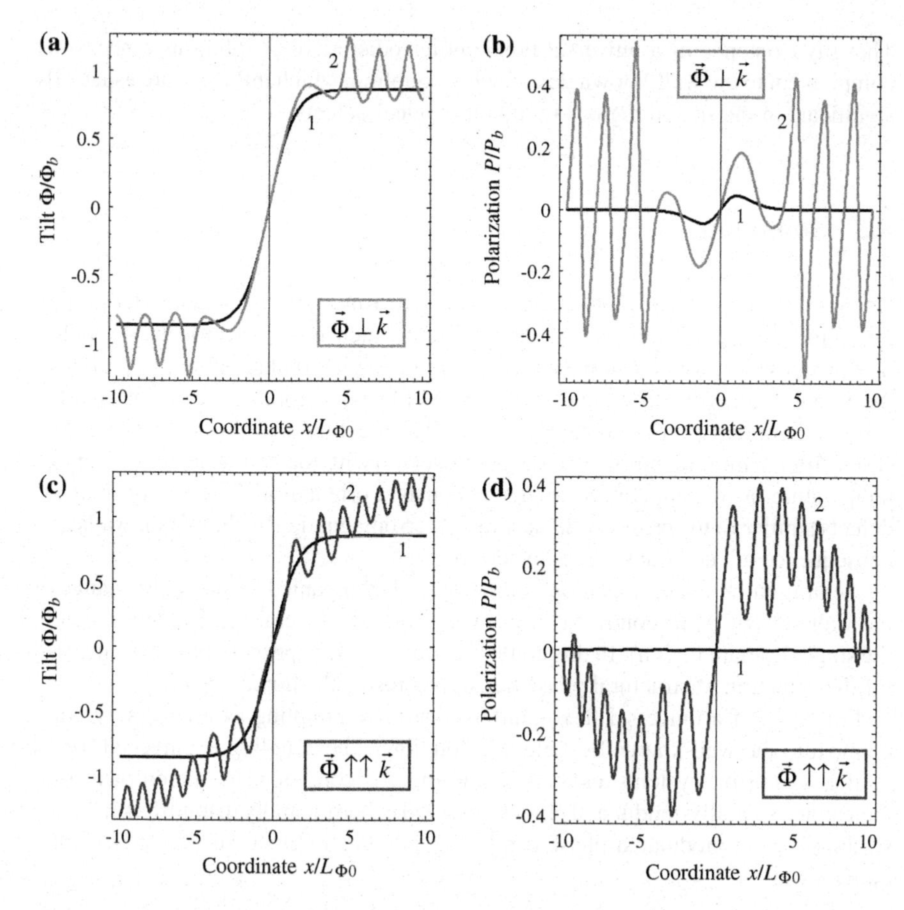

Fig. 8.4 Tilt (**a, c**) and polarization (**b, d**) spatial modulation originated near $EuTiO_3$ antiphase boundary between the domains with opposite orientations of tilt vector calculated for different values of flexoantiferrodistortive coupling coefficient. For orientation $\vec{k} \perp \vec{\Phi}$ (**a, b**) the value $\xi_{12} = 0$ and 3×10^{20} V/m^2 (curves 1 and 2 respectively). For orientation $\vec{k} \uparrow\uparrow \vec{\Phi}$ (**c, d**) the value $\xi_{11} = 0$ and 2×10^{20} V/m^2 (curves 1 and 2 respectively). Temperature $T = 200$ K, scales for the tilt, polarization and x-coordinate are Φ_b, and P_b and $L_{\Phi 0}$. (Adapted from [93])

then the critical value dependent on the modulation vector $\vec{k}$ orientation with respect to the tilt $\vec{\Phi}$ (compare dashed curves for $\vec{k} \uparrow\uparrow \vec{\Phi}$ and $\vec{k} \perp \vec{\Phi}$). Finally the polarization modulation amplitude increases and its period decreases with the coupling value increase. Note, that in agreement with experimental observations [29] the spatial modulation of polarization can acquire antiferroelectric features for the case $\vec{k} \perp \vec{\Phi}$ and high values of $\xi = 3 \times 10^{20}$ V/m^2 (Fig. 8.4b, d).

The proposed description could be helpful for understanding of the modulated phases appearance around different topological defects in other antiferrodistortive ferroics. A promising candidate could be a $Eu_xSr_yBa_{1-x-y}TiO_3$ solid solution [95].

That say existence of a universal flexo-antiferrodistortive coupling as a necessary complement to the well-known flexoelectric coupling, which influence are especially significant in the spatial regions round topological defects.

8.5 Summary

Using LGD theory we explore some unusual electronic, structural and polar properties of the topological defects inherent in ferroics, such as ferroelectric and ferroelastic domain walls, which can have reach and tunable internal structure. In particular the domain walls with different (e.g. {211} and {110}) orientations are shown to have different symmetry in rhombohedral ferroelectrics, achiral and chiral, the phase transition from achiral to chiral state occurs under varying the wall orientation at fixed temperature or by temperature change at constant orientation. The existence of 2D defects similar to the cross-tie defects in the ferromagnetic Bloch domain walls as a consequence of such transition is expected.

Achiral-chiral phase transition in the wall is accompanied by the rapid change of the wall current-AFM contrast (from several times to several orders higher than in the single-domain region). In this context current-AFM appears to be promising tool for the detection of structural phase transitions inside the domain walls.

Existence of a universal flexo-antiferrodistortive coupling as a necessary complement to the well-known flexoelectric coupling. The coupling is universal for all antiferrodistortive systems and can lead to the self-consistent formation of incommensurate, spatially-modulated phases observed experimentally in multiferroics. The seeding for the modulated phase can be a topological defect, such as a structural domain wall.

Acknowledgments E.A.E. and A.N.M. acknowledge Center for Nanophase Materials Sciences (CNMS), user projects CNMS 2013-293 and CNMS 2014-270, and National Academy of Sciences of Ukraine (grant 35-02-14). S.V.K. acknowledges Office of Basic Energy Sciences, U.S. Department of Energy.

References

1. N.A. Spaldin, M. Fiebig, Science **309**, 391–392 (2005)
2. J.M. Rondinelli, N.A. Spaldin, Adv. Mater. **23**, 3363–3381 (2011)
3. M. Fiebig, T. Lottermoser, D. Frohlich, A.V. Goltsev, R.V. Pisarev, Nature **419**, 818 (2002)
4. R. Ramesh, N.A. Spaldin, Nat. Mater. **6**, 21 (2007)
5. M.J. Haun, E. Furman, S.J. Jang, L.E. Cross, Thermodynamic theory of the lead zirconate-titanate solid solution system, part I: phenomenology. Ferroelectrics **99**, 13–25 (1989)
6. M.J. Haun, E. Furman, T.R. Halemane, L.E. Cross, Ferroelectrics **99**, 55 (1989)
7. B. Houchmanzadeh, J. Lajzerowicz, E. Salje, J. Phys.: Condens. Matter **3**, 5163 (1991)
8. B. Houchmanzadeh, J. Lajzerowicz, E. Salje, Phase Trans. **38**, 77 (1992)
9. E.V. Balashova, A.K. Tagantsev, Phys. Rev. B **48**, 9979 (1993)

10. M. Fiebig, J. Phys. D: Appl. Phys. **38**, R123–R152 (2005)
11. P.J. Ryan, J.-W. Kim, T. Birol, P. Thompson, J.-H. Lee, X. Ke, P.S. Normile, E. Karapetrova, P. Schiffer, S.D. Brown, C.J. Fennie, D.G. Schlom, Nat. Commun. **4**, 1334 (2013)
12. A.K. Tagantsev, L.E. Cross, J. Fousek, *Domains in Ferroic Crystals and Thin Films* (Springer, Dordrecht, 2010), 827 pp
13. M. Daraktchiev, G. Catalan, J.F. Scott, Phys. Rev. B **81**, 224118 (2010)
14. O. Dieguez, P. Aguado-Puente, J. Junquera, J. Iniguez, Phys. Rev. B **87**, 024102 (2013)
15. M. Imada, A. Fujimori, Y. Tokura, Rev. Mod. Phys. **70**(4), 1039–1263 (1998)
16. E. Dagotto, Science **309**(5732), 257–262 (2005)
17. E. Dagotto, T. Hotta, A. Moreo, Phys. Rep.-Rev. Sec. Phys. Lett. **344**(1–3), 1–153 (2001)
18. K.H. Ahn, T. Lookman, A.R. Bishop, Nature **428**, 401 (2004)
19. A.A. Bokov, Z.G. Ye, J. Mat. Sci. **41**, 31 (2006)
20. Z. Kutnjak, J. Pelzelt, R. Blinc, Nature **441**, 956 (2006)
21. D.I. Woodward, J. Knudsen, I.M. Reaney, Phys. Rev. B **72**, 104110 (2005)
22. B. Noheda, Curr. Opin. Solid State Mater. Sci. **6**, 27 (2002)
23. C.J. Cheng, D. Kan, S.H. Lim, W.R. McKenzie, P.R. Munroe, L.G. Salamanca-Riba, R.L. Withers, I. Takeuchi, V. Nagarajan, Phys. Rev. B **80**, 014109 (2009)
24. V. Wadhawan, *Introduction to Ferroic Materials* (CRC Press, Boca Raton, 2000)
25. A. Khachaturyan, D. Viehland, Mater. Trans. **38A**, 2308; **38A**, 2317 (2007)
26. R. Pirc, R. Blinc, J.F. Scott, Phys. Rev. B **79**, 214114 (2009)
27. A.G. Khachaturyan, Phil. Mag. **90**, 37 (2010)
28. A.Y. Borisevich, E.A. Eliseev, A.N. Morozovska, C.-J. Cheng, J.-Y. Lin, Y.H. Chu, D. Kan, I. Takeuchi, V. Nagarajan, S.V. Kalinin, Nat. Commun. **3**, 775 (2012)
29. J.-W. Kim, P. Thompson, S. Brown, P.S. Normile, J.A. Schlueter, A. Shkabko, A. Weidenkaff, P.J. Ryan, Phys. Rev. Lett. **110**, 027201 (2013)
30. A.K. Tagantsev, E. Courtens, L. Arzel, Phys. Rev. B **64**, 224107 (2001)
31. B. Houchmandzadeh, J. Lajzerowicz, E.K.H. Salje, J. Phys. Condens. Matter **4**, 9779 (1992)
32. M. Daraktchiev, G. Catalan, J.F. Scott, Ferroelectrics **375**, 122 (2008)
33. V.G. Bar'yakhtar, V.A. L'vov, D.A. Yablonskii, JETP Lett. **37**, 673 (1983)
34. A.P. Pyatakov, A.K. Zvezdin, Eur. Phys. J. B **71**, 419–427 (2009)
35. V.A. Zhirnov, Zh. Eksp. Teor. Fiz. **35**, 1175 (1958), [Sov. Phys. JETP **35**, 822 (1959)]
36. W. Cao, L.E. Cross, Phys. Rev. B **44**, 5 (1991)
37. B.M. Darinskii, V.N. Fedosov, Sov. Phys.-Solid State **13**, 17 (1971)
38. B.M. Vul, G.M. Guroa, I.I. Ivanchik, Ferroelectrics **6**, 29 (1973)
39. J. Seidel, L.W. Martin, Q. He, Q. Zhan, Y.-H. Chu, A. Rother, M.E. Hawkridge, P. Maksymovych, P. Yu, M. Gajek, N. Balke, S.V. Kalinin, S. Gemming, F. Wang, G. Catalan, J.F. Scott, N.A. Spaldin, J. Orenstein, R. Ramesh, Nat. Mater. **8**, 229–234 (2009)
40. J. Seidel, P. Maksymovych, Y. Batra, A. Katan, S.-Y. Yang, Q. He, A.P. Baddorf, S.V. Kalinin, C.-H. Yang, J.-C. Yang, Y.-H. Chu, E.K.H. Salje, H. Wormeester, M. Salmeron, R. Ramesh, Phys. Rev. Lett. **105**, 197603 (2010)
41. S. Farokhipoor, B. Noheda, Phys. Rev. Lett. **107**, 127601 (2011)
42. N. Balke, B. Winchester, W. Ren, Y.H. Chu, A.N. Morozovska, E.A. Eliseev, M. Huijben, R.K. Vasudevan, P. Maksymovych, J. Britson, S. Jesse, I. Kornev, R. Ramesh, L. Bellaiche, L.Q. Chen, S.V. Kalinin, Nat. Phys. **8**, 81–88 (2012)
43. D. Meier, J. Seidel, A. Cano, K. Delaney, Y. Kumagai, M. Mostovoy, N.A. Spaldin, R. Ramesh, M. Fiebig, Nat. Mater. **11**, 284 (2012)
44. V. Ya, Shur, A.V. Ievlev, E.V. Nikolaeva, E.I. Shishkin, M.M. Neradovskiy, J. Appl. Phys. **110**, 052017 (2011)
45. J. Guyonnet, I. Gaponenko, S. Gariglio, P. Paruch, Adv. Mater. **23**, 5377–5382 (2011)
46. P. Maksymovych, A.N. Morozovska, P. Yu, E.A. Eliseev, Y.H. Chu, R. Ramesh, A.P. Baddorf, S.V. Kalinin, Nano Lett. **12**, 209–213 (2012)
47. Y.-H. Chu, Q. Zhan, L.W. Martin, M.P. Cruz, P.-L. Yang, G.W. Pabst, F. Zavaliche, S.-Y. Yang, J.-X. Zhang, L.-Q. Chen, D.G. Schlom, I.-N. Lin, T.-B. Wu, R. Ramesh. Adv. Mater. **18**, 2307 (2006)

48. R.K. Vasudevan, A.N. Morozovska, E.A. Eliseev, J. Britson, J.-C. Yang, Y.-H. Chu, P. Maksymovych, L.Q. Chen, V. Nagarajan, S.V. Kalinin, Domain wall geometry controls conduction in ferroelectrics. Nano Lett. **12**(11), 5524–5531 (2012)
49. G.I. Guro, I.I. Ivanchik, N.F. Kovtoniuk, Sov. Sol. St. Phys. **11**, 1956–1964 (1969)
50. M.Y. Gureev, A.K. Tagantsev, N. Setter, Phys. Rev. B **83**, 184104 (2011)
51. E.A. Eliseev, A.N. Morozovska, G.S. Svechnikov, Venkatraman Gopalan, V.Ya. Shur, Phys. Rev. B **83**, 235313 (2011)
52. E.A. Eliseev, A.N. Morozovska, G.S. Svechnikov, P. Maksymovych, S.V. Kalinin, Phys. Rev. B. **85**, 045312 (2012)
53. A.N. Morozovska, R.K. Vasudevan, P. Maksymovych, S.V. Kalinin, E.A. Eliseev, Anisotropic conductivity of uncharged domain walls in BiFeO3. Phys. Rev. B. **86**, 085315 (2012)
54. M. Mostovoy, Phys. Rev. Lett. **106**, 047204 (2011)
55. E.A. Eliseev, A.N. Morozovska, A.Y. Borisevich, Y. Gu, L.-Q. Chen, V. Gopalan, S.V. Kalinin, Phys. Rev. B **86**, 085416 (2012)
56. J. Hlinka, P. Márton, Phys. Rev. B **74**, 104104 (2006)
57. P. Marton, I. Rychetsky, J. Hlinka, Phys. Rev. B **81**, 144125 (2010)
58. D. Lee, R.K. Behera, W. Pingping, X. Haixuan, S.B. Sinnott, S.R. Phillpot, L.Q. Chen, V. Gopalan, Phys. Rev. B **80**, 060102(R) (2009)
59. S.M. Kogan, Sov. Phys. Solid State **5**, 2069 (1964)
60. A.K. Tagantsev, Phys. Rev B **34**, 5883 (1986)
61. A.K. Tagantsev, Phase Transitions **35**, 119 (1991)
62. V.S. Mashkevich, K.B. Tolpygo, Zh. Eksp. Teor. Fiz. **31**, 520 (1957) [Sov. Phys. JETP **4**, 455 (1957)]
63. G. Catalan, L.J. Sinnamon, J.M. Gregg, J. Phys.: Condens. Matter **16**, 2253 (2004)
64. G. Catalan, B. Noheda, J. McAneney, L.J. Sinnamon, J.M. Gregg, Phys. Rev B **72**, 020102 (2005)
65. R. Maranganti, N.D. Sharma, P. Sharma, Phys. Rev. B **74**, 014110 (2006)
66. S.V. Kalinin, V. Meunier, Phys. Rev. B **77**, 033403 (2008)
67. M.S. Majdoub, P. Sharma, T. Cagin, Phys. Rev. B **77**, 125424 (2008)
68. A.K. Tagantsev, V. Meunier, P. Sharma, MRS Bull. **34**, 643 (2009)
69. M.S. Majdoub, R. Maranganti, P. Sharma, Understanding the origins of the intrinsic dead-layer effect in nanocapacitors. Phys. Rev. B **79**, 115412 (2009)
70. E.A. Eliseev, A.N. Morozovska, M.D. Glinchuk, R. Blinc, Phys. Rev. B. **79**, 165433 (2009)
71. A.N. Morozovska, E.A. Eliseev, G.S. Svechnikov, S.V. Kalinin, Phys. Rev. B **84**, 045402 (2011)
72. W. Ma, L.E. Cross, Appl. Phys. Lett. **79**, 4420 (2001)
73. W. Ma, L.E. Cross, Appl. Phys. Lett. **81**, 3440 (2002)
74. W. Ma, L.E. Cross, Appl. Phys. Lett. **82**, 3293 (2003)
75. P. Zubko, G. Catalan, A. Buckley, P.R.L. Welche, J.F. Scott, Phys. Rev. Lett. **99**, 167601 (2007)
76. P.V. Yudin, A.K. Tagantsev, E.A. Eliseev, A.N. Morozovska, N. Setter, Bichiral structure of ferroelectric domain walls driven by flexoelectricity. Phys. Rev. B **86**, 134102 (2012)
77. E.A. Eliseev, P.V. Yudin, S.V. Kalinin, N. Setter, A.K. Tagantsev, A.N. Morozovska, Structural phase transitions and electronic phenomena at 180-degree domain walls in rhombohedral BaTiO3. Phys. Rev. B **87**, 054111 (2013)
78. A.K. Tagantsev, G. Gerra, J. Appl. Phys. **100**, 051607 (2006)
79. Supplementary materials to Ref. [51]
80. A.N. Morozovska, E.A. Eliseev, A.K. Tagantsev, S.L. Bravina, L.-Q. Chen, S.V. Kalinin, Phys. Rev. B **83**, 195313 (2011)
81. C. Gahwiller, Zeitschrift fur Physik B Condens. Matter **6**(4), 269–289 (1967). doi:10.1007/BF02422508
82. V.M. Fridkin, Ferrolectrics-semiconductors, Consultant Bureau, New-York and London, Eqs. (1.46) on p. 25, and parameters on p. 162 (1980)
83. S.H. Wemple, Phys. Rev B **2**, 2679 (1970)
84. Y. Sun, S.E. Thompson, T. Nishida, J. Appl. Phys. **101**, 104503 (2007)
85. R.F. Berger, C.J. Fennie, J.B. Neaton, Phys. Rev. Lett. **107**, 146804 (2011)

86. G.L. Bir, G.E. Pikus, *Symmetry and Deformation Effects in Semiconductors* (Moscow, 1972)
87. R.K. Behera, C.-W. Lee, D. Lee, A.N. Morozovska, S.B. Sinnott, A. Asthagiri, V. Gopalan, S.R. Phillpot, J. Phys.: Condens. Matter **23**, 175902 (2011)
88. R. Maranganti, P. Sharma, Phys. Rev. B **80**, 054109 (2009)
89. P. Zubko, G. Catalan, A.K. Tagantsev, Flexoelectric Eff. Solids Rev. (2013)
90. M. Mostovoy, Phys. Rev. Lett. **96**, 067601 (2006)
91. A. Sparavigna, A. Strigazzi, A. Zvezdin, Phys. Rev. B **50**, 2953 (1994)
92. V. Gopalan, D.B. Litvin, Nat. Mater. **10**, 376–381 (2011)
93. E.A. Eliseev, S.V. Kalinin, G. Yijia, M.D. Glinchuk, V. Khist, A. Borisevich, V. Gopalan, L.-Q. Chen, A.N. Morozovska, Phys. Rev. B **88**, 224105 (2013)
94. A.N. Morozovska, E.A. Eliseev, M.D. Glinchuk, L.-Q. Chen, V. Gopalan, Phys. Rev. B. **85**, 094107 (2012)
95. V. Goian, S. Kamba, P. Vaněk, M. Savinov, C. Kadlec, J. Prokleška, Phase Trans.: A Multinatl. J. **86**(2–3), 191 (2013)

Chapter 9
Topological Defects in Nanostructures—Chiral Domain Walls and Skyrmions

Benjamin Krüger and Mathias Kläui

Abstract In this chapter, spin structures with particular topologies in confined geometries are presented. Domain walls in nanowires exhibit a spin structure that depends on the material and geometry while in discs Skyrmions can be stabilized by different competing interactions. The topologies of these spin structures can be characterized by a Skyrmion or Winding number that governs the dynamics and stability.

9.1 Introduction to Topological Spin Structures in Confined Geometries

In ferromagnetic materials different interaction energy terms compete with one other and the resulting equilibrium spin structures minimize the appropriate thermodynamic potential that includes all the contributing terms [1, 2]. The most obvious energy term is the exchange interaction, which is of quantum mechanical origin. It is a short range interaction that favors a parallel alignment of neighboring spins. This is competing with the long range dipolar interaction of the stray field of the magnetic moments. This stray field energy favours the minimization of surface or bulk magnetic "charges" and leads in thin films to a preferential alignment of magnetization with the film plane. Additional energy terms include the anisotropy term that favours certain magnetization directions and the Zeeman energy that favours alignment of the magnetization with the magnetic field. There are different sources of anisotropy: One exists similarly in bulk and patterned thin films and originates from the spin-orbit coupling between the magnetic spin system and the crystalline

B. Krüger (✉)
Institute of Physics, Johannes Gutenberg University Mainz, 55128 Mainz, Germany
e-mail: bkrueger@uni-mainz.de

M. Kläui
Institute of Physics and Materials Science in Mainz, Johannes Gutenberg University Mainz, 55128 Mainz, Germany
e-mail: Klaeui@Uni-Mainz.de

J. Seidel (ed.), *Topological Structures in Ferroic Materials*, Springer Series in Materials Science 228, DOI 10.1007/978-3-319-25301-5_9

lattice of the material. Another source of anisotropies is the symmetry breaking at interfaces and surfaces. Details of the energy terms and the calculations can be found in the literature [1–3].

The minimization of the sum of these energies leads in confined geometries to the creation of generally inhomogeneous magnetization configurations [4]. These configurations usually contain some areas with uniform magnetization, the domains. These regions are separated by domain walls in which the magnetic moments change their orientation [4].

To characterize the topology of these spin structures, certain characteristic numbers have been introduced. For the thin film geometries that we are dealing with in this chapter, one can introduce a topological charge, called Skyrmion number or Skyrmion charge, which is given by [5]:

$$q = \frac{1}{4\pi} \int d^2 r \boldsymbol{m} \left(\frac{\partial \boldsymbol{m}}{\partial x} \times \frac{\partial \boldsymbol{m}}{\partial y} \right) \tag{9.1}$$

where the z direction is the out-of-plane direction and $\boldsymbol{m}$ is a unit vector in the direction of the magnetization. This charge counts how often the magnetization field covers the unit sphere. Assuming a continuous vector field that is fixed at the boundaries of the film, this charge is conserved. Hence, configurations with non-vanishing Skyrmion charge cannot be transformed to a state with homogeneous magnetization by continuous transformations. Due to the lattice structure of real matter a transformation between configurations with different Skyrmion charges is possible but often energetically costly and thus suppressed. This explains why the Skyrmion charge is an important quantity that allows one to describe the topology as this is usually conserved.

For domain walls, the winding number has been introduced in [6]. This number counts how often the magnetization curls in-plane. The winding number can be split into one part n_{edge} that counts for defects that are located at the edges and another that part n_{bulk} containing bulk defects. These numbers then read

$$n_{\text{edge}} + n_{\text{bulk}} = \left(1 - \frac{1}{2\pi} \int d\boldsymbol{S}\, \nabla\theta \right) + \frac{1}{2\pi} \int d^2 r \left(\frac{\partial^2 \theta}{\partial x \partial y} - \frac{\partial^2 \theta}{\partial y \partial x} \right) = 1 \tag{9.2}$$

with the in-plane angle θ. The line integral is the integration around the film element while the area in the second integral is given by the film element. The second equality in (9.2) can be derived from Stoke's theorem. Here, it is worth noting that according to Schwarz' theorem n_{bulk} is zero if θ is continuous, that is, there is no region with complete out-of-plane magnetization.

As an example, we first discuss square-shaped thin-film elements without any bulk anisotropy. For this geometry, depending on the materials properties and dimensions, different stable magnetic configurations are possible. The S-state, the C-state, and the flower state (see Fig. 9.1) share their topology with a homogeneous magnetized sample [4]. However, other stable magnetization configurations exist that possess different topologies.

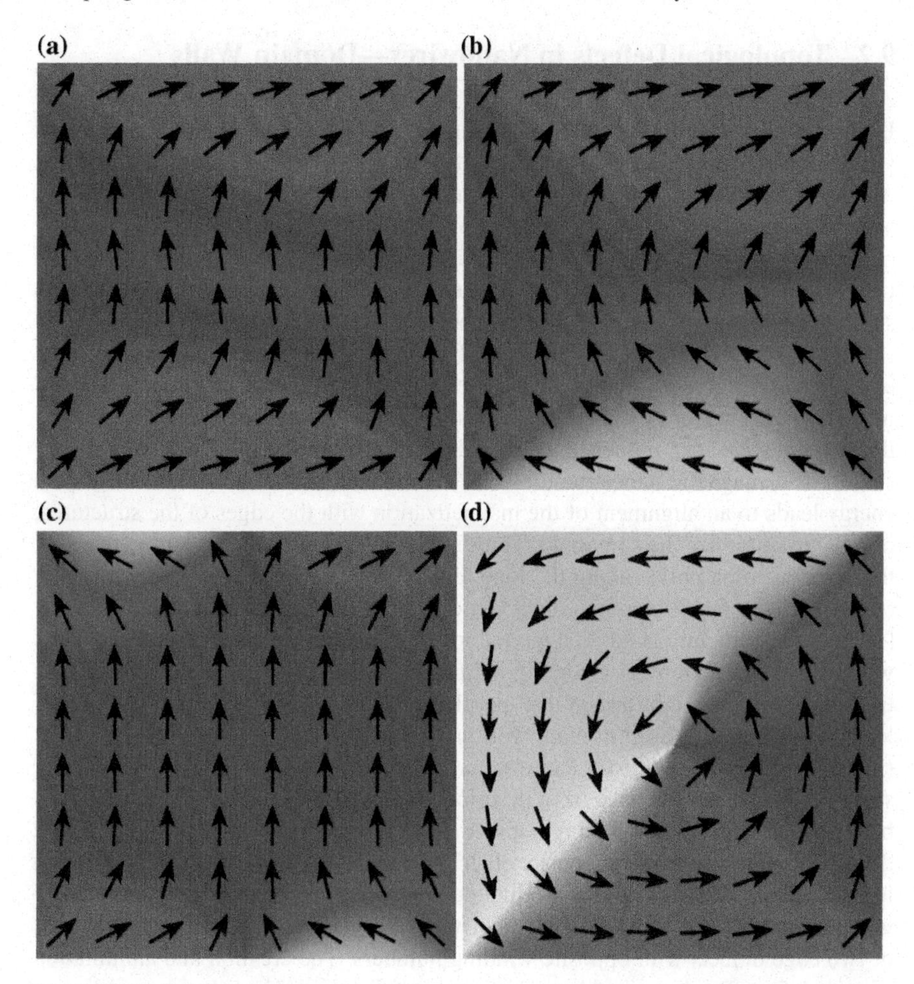

Fig. 9.1 Different magnetization configurations that can be stable in ferromagnetic thin-film elements. The S-state (**a**), C-state (**b**), and flower state (**c**) have the same topology as a homogeneous magnetization pattern. In contrast the vortex state (**d**) has a non-trivial topology. This state cannot be transformed to a homogeneous magnetization by continuous transformations

The simplest example of such a configuration is the magnetic vortex, as shown in Fig. 9.1d, that has a Skyrmion charge of 1/2. In a vortex state the magnetization curls in the film plane around a center region. In this region, called vortex core, the exchange interaction forces the magnetization to point out-of-plane. In these states there can be two different senses of the curling of the in-plane magnetization. This sense is called chirality of the vortex. In addition to the chirality there are two different orientations of the out-of-plane magnetization in the vortex core. This is normally referred to as the polarization of the vortex. Due to its symmetry and spatial confinement, the magnetic vortex is a very suitable system for investigations and makes the vortex one of the most extensively investigated magnetic states [7, 8].

9.2 Topological Defects in Nanowires—Domain Walls

Domain walls, which constitute the boundary between domains, have been intensively researched in the past, though with a focus on the domain wall types that occur in the bulk or in continuous films. The most prominent examples are the Bloch and the Néel wall types, which occur in continuous thin films. A thorough overview of such domain walls in particular in the bulk can be found in the literature [9–12]. In patterned magnetic structures, novel domain wall types emerge, when the wall spin structures start to be dominated by the geometry rather than by the intrinsic materials properties. This is particularly true for soft magnetic materials (and to some extent also for polycrystalline hard magnetic structures), where the effects of magnetocrystalline anisotropy and other material-dependent anisotropies are small and the element shape can influence the domain wall spin structures. The lowest energy state in soft magnetic nanostructures is a monodomain state, where the stray field energy leads to an alignment of the magnetization with the edges of the structures, in order to minimize the stray field. Thus, in elongated elements, such as wires, etc., the magnetization points along the long axis of the element and most often such a (quasi-)monodomain state constitutes the lowest energy magnetization configuration [4]. If such a structure is not in a monodomain state, say, for instance, two domains with opposite direction exist (see Fig. 9.2a where in the "green" domain the magnetization is pointing right and in the "purple" domain the magnetization is pointing left), a 180° head-to-head domain wall has to be present in between the domains (interfaces between the green and the purple domains). The domain walls in such soft magnetic materials have been described in detail in [13] and the topology has been analyzed using the winding number concept in [6]. In particular it was found that the domain walls are composite objects made of two or more elementary topological defects. Vortices have a winding number of $n = \pm 1$ and edge defects exhibit winding numbers of $n = \pm 1/2$. Domain walls in the simplest case are composed of two edge defects with opposite winding numbers. The creation and annihilation of such defects is constrained by conservation of the topological charge (winding or Skyrmion number). This description of the topology provides a basic understanding of the complex switching processes observed in ferromagnetic nanostructures, such as wires. Note that the Skyrmion charge of a vortex is $q = 1/2$ since there is only positive or negative out-of-plane magnetization for vortices with positive or negative polarization, respectively. This means that only the upper or lower hemisphere of the unit sphere is covered by the magnetization vector field and therefore the winding number and the Skyrmion charge yield different quantities for the same spin structure. However both numbers describe the topology equally well and are similarly conserved quantities.

While such domain walls in soft magnetic materials have been discussed in great detail [13], there are high anisotropy materials, in which the magnetization can point perpendicularly to the plane (see Fig. 9.2b). In this case the domain wall is a transition between the domain with the magnetization pointing up and the domain with the magnetization pointing down. Here the possible domain walls types are shown in

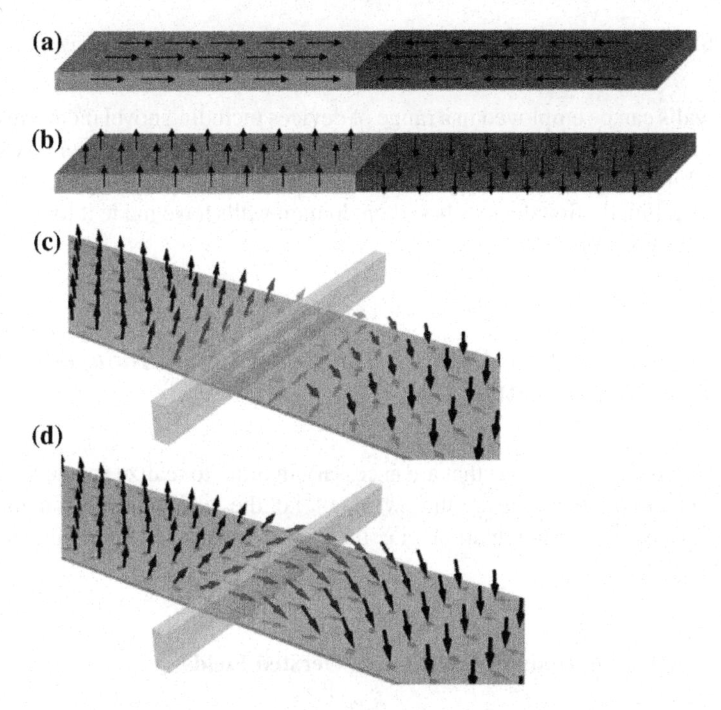

Fig. 9.2 **a** Schematic depiction of a multi-domain state in a wire made of a soft magnetic material where the magnetization in the domains is pointing along the wire axis. **b** Schematic depiction of a multi-domain state in a wire made of a high anisotropy material where the magnetization in the domains is pointing perpendicularly to the wire axis. **c** Bloch domain wall and **d** Néel domain wall spin structure

Fig. 9.2 c, d. In Fig. 9.2c a Bloch wall is shown, where the magnetization in the domain wall rotates perpendicularly to the wire edge. In the Néel domain wall in Fig. 9.2d, the magnetization rotates along the wire. Here the important concept of domain wall chirality can be introduced as there are two domain walls realizations that are energetically degenerate if no symmetry breaking interaction term is introduced. For the Bloch wall the magnetization can rotate either clockwise or counter-clockwise (when looking along the wire) while for the Néel wall the magnetization can rotate clockwise or counter-clockwise (when looking perpendicularly to the wire as shown in Fig. 9.2c, d). In Fig. 9.2d a clockwise rotating Néel wall is visible when looking from left to right.

A symmetry breaking interaction that favours one domain wall chirality is the Dzyaloshinskii-Moriya interaction (DMI) (see [14] and references therein). The DMI interaction is resulting from bulk inversion asymmetry or structural inversion asymmetry as in multilayer stacks where it is governed by the interface between a heavy metal and a magnetic thin film layer [15, 16]. In this scenario the DMI results in a preferential chirality of the domain walls. In particular for structural inversion asymmetry, the resulting DMI favours Néel walls with a given chirality.

9.3 Using Magnetic Domain Walls in Devices

Domain walls can be employed in a range of devices including novel memories, such as the racetrack memory device [17] and a field driven domain wall memory device based on chiral walls [18]. Furthermore domain walls have been suggested for logic devices [19]. Finally first devices based on domain walls have made it to the market like multi-turn sensors [20].

9.3.1 Operations for a Devices Based on Current Induced DW Motion in Wires

We next discuss the operations that are necessary in order to realize any device based on DWs. The key operations are the "writing", i.e. the nucleation of domain walls and domains and the "manipulation", i.e. the displacement of the domain walls and thereby the domains.

9.3.1.1 Nucleating Domain Walls Using Oersted Fields

There are different approaches to nucleating domains and domain walls. The simplest is the conventional approach using Oersted fields generated by current passing through striplines. As seen in Fig. 9.3a, b, this can be achieved by combining the magnetic wires with a low resistivity and thus low heating and ohmic losses Oe field line (stripline). Here Au is used in the geometry shown in Fig. 9.3.

To calculate the magnetic fields resulting from the current pulses, numerical integration of the current paths is carried out. Figure 9.4a shows a small rectangle, whose

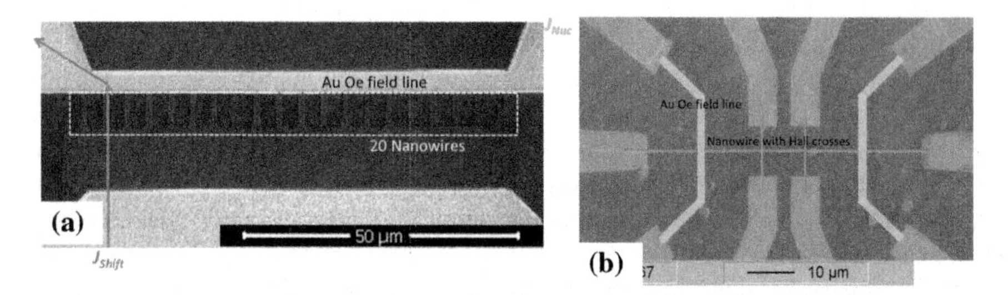

Fig. 9.3 **a** Scanning electron micrograph of a device that operates by domain wall motion. The figure shows 20 nanowires of Ta/CoFeB/MgO with gold Oe lines (*yellow region*) across the wires. The current path for the DW nucleation is indicated as a *green arrow* and the current path for DW shifting as an *orange arrow*. **b** Device based on the same concept but including Hall crosses for electrical detection. The Oe field lines are indicated in *yellow*, while the magnetic nanowire is shown in *blue* (partly from [32])

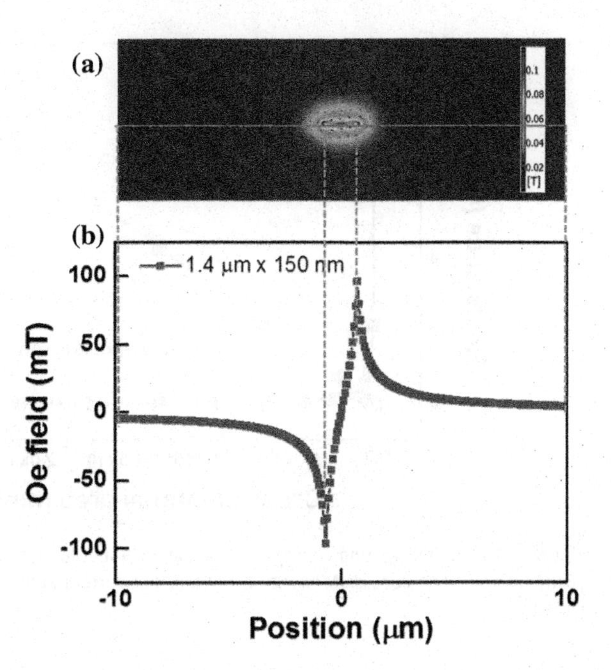

Fig. 9.4 **a** Simulation of a homogeneous current density (10^{12} A/m^2) creating a circular magnetic field. The current density is perpendicular to the cross sectional area of a gold Oe field line. The width and thickness (height) of the current line (*small rectangle*) are 1.4 μm (W) × 150 nm (H), which is the same as the sample dimensions. **b** Component of the magnetic field, which is perpendicular to the wire plane, as a function of the distance from the centre of the Au Oe field line (0)

width and height is the same as in the experiment representing a cross sectional area of a gold Oe field line (1.4 μm (width W) × 150 nm (thickness H)). For the simulation of a local Oe field, we assume a fixed current density of 1×10^{12} A/m^2 being perpendicular to the nanowire, which is a reasonable density that is compatible with reliable operation based on our experimental results. In particular using these values, we obtain 100 % DW nucleation probability using this current density (details, see further below). The arrows in Fig. 9.4a indicate the direction of the Oe field and the color indicates the absolute value of the magnetic field. A line scan along the top of the wire (a red line) shows the magnetic field component perpendicular to the wire in Fig. 9.4b as a function of its position along the wire. A sufficient high magnetic field for magnetization reversal is created in the surrounding of the gold Oe field line. The resulting localized Oe field has field strengths going up from 5 to 90 mT for distances from 10 μm from the gold Oe field line to the edge of the line. Therefore a localized nucleation of a single domain wall in an out-of-plane magnetized wire is possible for soft materials such as CoFeB based structures with low coercivity.

To gauge the scalability of this writing approach using the Oe field line, we calculate the writing energy and the generated Oe fields for various dimensions as shown in Fig. 9.5. Assuming a length of a gold Oe field line of 160 nm (for applications, we assume a design rule of 32 nm and the length is thus 5 times this design rule), we can calculate the power necessary for writing a single bit. Figure 9.5 shows the calculated resistance for a gold Oe line and the calculated energy for a constant current density of 1×10^{12} A/m^2 as a function of the cross sectional area of the gold line. The writing energy is a few pJ at $j = 1 \times 10^{12}$ A/m^2 and is reduced as the cross sectional area of a gold Oe field line is reduced. The Oe fields that

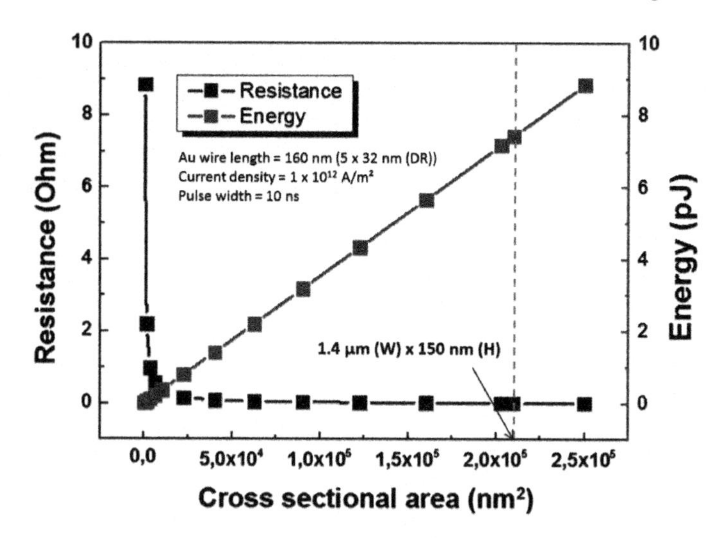

Fig. 9.5 Calculated writing energy and resistance of a gold Oersted line as a function of the cross sectional area. The necessary energy for a constant current density of 1×10^{12} A/m^2 is in the few pJ range

can be generated are up to 15 mT for 100 nm wire width and 2 mT for a 20 nm wire width, which is compatible with future design rules that continuously scale to smaller dimensions. This field is still sufficient to realize magnetization switching in specially designed nanowires when a sufficiently low coercivity is achieved for a tailored material.

Next we study this DW nucleation experimentally. As shown in Fig. 9.3, we can either determine the necessary current densities by directly imaging the nucleated domains and domain walls (device from Fig. 9.3a) or by electrically detecting the walls (Fig. 9.3b) [21]. In the latter case, the DW is detected by the extraordinary Hall effect (EHE) resistance that shows changes when the DW is pinned at the Hall cross [22]. Figure 9.6 shows the EHE hysteresis of a Ta/CoFeB/MgO multilayer nanowire with and without a DW, and indicates schematically the resulting magnetization configuration in the nanostructure. The black curve is the normal hysteresis of the Ta/CoFeB/MgO nanowire without a DW, and the blue one shows the hysteresis of Ta/CoFeB/MgO when a DW is nucleated using the Oe field. The coercivity changes drastically from $H_c \approx 25 \sim 30$ mT (normal hysteresis loop without DW) to $H_c^{DW} \approx$ 13 mT in the case where a DW is nucleated. After nucleating the domain wall close to the Oe field line, we see that the domain walls moves to the Hall cross at a propagation field H_{prop} of 3 mT. It then takes 13 mT to move the DW across the Hall cross to reverse the magnetization. The increased field of 25–30 mT without a nucleated domain wall means, that the nucleation field for a DW is much higher than the propagation field and even the field needed to push the domain wall across the Hall cross, which allows us to unambiguously determine these fields.

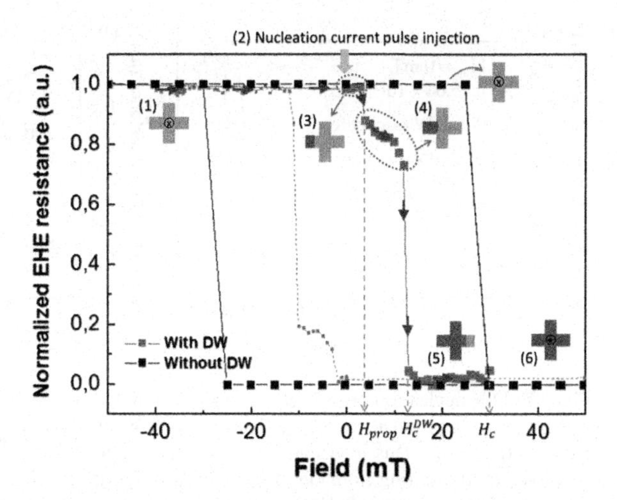

Fig. 9.6 Normalized EHE hysteresis of Ta/CoFeB/MgO nanowire with (*blue*) and without (*black*) DW. The coercivity is indicated $H_c \approx$ 25–30 mT for a normal hysteresis loop without DW, and $H_c^{DW} \approx 13$ mT for hysteresis loop with DW in the nanowire. We find the DW propagation field is around 3 mT. We indicate the magnetization direction is up/down as *red*/*sky blue* in the schematic depictions as insets

The protocol for this measurement is that we first saturate the sample in one direction with a negative field (here marked as down magnetization indicated in blue in the schematic representations of the magnetization in Fig. 9.6). Then the current pulse is injected through the gold Oe field line for DW nucleation at point (2) in Fig. 9.6. The DW is nucleated in the nanowire using the local Oe field generated by a current pulse with a current density of 8×10^{11} A/m^2 and pulse length of 100 μs without any external field. Then there is a DW present the wire close to the Oe field line (Fig. 9.6 (3)) and this DW is propagated to the Hall cross at 3 mT external field. Figure 9.6 (4) shows the DW at the Hall cross, and the hysteresis shows small changes in the signal as the DW moves in the Hall cross and then when the coercivity field is H_c^{DW} reached the DW moves completely across the Hall cross (Fig. 9.6 (5)), and finally the samples is saturated in the up magnetization direction (Fig. 9.6 (6)).

Next we study this nucleation systematically. We determine the DW nucleation probability, for two different current densities as a function of pulse length. All measurements shown in Figs. 9.7 and 9.8 were repeated 10 times. At a current density of 5.64×10^{11} A/m^2, only 10 % of the time DW nucleation is achieved without any external field. At a higher current density of 6.3×10^{11} A/m^2, we can obtain 100 % DW nucleation probability without any external field as shown in Fig. 9.7b. Figure 9.8 finally shows the DW nucleation probability for different current pulse widths from 1 ns to 10 ns. The DW nucleation probability increases with increasing the current pulse width. In particular we can obtain 100 % for pulse lengths down to 2 ns and possibly lower (the limitation of the pulse shape due to the equipment used, did not allow us to obtain results for 1 ns for higher current densities).

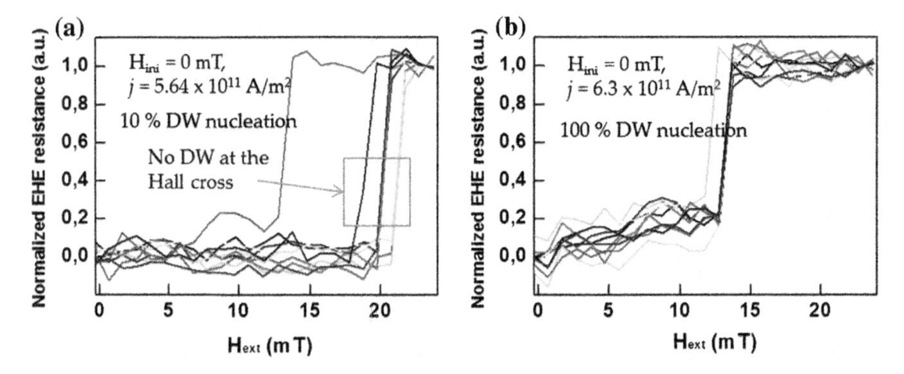

Fig. 9.7 DW nucleation probability for an Au Oe field line on top of a Ta/CoFeB/MgO nanowire using a 10 ns long pulse. **a** 10 % probability for DW nucleation is found for a current density of 5.64×10^{11} A/m^2. This is visualized by the number of traces, where one trace shows the lower coercivity field resulting from the successful DW nucleation while the other traces show the higher field if no DW is present. **b** 100 % DW nucleation is possible at a current density of 6.3×10^{11} A/m^2 without any external field a visible by all traces jumping at the lower field

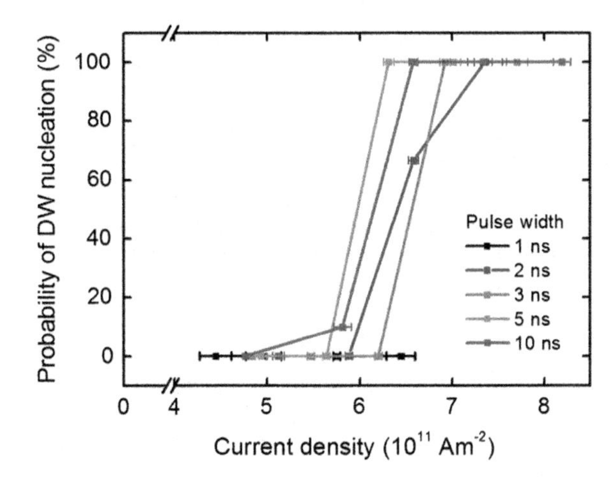

Fig. 9.8 DW nucleation probability for an Au Oe field line on a Ta/CoFeB/MgO nanowire for various current pulse widths. The DW nucleation probability increase with increasing the current pulse width

9.3.1.2 Manipulating Magnetization by Spin Orbit Torques

Next the domain walls need to be displaced. Here we study current-induced domain wall motion, which exhibits more favourable scaling than field-induced wall motion [23]. Recently it was shown that more efficient current-induced domain wall motion can be obtained in asymmetric magnetic multilayers due to novel spin orbit torques compared to conventional spin transfer torques [23]. A first observation of spin-orbit torque resulting in DW motion in a Pt/Co/AlOx nanowire was reported by Miron et al. [24] as the presence of structure inversion asymmetry (SIA) gives rise to an effective field (due to the Rashba effect [25] perpendicular to both the current flow direction and the magnetic easy-axis, which makes it energetically easier to rotate the magnetization inside the magnetic layer so that DW velocities have been detected

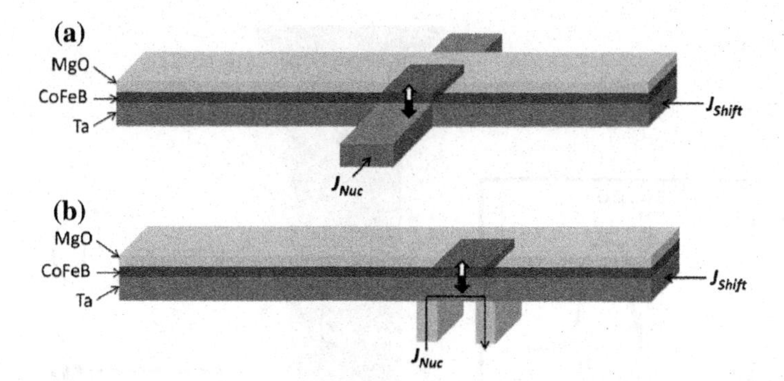

Fig. 9.9 Schematic of DW writing using SHE—induced torques. The magnetization in CoFeB is locally switched by SHE in *red region* by current perpendicular to nanowire using cross shaped Ta layer (**a**) or using two contacts under the Ta layer (**b**). The *black* and *white arrows* indicate the magnetization direction in CoFeB

up to 400 m/s. Another possible origin for such spin-orbit torque was suggested by Manchon [26], who suggested as the origin of the high DW velocity the spin Hall effect (SHE) occurring in the SIA-stack when an electric current is injected through it.

To obtain efficient domain wall motion of multiple domain walls in the same direction by spin orbit torques, chiral walls are needed. Therfore, recently the other contribution necessary for the fast current-induced DW motion has been explored theoretically and experimentally: the DMI (see [15, 16, 27–29] and references therein). It was realized that the spin orbit torques are not sufficient to move a domain wall that is usually a Bloch wall but one needs Néel DWs all with the same chirality, so that the SHE torque efficiency is maximized and synchronous motion is obtained. This combination of SHE and DMI is thus the real trigger of the fast current-induced DW motion in magnetic nanostructures with SIA.

Furthermore, switching of the magnetization by spin orbit torques has been observed [30, 31]. Such switching can be used to reverse the magnetization and thus also to "write" information by nucleating new domains and domain walls.

Figure 9.9 shows two different geometries to generate a spin current through Ta and induce DW nucleation in the adjacent ferromagnetic layer. In order to investigate this approach to initiate locally DW nucleation, we have used 1 μm wide and 10 μm long wires as shown in Fig. 9.10. We have measured the combination of longitudinal fields H_x along the wire and current to induce magnetization switching. The longitudinal fields are used to lower the switching barrier by tilting the magnetization. Figure 9.10 shows a typical result of magnetization switching at a current density of 4.02×10^{11} A/m^2 flowing in the wire with 100 ns pulse width and a longitudinal field of 200 Oe.

Furthermore we find that we can achieve switching even at zero external field due to thermal fluctuations that lead to transient magnetization components along the wire onto which the spin orbit torques act that lead to switching [31].

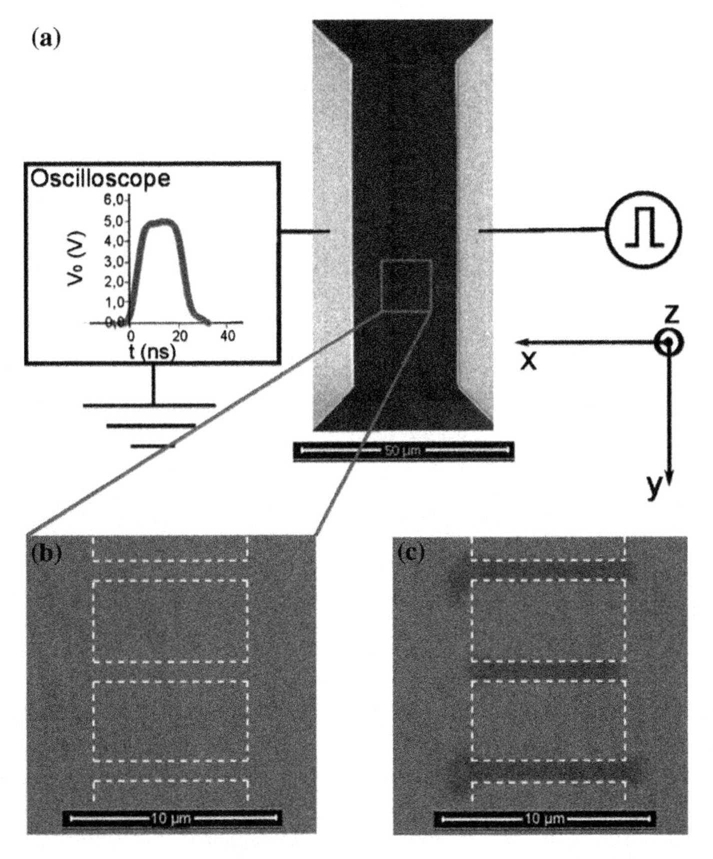

Fig. 9.10 (From [31]) **a** Schematic of the experimental set-up for current pulse injection, including an SEM micrograph of the Ta\CoFeB\MgO nanowires. The inset shows the shape of one of the voltage pulses applied to the device, measured with the oscilloscope (across the 50 Ω internal resistance). **b** Differential Kerr microscopy image of the initialized nanowires with the magnetization pointing down (−z) everywhere. **c** Differential Kerr microscopy image of the same nanowires in (**b**), after their magnetization has been switched up (+z) by a current pulse in the presence of an in-plane magnetic field collinear with the current-flow

9.3.1.3 Displacing Domain Walls Using Spin Orbit Torques

Having established the generation (writing) of domain walls, we next turn to shifting (displacing) them. As discussed above, the combination of the spin orbit torques and the DMI leads to efficient displacement of domain walls. We determine the key properties, which is the strength of the DMI by measuring the DW velocity as a function of an applied magnetic field along the wire axis (x-direction) for fixed current densities (see Fig. 9.11). First of all, both types of DW (↑↓ and ↓↑) are nucleated in the pre-saturated nanostructures by current-induced magnetization switching (see Fig. 9.11a). Typical nucleation pulses used in the experiment have a current density

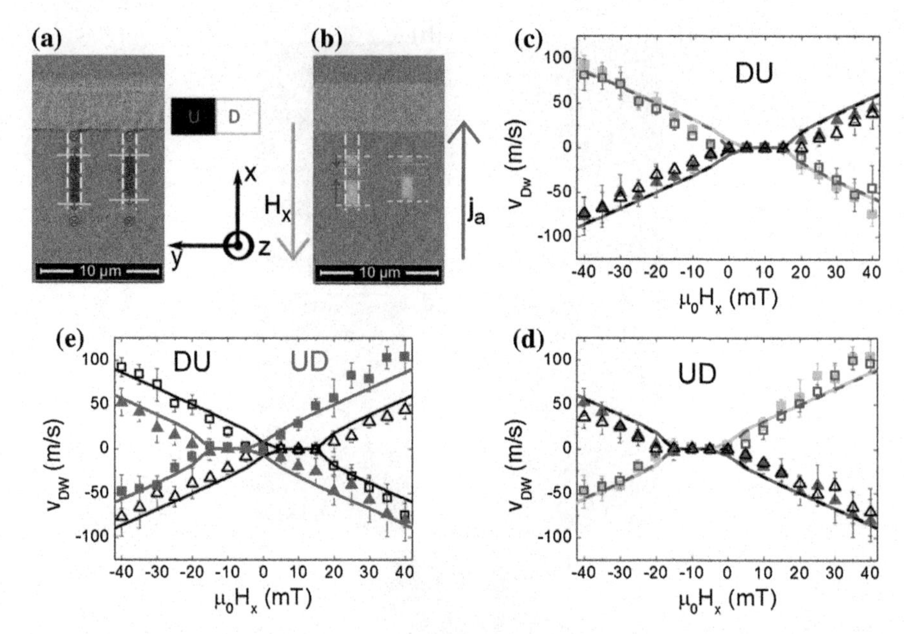

Fig. 9.11 (From [32]): Effect of a longitudinal magnetic field on the current-induced DW motion. **a** Differential Kerr microscopy image of nucleated magnetic domains in pre-saturated nanowires. The magnetization in the reversed domains points in the +z direction (*black areas*). The *green lines* indicate the position of the DWs. The *red arrows* describe the DWs magnetization configuration. **b** Differential Kerr microscopy image of the domain walls moved due to current pulse injection (j_a = +3.6 × 10^{11} A/m^2), when a longitudinal field is applied ($\mu_0 H_x$ = −35 mT). The dashed green lines indicate the starting position of the DWs, while the *solid orange lines* indicate their final position. The *blue arrows* show the DW motion. Down-up (DU, ↓↑) and up-down (UD, ↑↓) DWs move in opposite direction. **c** Average velocity of ↓↑- and **d** ↑↓-DWs as a function of the longitudinal field ($\mu_0 H_x$), for two different current densities. Solid symbols refer to j_a = 3.6 × 10^{11} A/m^2, while empty symbols refer to j_a = 2.8 × 10^{11} A/m^2. *Squares* refer to positive j_a, while *triangles* refer to negative j_a. The *solid* (*dashed*) *lines* are the 1D-model fitting-curves for $j_a = \pm 3.6 \times 10^{11}$ A/m^2 ($j_a = \pm 2.8 \times 10^{11}$ A/m^2) (see text for details). **e** Average velocity of ↓↑-(empty symbols) and ↑↓-(*solid symbols*) domain walls as a function of $\mu_0 H_x$, for a current density of $j_a = +3.6 \times 10^{11}$ A/m^2 (*squares*), and j_a = −3.6× 10^{11} A/m^2 (*triangles*). *Lines* represent the 1D-model fitting-curves

j_a ~10^{12} A/m^2 and a duration Δt = 20 ns. Once the DWs are generated, they are displaced by injecting a burst (*n* = 1–20) of 20 ns-long current pulses with lower current densities (2.8 − 3.6 × 10^{11} A/m^2), as shown in Fig. 9.11b. In order to calculate the DW velocity, the full width at half maximum of the current pulse is used as the time duration of a single pulse. For each combination of current-density and field-amplitude, the measurement is repeated five times. The DW velocity as a function of the longitudinal field $\mu_0 \mathbf{H_x}$ is shown in Fig. 9.11c, d (symbols), respectively for the ↓↑- and ↑↓-DWs. The graphs show that the DW velocity is strongly influenced by the presence of the longitudinal field.

While at zero field the velocity of both types of DWs is the same, in the presence of the field the two DWs move at different velocities. We observe a symmetric

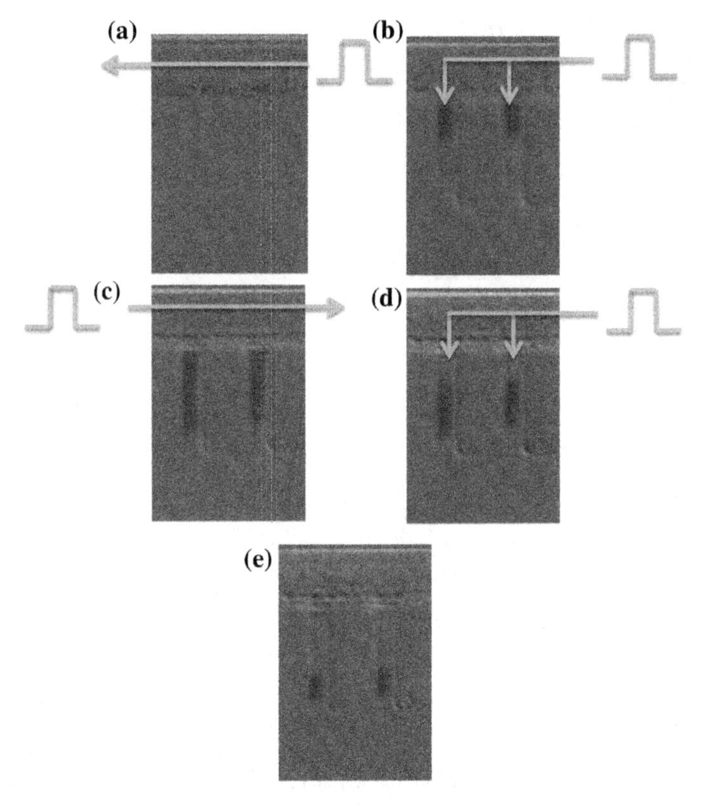

Fig. 9.12 Demonstration of domain wall writing and shifting. Initially the wire is uniformly magnetized (**a**). A pair of domain walls is generated by an Oe field pulse (current pulse direction indicated) (**b**). The two domain walls are shifted by a current pulse along the wire (**c**), and an opposite current pulse along the Oe field line generates a second pair of domain walls thus "writing" a domain with two domain walls in both wires (**d**), which is then shifted by a current pulse along the wires (**e**)

behavior of the DW velocity as a function of $\mu_0\mathbf{H_x}$ for the two types of DW, as shown in Fig. 9.11e (solid symbols for the $\uparrow\downarrow$-domain wall, empty symbols for the $\downarrow\uparrow$-domain wall). The field at which the SOT is minimized, resulting in a stationary DW, is the so-called DMI effective field $\mu_0 H_{DMI} = D/(M_S\Delta)$ [28], where D is the DMI coefficient, and Δ is the DW width. Figure 9.11c–e show that there is a range of in-plane longitudinal fields $\mu_0\mathbf{H}_x$ where the DW remains stationary (with zero or very small DW velocity compared to the velocities measured for larger longitudinal fields). This zero motion field range is not reproduced by the standard SOT-DWM model and was not discussed in some other experiments. In order to properly analyze the experimental data a more accurate model is needed, where this "pinning" effect is taken into account.

Since the reversal of the direction of the DW motion occurs for the low-velocity field range, a more detailed analysis of this behavior follows. The DMI-field is extracted by linearly fitting the experimental data in Fig. 9.11e, for both types of

DW and for both positive and negative current. Considering only the high velocity experimental data, the crossing of the two best fitting lines for the ↑↓-DW data occurs at a longitudinal field value $\mu_0 H_x^{\uparrow\downarrow} = -8.5 \pm 1.8$ mT. While, for the ↓↑-DW the crossing occurs at $\mu H_x^{\downarrow\uparrow} = +7.0 \pm 1.5$ mT. Assuming the amplitude of the DMI field to be the average of the two fields (in absolute values) we obtain $|\mu_o H_{DMI}| = 7.8 \pm 1.2$ mT. All the errors correspond to one standard deviation. Since the DW width is $\Delta = 7$ nm $\Delta = (A/K_{eff})^{1/2}$, where we use a strength of the exchange interaction of $A = 10^{-11}$ J/m and K_{eff} is the effective anisotropy [28]), $\mu H_x^{\uparrow\downarrow} < 0$ and $\mu_o H_x{}^{\downarrow\uparrow} > 0$, and knowing that Ta-$\theta_{SH}$ has a negative sign [31] we obtain a DMI constant $D = +0.06 \pm 0.01$ mJ/m^2.

9.3.1.4 Combined Operation of Writing and Shifting Domain Walls

Finally we combine the writing by the Oe field and the shifting to demonstrate the operation of both together. As seen in Fig. 9.12, we can write and shift domain walls thus demonstrating the functionality of a shift register memory device with domain wall writing by Oe fields and domain wall displacement by spin orbit torques. Note that in the last displacement (e), the bottom two walls reach the end of the wire and thus stop (due to the reduced current density as the wire widens into the pad) leading to a change in the domain length.

9.4 Topological Defects in Discs—Skyrmions

We next change the geometry from a wire to a disc. We use similar materials with out of plane magnetization as discussed above for current-induced domain wall motion of the Bloch or Néel walls in wires. If we however structure the films in discs, we find under the right conditions bubble Skyrmion spin structures (Fig. 9.13b). These structures gain their stability from the dipolar interaction of the magnetization in the disk. Compared to a state with a homogeneous out-of-plane magnetization the state containing a bubble has a much lower stray field outside the sample. The reduction of the stray-field energy can be larger than the additional energy that emerges from the domain wall that occurs in the bubble state. We start by comparing the dynamics of the well-known vortex (Fig. 9.13a) with that of the bubble Skyrmion.

A magnetic field pulse, that is aligned parallel to the film plane displaces the magnetic vortex core from its equilibrium position, since a domain parallel to the field is energetically favored during the pulse. After the excitation, the core does not move back to its original position on a straight line but performs a spiral gyration as shown in Fig. 9.13c. The sense of this gyration depends on the polarization of the vortex and has been experimentally observed using x-ray microscopy [33]. A similar excitation can be achieved by sending an in-plane current pulse through the film [34, 35].

Analytically, the dynamics of the vortex can be described by the so called, Thiele equation [38, 39]

$$0 = \boldsymbol{F} + \boldsymbol{G} \times (\dot{\boldsymbol{X}} + \boldsymbol{u}) + D(\alpha \dot{\boldsymbol{X}} + \xi \boldsymbol{u}) \tag{9.3}$$

where $\boldsymbol{X}$ is the position of the vortex core and $\boldsymbol{u} = P\mu_{\mathrm{B}}\boldsymbol{j}/(e\ M_s)$ introduces the action of a current density $\boldsymbol{j}$ [40]. Here P is the polarization of the current, M_s is the saturation magnetization, e is the elementary charge, and μ_B is the Bohr magneton. The constant $\boldsymbol{G}$ is the gyro vector which direction depends on the orientation of the vortex core. The tensor $D\ \alpha$ describes the damping of the excitation. The strength of the non-adiabatic spin torque ξ is a parameter that describes the interaction between the current and the vortex core [41]. Using (9.3) it has been found that a vortex that is excited by alternating in-plane fields or currents moves on an elliptic trajectory [36, 42]. The ratio of the two semi axis depends on the ratio of the frequency of the excitation and the resonance frequency of the potential. In the case of a resonant excitation the trajectory becomes circular.

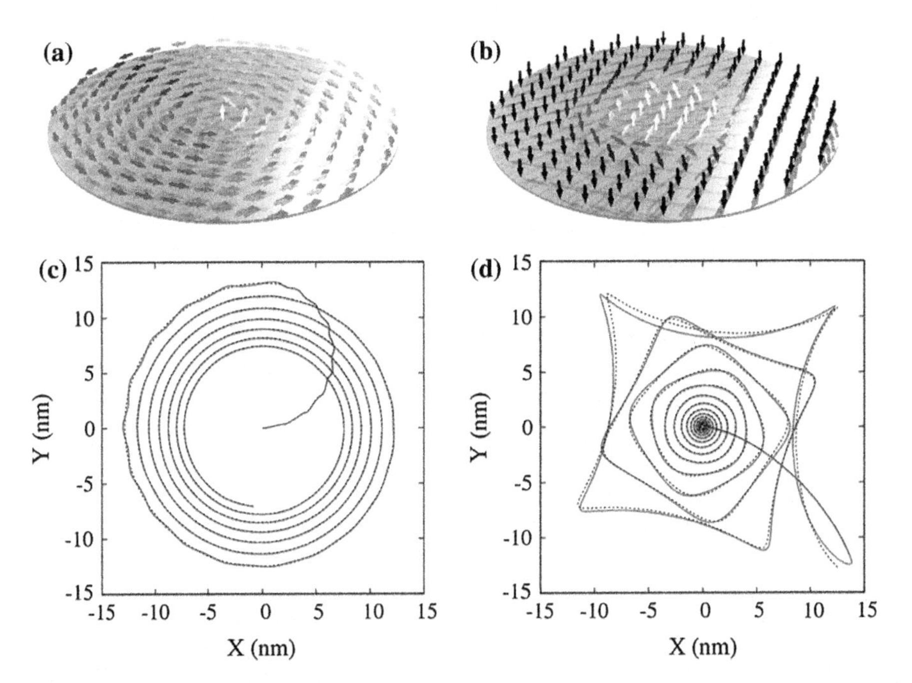

Fig. 9.13 **a** and **b** Schematic representation of the magnetization of (**a**) a vortex state and (**b**) a magnetic bubble Skyrmion. **c** Trajectory of a vortex in a $Ni_{81}Fe_{19}$ disc with a diameter of 200 nm and a thickness of 10 nm. The *blue line* denotes the trajectory of the maximum out-of-plane magnetization [36], that is, the center of the vortex core, during the application of the pulse. The *red* one shows the free dynamics after the field is switched off. The *dashed black line* is a fit with the analytical result in (9.3). **d** Trajectory of the center of mass of a bubble Skyrmion [37] in a FePt disc with a diameter of 200 nm and a thickness of 30 nm. The lines have the same meaning as in (**c**). In the case of the bubble Skyrmion the analytical fit was performed using (9.4)

It can be seen that the chirality and polarization both provide a binary state that can be used to store data. Therefore, it is in principle possible to store two bits in a vortex state. To change the polarization of a vortex, a very high static field perpendicular to the film is necessary. However, it has been found that a vortex excited by an in-plane field that is above a critical amplitude is able to change the orientation of the vortex core as well [33]. Here, the critical fields are much smaller than for the perpendicular field.

While vortices exist in materials without bulk anisotropy, there is a similar magnetization configuration, called magnetic bubble Skyrmion, that exists in materials with an out-of-plane easy axis. In the bubble the magnetization points out-of-plane while in the outer region the magnetization points in the opposite direction. Both domains are separated by a domain wall. A view on (9.1) shows that a bubble Skyrmion possesses a Skyrmion charge of one as the magnetization covers a sphere once. This can be seen by mapping the inner region of the bubble Skyrmion and the boundary of the film to the north and south poles of the sphere, respectively. The domain wall then moves to the equator, as shown in Fig. 9.14.

In contrast to a vortex, a bubble Skyrmion cannot be displaced by in-plane fields since there is no domain with a magnetization that is parallel to such a field. A homogeneous field that points perpendicular to the film causes the bubble to change its size but would not cause any movement due to the rotational symmetry of the system. This rotational symmetry can be broken by applying a gradient field. In this case, the bubble moves in the direction where the field parallel to the magnetization in the bubble Skyrmion increases.

From micromagnetic simulations it has been found that a bubble that is excited by a short pulse of a gradient field, is displaced from its equilibrium position. But in contrast to the vortex the bubble is not moving on a spiral trajectory but on a more complex one [37]. This trajectory has been identified to be a hypocycloid [43]. The hypocyclodic trajectory of the bubble can be analytically explained by having a closer look on the domain wall that is surrounding the bubble. One then finds that the excitations of the bubble can be described by waves that travel along the domain wall, where waves with the same wavelength that travel in different direction, that is, clockwise or counter clockwise, show different velocities [43]. It can be seen that the two modes for which the wavelength is exactly the circumference of the bubble are similar to a displacement of the bubble. A superposition of these two waves then resembles the hypocyclodic trajectory of the bubble.

From this observation it follows that (9.3) has to be extended by an additional term that is proportional to the acceleration of the vortex core, introducing some inertia of the bubble. The equation for the motion of the bubble Skyrmion reads [43]

$$0 = M\ddot{\boldsymbol{X}} + \boldsymbol{F} + \boldsymbol{G} \times (\dot{\boldsymbol{X}} + \boldsymbol{u}) + D(\alpha\dot{\boldsymbol{X}} + \xi\boldsymbol{u}) \tag{9.4}$$

with the inertial mass M.

Such a trajectory has been found experimentally by x-ray holography [44, 45]. The advantage of this technique is that it is drift-free allowing a measurement of the bubble position with a high accuracy. In the experiment [46] a multilayer film of

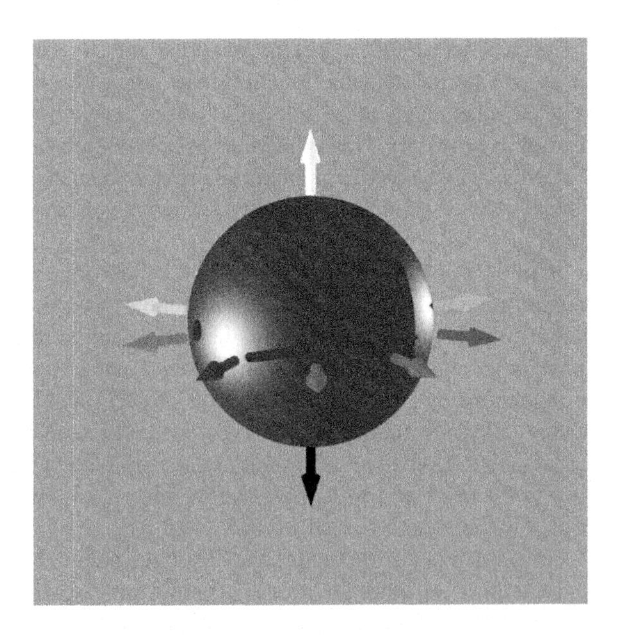

Fig. 9.14 Schematic view of the vector field from Fig. 9.13b mapped on a sphere. The field covers the sphere once

CoB/Pt has been grown on a thin SiN membrane to allow the photons to be transmitted through the sample. The magnetic layer, that is CoB, grows amorphous, reducing the pinning compared the pinning at crystalline imperfections in a pure Co film. The CoB/Pt film is patterned to a disc element. This element is surrounded by a gold microcoil as shown in Fig. 9.15a. The system is then excited by sending the current pulse shown in Fig. 9.15b through the microcoil. The Oersted field then excites the magnetic structure. As it can be seen in Fig. 9.15a there are two bubbles, however, one of them turns out to be pinned. After the current pulse the bubble performs a motion on a hypocycloidic trajectory. A part of this trajectory is shown in Fig. 9.15c. From the shape of the trajectory the Skyrmion number is deduced to be 1 and we extract the effective mass [46].

Currently, the focus of the research includes chiral Skyrmions that exist in systems where DMI is present. In contrast to bubble Skyrmions that are stabilized by the dipolar interaction, the chiral Skyrmion obtain its stability in thin films additionally from the DMI. The DMI favors a continuous rotation of the magnetization with a fixed sense of rotation. This results in the fact that Skyrmions with different senses of the in-plane magnetization are not energetically degenerate. If the strength of the DMI is large enough the Skyrmions with one sense of gyration become unstable. Therefore, the sense of gyration is determined by a material parameter, that is, the sign of the DMI, making the magnetization pattern chiral. In addition to the fixed

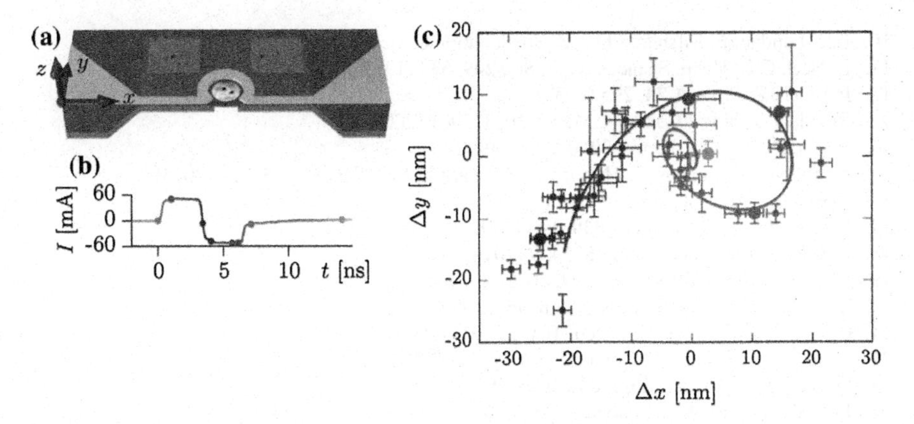

Fig. 9.15 (From [46]) **a** Schematic depiction of the experimental setup. The CoB/Pt disc is surrounded by a *gold stripline* that generates a field. **b** The shape of the bipolar current pulse. **c** Measured positions of the bubble as a function of time (*colour code* depicts the time and is the same as in (**b**)). The line is a fit with (9.4) to extract the Skyrmion properties

sense of rotation a strong DMI is also capable of decrease the size of a Skyrmion, making these topological spin structures apt for novel devices, such as the proposed Skyrmion racetrack [47].

Acknowledgments A number of students, postdocs and colleagues have contributed to these experiments to all of which we are indebted. In particular we would like to acknowledge T. Schulz, S.-J. Noh, T. Zacke, R. Lo Conte and G. Karnad for much of the experimental and simulations work carried out.

A large part of the work was performed within the EU projects MAGWIRE (FP7-ICT-2009-5 257707), WALL (FP7-PEOPLE-2013-ITN 608031) the European Research Council Starting Independent Researcher Grant MASPIC (No. ERC-2007-StG 208162) and furthermore the Graduate School of Excellence "Materials Science in Mainz" (No. GSC266), the Deutsche Forschungsgemeinschaft (DFG) and the State Research Center of Innovative and Emerging Materials. We thank O. Tchernychyov and E. Martinez for valuable discussions.

References

1. A. Aharoni, *Introduction to the Theory of Ferromagnetism* (Clarendon Press, Oxford, 1996)
2. H. Kronmüller, M. Fähnle, *Micromagnetism and the Microstructure of Ferromagnetic Solids* (Cambridge University Press, Cambridge, 2003)
3. S. Chikazumi, *Physics of Ferromagnetism*, 2nd edn. (Clarendon Press, Oxford, 1997)
4. M. Kläui, C.A.F. Vaz, in *Handbook of Magnetism and Advanced Magnetic Materials*, vol. 2, ed. by H. Kronmüller, S.S.P. Parkin (John Wiley and Sons, Chichester, 2007)
5. A.A. Belavin, A.M. Polyakov, JETP Lett. **22**, 245 (1975)
6. O. Tchernyshyov, G.-W. Chern, Phys. Rev. Lett. **95**, 197204 (2005)
7. R.P. Cowburn et al., Phys. Rev. Lett. **83**, 1042 (1999)
8. T. Shinjo et al., Science **289**, 930 (2000)
9. A. Hubert, R. Schäfer, *Magnetic Domains-The Analysis of Magnetic Microstructures* (Springer, Berlin, 1998)

10. L.D. Landau, E. Lifshitz, Phys. Z. Sowjetunion **8**, 153 (1935)
11. L. Néel, C.R. Hebd, Séances Acad. Sci. **249**, 533 (1955)
12. F. Bloch, Z. Phys. A **74**, 295 (1932)
13. M. Kläui, J. Phys.: Condens. Matter **20**, 313001 (2008)
14. A. Fert, Mater. Sci. Forum **50–60**, 439 (1990)
15. K.-S. Ryu, L. Thomas, S.-H. Yang, S.S.P. Parkin, Nat. Nanotech. **8**, 527 (2013)
16. S. Emori et al., Nat. Mater. **12**, 611 (2013)
17. S.S.P. Parkin et al., Science **320**, 190 (2008)
18. J.-S. Kim et al., Nat. Comm. **5**, 3429 (2014)
19. D.A. Allwood et al., Science **309**, 1688 (2005)
20. M. Diegel et al., IEEE Trans. Magn. **45**, 3792 (2009)
21. O. Boulle et al., Phys. Rev. Lett. **101**, 216601 (2008)
22. J. Heinen et al., J. Phys. Cond. Matter **24**, 024220 (2012)
23. O. Boulle, Mater. Sci. Eng. R **72**, 159 (2011)
24. I.M. Miron et al., Nat. Mater. **9**, 230 (2010)
25. P. Gambardella et al., Phil. Trans. R. Soc. A **369**, 3175 (2011)
26. A. Manchon, arXiv:1204.4869v1 (2012)
27. A. Thiaville et al., Europhys. Lett. **100**, 57002 (2012)
28. S. Emori et al., Phys. Rev. B **90**, 184427 (2014)
29. O. Boulle et al., Phys. Rev. Lett. **111**, 217203 (2013)
30. L. Liu et al., Science **336**, 555 (2012)
31. R. Lo Conte et al., Appl. Phys. Lett. **105**, 122404 (2014)
32. R. Lo Conte et al., Phys. Rev. B **91**, 014433 (2015)
33. B. Van Waeyenberge et al., Nature **444**, 461–464 (2006)
34. J. Shibata et al., Phys. Rev. B **73**, 020403 (2006)
35. M. Bolte et al., Phys. Rev. Lett. **100**, 176601 (2008)
36. B. Krüger et al., Phys. Rev. B **76**, 224426 (2007)
37. C. Moutafis et al., Phys. Rev. B **79**, 224429 (2009)
38. A. Thiele, Phys. Rev. Lett. **30**, 230–233 (1973)
39. A. Thiele, J. Appl. Phys. **45**, 377 (1974)
40. Y.B. Bazaliy et al., Phys. Rev. B **57**, R3213 (1998)
41. S. Zhang, Z. Li, Phys. Rev. Lett. **93**, 127204 (2004)
42. K.-S. Lee et al., Appl. Phys. Lett. **92**, 192513 (2008)
43. I. Makhfudz et al., Phys. Rev. Lett. **109**, 217201 (2012)
44. S. Eisebitt et al., Nature **432**, 885 (2004)
45. F. Büttner et al., Phys. Rev. B **87**, 134422 (2013)
46. F. Büttner et al., Nat. Phys. **11**, 225 (2015)
47. A. Fert et al., Nat. Nano. **8**, 152 (2013)

Chapter 10
Magnetic Solitons in Superlattices

Amalio Fernández-Pacheco, Rhodri Mansell, JiHyun Lee, Dishant Mahendru, Alexander Welbourne, Shin-Liang Chin, Reinoud Lavrijsen, Dorothee Petit and Russell P. Cowburn

10.1 Introduction

The injection and controlled motion of domain walls in nanowires has been extensively investigated in the last few years by the spintronics community, due to the great potential of these systems for storage and logic operations [1–7], as well as for the interesting fundamental properties of the walls [8–15]. By using planar nanowires, the result of lithography-patterned sputtered or evaporated films, a high level of control can now be achieved in injecting walls into wires [16–18], and trapping them using local alterations in the shape of the wires and other mechanisms [19–27]. Moreover, it is now well understood that the type of domain wall present depends on the wire size and material. [9, 12, 13, 28] and the torques present on the walls when applying external magnetic fields, spin-polarized currents and spin currents [29–32]. As well as their possible use in nanoelectronics, this technology is now proposed for biosensing applications [33, 34], and multi-turn sensors based on NOT-gate nanowires are currently used in the automotive industry [35].

Regarding a possible application of domain wall technology in nanoelectronics, in spite of possessing great advantages with respect to commercial products, that is they are non-volatile data storage devices, have high access speeds and do not have mobile parts, their storage capacities are not high enough to compete against CMOS counterparts, constituting a great limitation for this technology. As a possible solution, schemes based on stacking multiple planes of functional nanowires, separated

A. Fernández-Pacheco (✉) · R. Mansell · J. Lee · D. Mahendru · A. Welbourne · S.-L. Chin · R. Lavrijsen · D. Petit · R.P. Cowburn
Cavendish Laboratory, University of Cambridge, JJ Thomson Avenue, Cambridge CB3 0HE, UK
e-mail: af457@cam.ac.uk

R. Lavrijsen
Department of Applied Physics, Center for NanoMaterials, Eindhoven University of Technology, P.O. Box 513, 5600 Eindhoven, MB, The Netherlands

J. Seidel (ed.), *Topological Structures in Ferroic Materials*, Springer Series in Materials Science 228, DOI 10.1007/978-3-319-25301-5_10

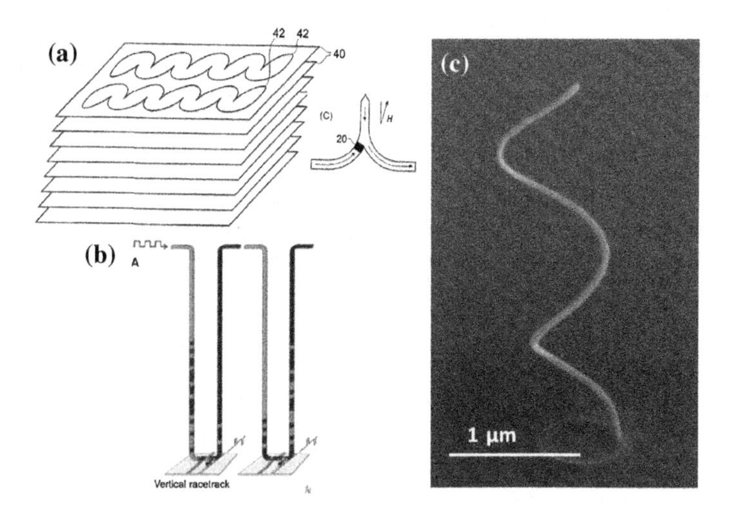

Fig. 10.1 Three-dimensional spintronic nanostructures based on magnetic nanowires. **a** Stack of planar NOT-gate nanowires, with information moving within planes [36]. **b** Vertical racetrack memory proposed by IBM [6]. **c** Suspended nano-spiral grown by focused electron beam induced deposition [37]

by non-magnetic spacers (Fig. 10.1a) [36], or the possibility to use vertical nanowires instead of horizontal ones (Fig. 10.1b) [6], have been proposed. However, these have not been implemented so far due to the great technical difficulty of fabricating these systems with the same degree of control as can be achieved for planar nanowires. It is clear that the possibility to create three-dimensional spintronic devices would be extremely interesting, which could result in a real revolution within nanoelectronics.

Recently, our group has been investigating ways to achieve this goal. For instance, we have developed new routes to pattern high aspect-ratio three dimensional nanowires using advanced lithography techniques (see Fig. 10.1c). Moreover, we have used a new concept to store and move information in the vertical direction, based on magnetic solitons in superlattices, which is the focus of this chapter.

10.2 Magnetic Solitons in Superlattices

We recently proposed [38] a new concept for the storage and motion of information along the vertical direction, based on magnetic solitons. This type of excitation is present in superlattices as the one shown in Fig. 10.2, formed by alternating magnetic/non-magnetic films. If the non-magnetic spacer is only a few nanometers thick, and is chosen appropriately, neighboring magnetic layers will preferentially align anti-parallel to each other due to Ruderman-Kittel-Kasuya-Yosida (RKKY) antiferromagnetic interactions [39]. In the case studied here, consisting of nanometer-thick magnetic layers with in-plane magnetization and uniaxial anisotropy, the simplest way to describe such a system is based on a macrospin approximation: each

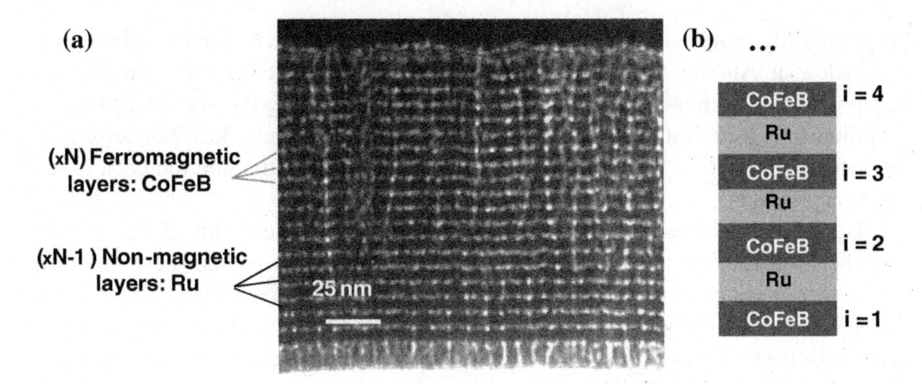

Fig. 10.2 **a** Transmission electron microscopy image of a sputtered magnetic superlattice formed by N CoFeB ferromagnetic layers and N − 1 Ru non-magnetic layers. **b** Scheme of the superlattice under investigation

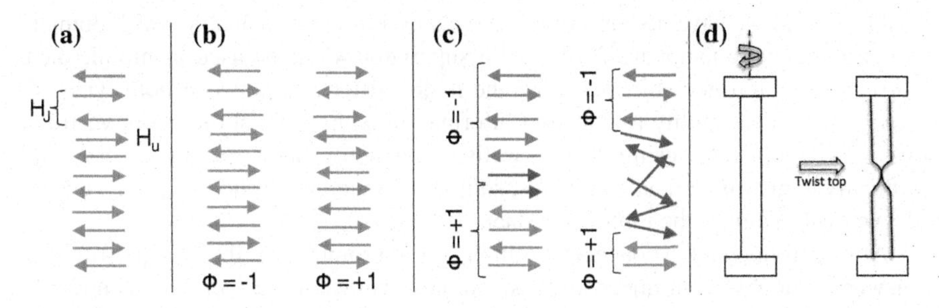

Fig. 10.3 Macrospin approximation for an antiferromagnetic superlattice. **a** 1D linear chain of spins with anti-parallel arrangement, where the anisotropy and coupling fields are indicated. **b** Two possible anti-parallel ground states of the system at remanence, with $\Phi = +1$ or -1. **c** Magnetic soliton (marked in *red*) in the *middle* of the system. The spin angles represent in-plane rotations. **d** The twist in a *top* forms a kink which can be topologically protected

layer is represented by a single spin and the superlattice by a linear 1D chain of spins, as shown in Fig. 10.3a. The fundamental fields which characterize this system are the coupling antiferromagnetic field $H_J = |J|/M_s t$ and the anisotropy field $H_u = 2K/M_s$, where J is the coupling surface energy density (erg/cm^2) between ferromagnetic layers, M_s is the saturation magnetization (emu/cm^3) of the layers, t is their thickness (cm) and K is the anisotropy energy density (erg/cm^3).

In this case, as shown in Fig. 10.3b, the ground level of the system is degenerate: defining the parameter $\Phi = (-1)^{i-1} \cos\theta_i$, with θ_i the angle formed by the magnetization of layer i with respect to the easy axis, there are two ground states corresponding to the two possible anti-parallel states ($\Phi = +1$ or -1). Analogously to a magnetic nanowire (Fig. 10.3c), when the two types of domains meet, a wall is formed separating the two domains. This anti-ferromagnetic wall, with the spins forming it marked in red in the figure, is a soliton or kink: an excitation with respect to the ground state, which is mobile along the vertical direction, and may be topologically protected: the only way to get remove it is to switch all the spins above (or below) it. Another way to think about it is by using a finite version of the Mobius strip [40]: as schematically shown in Fig. 10.3d: considering the system as a top, a topologically locked state is formed by twisting it: no matter how the top is deformed, it will retain the kink. The only way to remove it is by twisting the top back in the contrary manner.

Figure 10.3c shows two types of scenarios regarding the width of the soliton: on the left, a sharp soliton formed only by two layers is present between both antiferromagnetic domains, whereas on the right, this soliton is wider, being formed by several spins. The width of a soliton will be related to the ratio between coupling and anisotropy fields. It therefore means that on the left case $H_u > H_J$, whereas $H_J > H_u$ for the right one. This situation is analogous to a domain wall in a nanowire [28]; however, whereas in the case of nanowires the width is fixed by the properties of the material used, for a soliton in a superlattice the two coupling fields (especially H_J) can be finely tuned, making it possible to control its width.

From what we have just described, and again following an analogy with domain walls in magnetic nanowires, solitons in superlattices can be used as mobile data carriers for spintronic shift registers; the major difference between both types of systems is in their ability to transport information in the vertical direction: whereas it is very complicated using domain walls in magnetic nanowires, due to the great difficulty to grown vertical nanowires, solitons in superlattices do this easily, due to the vertical nature of the multilayer stack.

After introducing the concept of solitons for 3D spintronics, the first questions to answer are how we can inject solitons, and how can we move them in a controlled manner within a superlattice. In the following section, we will explain in detail the energetics of the system under consideration, as well as contextualize our work regarding the injection of solitons within superlattices.

10.3 Soliton Injection Using the Surface Spin-Flop Transition

Figure 10.4 shows a superlattice formed by N antiferromagnetic-coupled ferromagnetic layers. When a field H is applied along the easy axis (EA) direction, the total energy per unit area (u) will be:

$$u = M_s\left[\sum_{i=1}^{N} t_i\left(\frac{1}{2}H_{ui}\sin^2\theta_i - \mathrm{H}\cos\theta_i\right) + \sum_{i=1}^{N-1}\left(t_i H_{Ji}\cos(\theta_{i+1}-\theta_i)\right)\right] \quad (10.1)$$

where the index i refers to the ferromagnetic layer number and θ_i is the angle formed by the magnetization with the EA. The first term refers to the uniaxial anisotropy of the system, the second to the Zeeman energy when an external field is applied, and

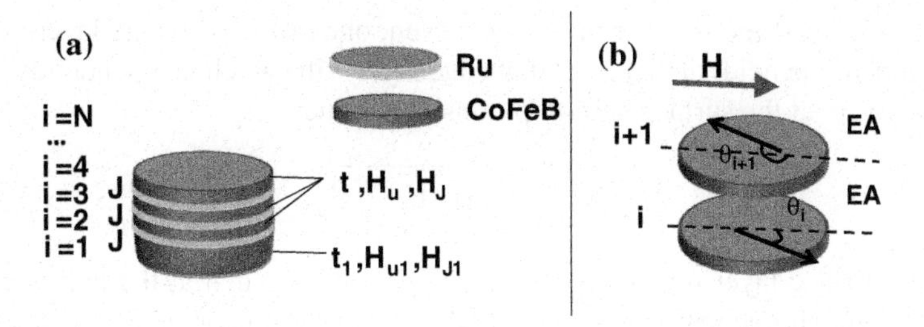

Fig. 10.4 **a** Schematic of a ferrimagnetic superlattice formed by alternating CoFeB-Ru layers, where the properties of one of the edge layers (the *bottom* one in this case) differ from the rest of the system. **b** Magnetization arrangement of two neighbouring layers under an external magnetic field applied along the easy axis direction. The angles of both layers with respect to that axis are indicated

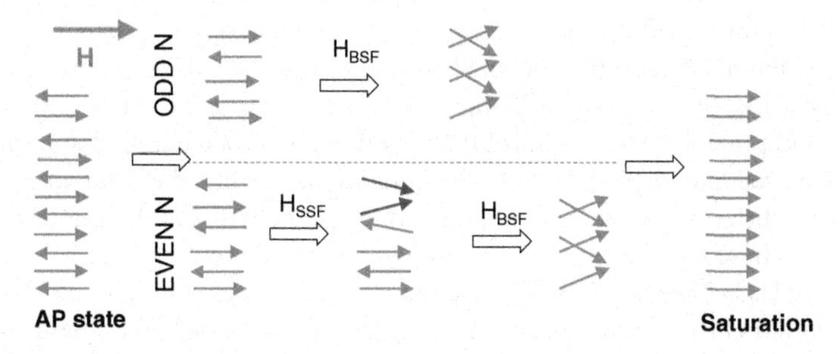

Fig. 10.5 Schematic of the evolution of a linear chain of spins will well-defined anisotropy and antiferromagnetic coupling under external magnetic fields along the easy axis, for chains with odd and even number of layers

the third to the RKKY antiferromagnetic coupling, which depends on the relative angle between first neighboring spins.

Taking this into account, it is well known [41, 42] that there exists an important asymmetry in the behavior of the system under external magnetic fields, if the number of layers constituting the superlattice is either even or odd, as shown in Fig. 10.5. Starting from an antiparallel state at remanence, and under an external positive magnetic field, the minimization of (10.1) gives a switching field for layer i:

$$H_{swi} = \sqrt{2H_{Ji}H_{ui} + H_{ui}^2} \tag{10.2}$$

If the number of layers is odd, as the two edge layers are already in the direction of the applied field, the bulk spins are those which will switch, at the so-called bulk spin-flop field:

$$H_{BSF} = \sqrt{4\mathrm{H_J H_u} + \mathrm{H_u^2}} \tag{10.3}$$

On the contrary, if the number of layers is even, one of the two edge layers will be antiparallel to the field; in that case, that edge layer will switch independently before the bulk layers at the surface spin-flop transition value:

$$H_{SSF} = \sqrt{2H_J H_u + H_u^2} \quad (10.4)$$

After the surface layer transition, the bulk layers will switch at the bulk spin-flop transition, which occurrs at higher field. Note that the factor of two between the prefactors multiplying $H_J H_u$ in (10.2) and (10.3) is due to the fact that a bulk layer has two neighboring layers and a surface layer only one. Importantly, in the field range between H_{SSF} and H_{BSF}, the system is incommensurate: part of it is still antiparallel, but at one edge the spins tend to be parallel to each other. We can therefore use the surface spin-flop transition as a method to inject solitons (marked in red in Fig. 10.5) at one edge of the superlattices. The concept presented here is not new. In fact there is a large amount of literature devoted to this effect, including experimental and theoretical works [41–52]. We will show in the next section the difference between our work and previous one to control the injection and propagation of solitons.

We have used macrospin Monte Carlo simulations to study the behavior of superlattices with different properties under external magnetic fields. Figure 10.6 shows snapshots of these simulations for the following parameters: N = 16, t = 1 nm, J/M_s = 1 kOe nm, H_u = 250 Oe. These correspond to superlattices similar to those studied by other authors [46, 51], where they investigated the behavior under field for large number of identical layers and observed both experimentally and via simulations the surface and bulk spin-flop transitions in these systems. The simulations shown here reproduce these results, with both bulk and surface transitions indicated. After the injection of a soliton at the bottom of the superlattice, when the field is increased further, the soliton is moved towards the middle of the superlattice, until the bulk spin-flop transition is reached, where the soliton is removed from the system. As in this case H_J is substantially larger than H_u, the formed soliton is broad, extending across several layers, which can be detrimental for applications where a large packing of data bits is required. Moreover, other evident problems for applications are the lack of control of where the soliton is formed (a system with all layers identical can end up in any of the two antiparallel states with equal probability, which defines what edge layer is antiparallel to the field at remanence) and the impossibility of propagating the nucleated soliton in a controlled manner using this field protocol.

In the system sketched in Fig. 10.4a we show the approach used here to solve these problems: our superlattices are synthetic ferrimagnets, i.e. one of the edge layers, in this case the bottom one, has different properties from the rest of the system: J is constant along the full superlattice, but t_1 and H_{u1} are different from the rest. Calling H_u and H_J the anisotropy and coupling fields for $i > 1$, $H_{u1} \neq H_u$ and $H_{J1} = (t/t_1)\, H_J$. This asymmetry in the system has a large influence in the nucleation and propagation of solitons, as will be shown in following sections.

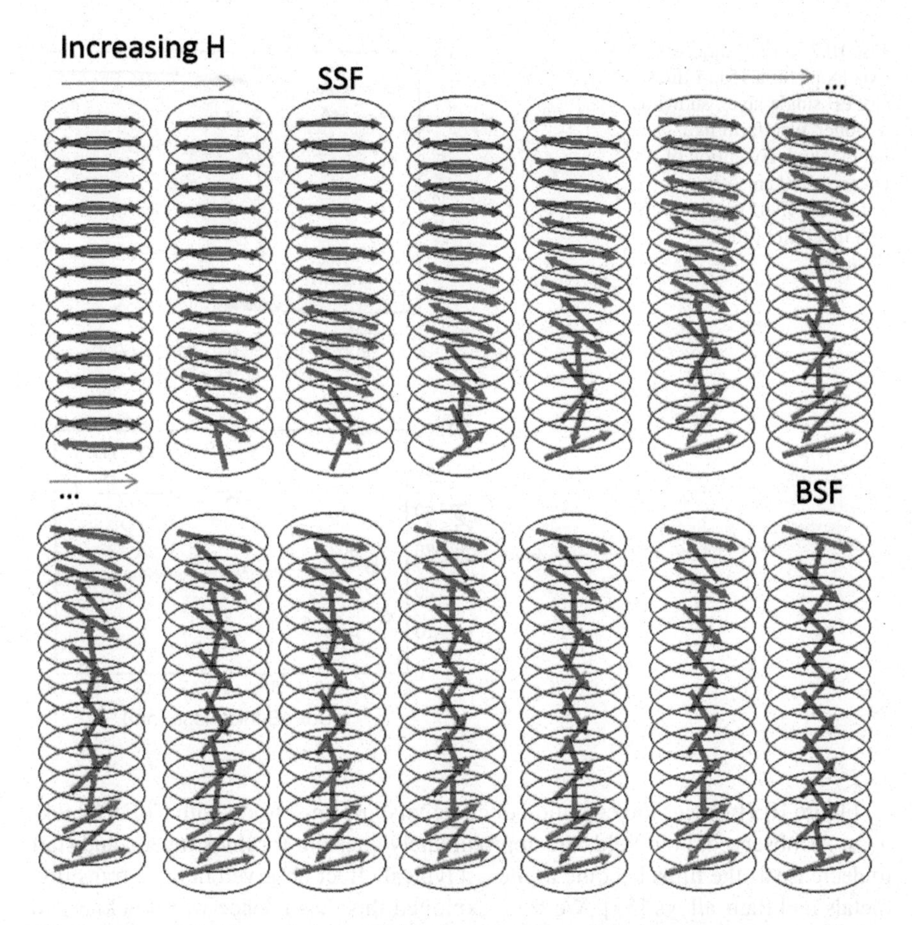

Fig. 10.6 Macrospin Monte Carlo simulations for a N = 16 superlattice similar to the one studied in [51]. Note that the magnetization of the layers is always in plane, due to the strong shape anisotropy in thin films

10.4 CoFeB/Ru Ferrimagnetic Superlattices

In this section, we show experimental results for CoFeB/Ru sputtered superlattices, which is the system that we have used for the injection and propagation of solitons. Figure 10.7a shows the hysteresis loop of 15 nm-thick CoFeB single layers measured by Kerr effect, with the field applied along the easy and hard axis. All samples presented here are grown by magnetron sputtering on Ta seedlayers (2–4 nm), with a Ta cap (4–5 nm). From these loops, we can infer the coercive field (H_c) and anisotropy field (H_u) of the layer. As observed, both fields have a similar value, indicating a negligible Brown's paradox; therefore, the use of macrospin simulations based on a Stoner-Wohlfarth analysis, as followed here, is in principle a good approximation.

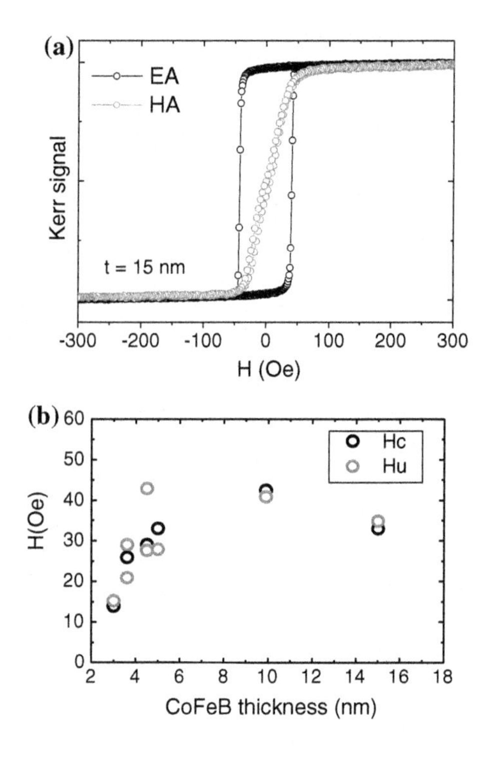

Fig. 10.7 **a** Easy and hard axis loops for a 15 nm-thick CoFeB single layer, showing a Stoner-Wohlfarth like behaviour. **b** Evolution of coercive and anisotropy fields of CoFeB single layers as a function of their thickness

Figure 10.7b shows the evolution of coercivity and anisotropy with CoFeB thickness. As observed, both fields are approximately constant for thick layers, and start to decrease as the films become thinner, a typical effect observed in most transition metals and their alloys [53]. We have exploited this dependence with thickness to have two well-defined anisotropy fields in our CoFeB films forming ferrimagnetic superlattices, since H_u (5 nm) $\approx$ 20 Oe, and H_u (15 nm) $\approx$ 40 Oe. Additionally, in order to control the RKKY coupling between layers, we have grown a series of bilayers, formed by two CoFeB layers separated by Ru, a non-magnetic interlayer with large RKKY interactions [39]. Figure 10.8a shows the experimental Kerr loop (black dots) of a ferrimagnetic bilayer formed by two CoFeB layers with thicknesses t_1 = 15 nm and t = 5 nm (notation as indicated before, "1" index for bottom layer), with the signal extracted from macrospin simulations superimposed (blue line) onto the experimental data. The behavior of the bilayer under the applied magnetic field is sketched by blue arrows. By making experimental data and simulations coincide, we can extract the parameters of the bilayer, in this case: H_{u1} = 40 Oe, H_u = 20 Oe, $J/M_s = -175$ Oe nm. The inset shows the experimental loop when the field is applied along the hard axis; the saturation field value of this loop can be used to extract the bilayer parameters independently [54], and confirm the validity of those obtained from the simulations.

We have varied the thickness of Ru for different bilayers, in a range around the third antiferromagnetic peak, located at 3.4 nm (see Fig. 10.8b). The maximum

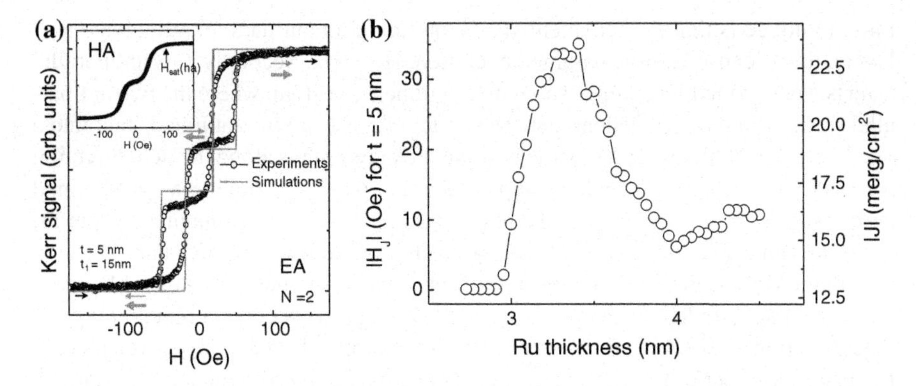

Fig. 10.8 **a** Easy axis experimental and simulation hysteresis Kerr loops for a ferrimagnetic bilayer, formed by two CoFeB layers with $t_1 = 15$ nm and $t = 5$ nm. The magnetic behaviour of the system during the field cycle is indicated by *blue arrows*. The *inset* shows a hard-axis loop. **b** RKKY third antiferromagnetic peak obtained by varying the interlayer Ru thickness around 3.5 nm. Both energy surface density and equivalent coupling field for a 5 nm-thick layer are included

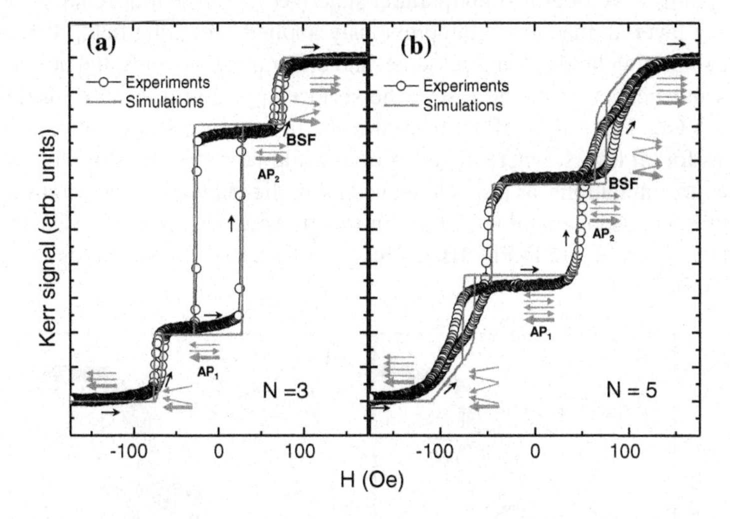

Fig. 10.9 Kerr hysteresis loops with the field along the easy axis, for superlattices with odd number of layers: N = 3 (**a**) and N = 5 (**b**)

corresponds to $J = -22.6\,\text{merg/cm}^2$, equivalent to $H_{J1} = 12.5$ Oe and $H_J = 37.6$ Oe, respectively. Once the properties of the single layers and the coupling between them have been fully characterized, the next objective is to grow ferrimagnetic superlattices with $N \geq 3$.

First, we investigated the behavior of ferrimagnetic superlattices with an odd number of layers. Figure 10.9 shows the cases of N = 3 (a) and N = 5 (b). As before, macrospin simulations are superimposed onto the experimental loops, with simulation parameters the same as those extracted from single layers and bilayers. The first thing to notice is that the remanent state consists of an antiparallel arrangement of layers; this is caused by the parameters chosen, $H_J > H_u$. Secondly, this antiparallel state is well-defined: in comparison with a symmetric system, where the two antiparallel states would be equivalent, here the system flops from the saturated state into a one which is well defined (AP1): the bottom thick layer stays along the field direction for longer than the others, which flop before, i.e. $\Phi = -1$. Having a well-defined antiparallel state at remanence is the first advantage of using ferrimagnetic superlattices. We can also notice that the system transits from that antiparallel state (AP1) to the other (AP2) as the field increases: that new state is energetically favorable under non-zero magnetic fields, since the bottom layer flips to become aligned along the field direction. After AP2, the system transits to a flopped state at H_{BSF}, as expected for an odd-number of layers system. Also notice that in a ferrimagnetic superlattice, the spins corresponding to thicker films (the edge bottom layer in this case) flop less than the thin ones, due to anisotropy and Zeeman energy contributions.

Several of the characteristics explained above also apply to systems with an even number of layers. Figure 10.10 shows the Kerr hysteresis loops for N = 4 and N = 6. Again, a well-defined antiparallel state (AP) is present at remanence, with the bottom layer aligned along the previously-applied saturating field, $\Phi = -1$ for negative saturating fields. After this state, and under positive fields, the bottom layer switches independently of the rest at the surface spin-flop transition, in this case $H_{SSF} \approx 45$ Oe. This value differs from the one in an equivalent system with same properties for all layers, where the transition would be ≈ 50 Oe. More importantly, according to simulations, as the field is increased, the magnetic configuration of the system does not change until the bulk spin-flop transition at $H_{BSF} \approx 65$ Oe, contrary to the case shown before in Fig. 10.6, where a soliton expanded towards the middle

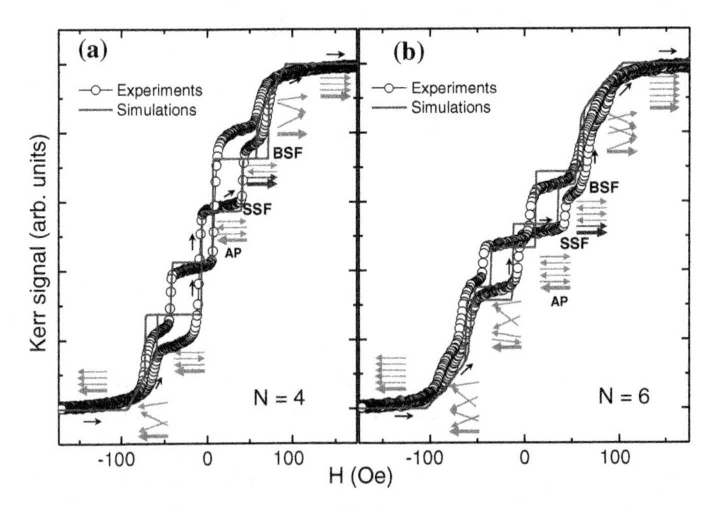

Fig. 10.10 Kerr hysteresis loops with the field along the easy axis, for superlattices with even number of layers: N = 4 (**a**) and N = 6 (**b**)

of the superlattice after being injected at one edge. The particular behavior described here is due to the value chosen for H_{J1}/H_{u1}, which is about 4 times smaller than H_J/H_u, keeping the soliton sharp and localized between layers 1 and 2 until the field reaches the bulk spin-flop transition. In order to use solitons in superlattices as mobile objects for possible data storage applications, a good control of the injection, direction of propagation and extension of that type of spin texture is required. According to what we describe here, using superlattices with an even number of layers, and with an edge layer different from the rest, is effective to inject sharp solitons in a well-defined position. We will see in the next section how this is also essential to controllably propagate them within superlattices using minor loops.

10.5 General Diagram for Injection and Propagation of Solitons in Ferrimagnetic Superlattices

As described in the previous section, the use of ferrimagnetic superlattices where one of the edge layers has different properties from the rest allows the creation of a sharp soliton at that edge after the surface spin-flop transition. We discuss now, by means of simulations, the behavior of the system under minor loops such as the one sketched on the right side of Fig. 10.11, i.e. field sequences as follows:

- The initial state (I) is a well-defined AP state at remanence, $\Phi = -1$, result of coming from negative saturation.
- The field is increased until the system switches at a field H_C (it may be H_{SSF}, but not necessarily).
- The final state (F) is the result of reducing the field, reaching a second remanent state.

For simplicity, we focus our discussion in a system with N = 6. The addition of more layers will be discussed later. The left part of Fig. 10.11 is a diagram generated by macrospin Monte Carlo simulations where the behavior of an N = 6 superlattice is studied under the type of minor loops described before. The properties of all layers in the superlattice (except for the bottom edge layer), are those described before:

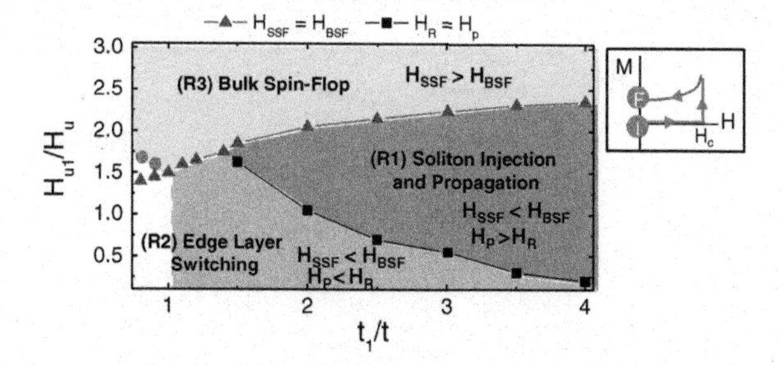

Fig. 10.11 Diagram generated by macrospin Monte Carlo simulations showing three different regions for the behaviour of the system under minor field loops. Adapted from [55]

$H_J = 37.6$ Oe and $H_u = 20$ Oe. The axes in the diagram correspond to the ratios between the anisotropy (H_{u1}/H_u, explored from 0.2 to 3) and thickness (t_1/t, explored from 0.8 to 4) of the edge layer with respect to all others of the superlattice; in other words, we are evaluating how different degrees of ferrimagnetism in the system affect the injection and propagation of solitons under minor loops. Note that the diagram is calculated for a fixed interaction energy density $J = -22.6$ merg/cm^2, so H_{J1} will change when t_1 is varied. As observed in that figure, the degree of ferrimagnetism of the superlattice is an essential ingredient for its behavior under minor loops. This had not been evaluated in previous works, which focused on systems with all layers having identical properties (both ratios = 1). Whilst the use of minor loops was briefly discussed in [52], here we extensively show how this type of field sequences can be used to control the injection and propagation of solitons.

Three main regions are observed in the diagram: soliton-injection-and-propagation (R1), edge-layer-switching (R2) and bulk spin-flop (R3). Simulation snapshots for the different regions at different parts of the field cycle are shown in Fig. 10.12.

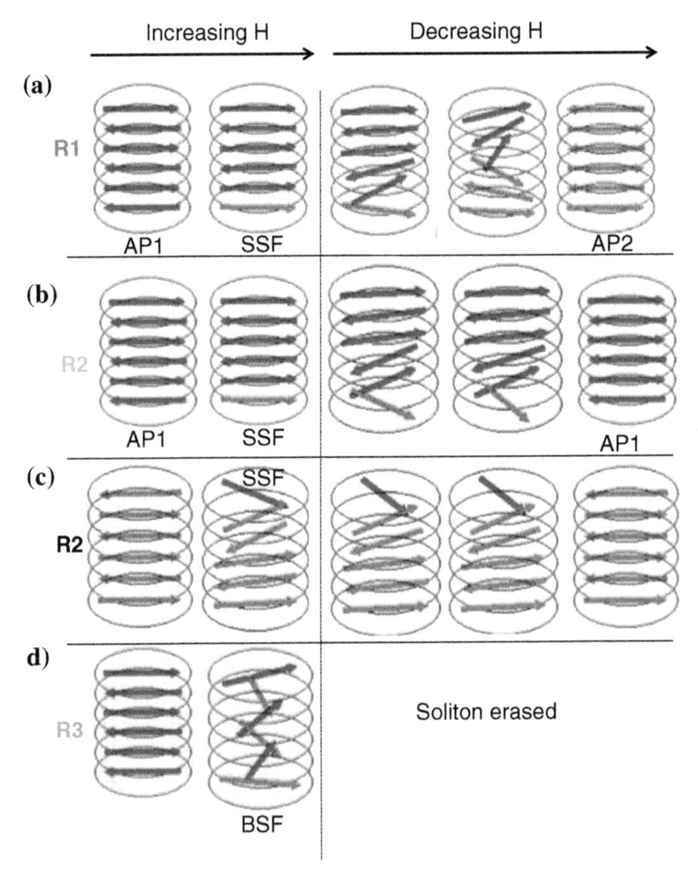

Fig. 10.12 Simulation snapshots for the different regions of the diagram shown in Fig. 10.11

Before describing in detail all of the regions, it is necessary to introduce an additional switching field in order to fully understand the diagram, the field necessary to reverse back layer 1 coming from positive saturation [56]:

$$H_R = \frac{2|H_{J1}| - H_{u1}}{\sqrt{1 + 2|H_{J1}|/H_{u1}}} \tag{10.5}$$

In regions R1 and R2-yellow, the first transition of the system under increasing magnetic fields corresponds to layer 1 switching at its corresponding H_{SSF} value. As discussed in the previous section, layer 1 reverses separately from the rest of the superlattice at that field, creating a sharp soliton at the bottom of the superlattice (second image in Fig. 10.12a, b). The behavior of these two regions, equal for increasing field, become different when decreasing it (third and fourth image in Fig. 10.12a, b): in R1, the field necessary to propagate the soliton H_p is such as $H_P > H_R$, with the contrary situation in R2; this means that when the field starts to decrease, a soliton is propagated through the superlattice for R1, whereas in R2 layer 1 switches back, ejecting the soliton downwards. As a result of an effective/non-effective propagation of the injected soliton, the second remanent state (F) will be the contrary as the initial one in R1 ($\Phi = +1$), whereas it will be the same in R2 ($\Phi = -1$). As expected, the snapshot corresponding to R1 (Fig. 10.12) show how the soliton broadens as is propagated, due to the increase of H_J/H_u for layers above the first one. Interestingly, R1 only exists for a critical ratio $t_1/t > 1.5$. As t_1/t increases from that value, the area of R1 is increased to the detriment of R2, becoming the mode of operation for lower H_{u1}/H_u ratios. It is therefore not necessary to have an edge layer with larger anisotropy than the others, as long as it is significantly thicker: Zeeman energy becomes dominant as t_1 increases, blocking a possible downwards expulsion of the soliton.

The case of R2-white is similar to R2-yellow: no soliton propagation is produced, but the injected soliton at one edge of the system is expelled instead. In the white case, however, $t_1/t < 1$, which means that the (I) state coming from negative saturation will be the contrary AP as the one before: now the top edge layer is thicker than the bottom one, which means that will stay along the field for longer, defining a remanent state with top (bottom) layer pointing left (right), i.e. $\Phi = +1$. The behavior of the superlattice for the rest of the field sequence is analogous to what it was described before for its yellow counterpart (see Fig. 10.12c): the edge layer which switched at H_{SSF} is not anisotropic or thick enough to avoid the expulsion of the soliton, resulting into the same (I) and (F) states, $\Phi = +1$. In the last region of the diagram, R3, H_{u1}/H_u becomes higher than before, resulting into $H_{SSF} > H_{BSF}$, due to the large anisotropy of the bottom layer. Therefore, in this case, the whole system transits directly from the initial AP state to a flopped state at H_{BSF}: no surface spin-flop transition is produced. This means that for large anisotropies of layer 1, the surface spin-flop transition is blocked, with an even number of layer system (N = 6) effectively behaving as a superlattice with an odd number of layers.

The vertical line $t_1/t = 1$ in the diagram requires special attention, since previous works mainly focused on this type of system. Point (1,1) of the diagram, with all layers having equal properties, is located inside R2 region, where no effective propagation of soliton occurs. Either of the two AP states ($\Phi = +1$ or -1) is possible, but in both cases the edge layer which switched at H_{SSF} switches back when reducing the field. By varying H_{u1}/H_u, the state initially present (I) is controlled: If $H_{u1}/H_u < 1$, the top layer points left at (I), i.e. $\Phi = -1$, and the top surface flops independently from the other layers, whereas if $1 < H_{u1}/H_u < 1.5$, (I) will be such as the bottom layer points left, i.e. $\Phi = -1$, and therefore it is the bottom layer which now switches independently of the others under increasing fields. However, in both cases these spin arrangements are always reversed when the field starts to decrease (R2). Still along the same line of the diagram, if $H_{u1}/H_u > 1.5$, the layers directly transit into a bulk-flopped state (R3).

The diagram shown here is a generalization of the surface-spin flop transition for ferrimagnetic systems, which shows a rich behavior just by introducing a slight asymmetry to the system. From it, we can conclude that the presence of a thicker edge layer is essential for a successful nucleation and propagation of solitons using the mechanisms explained here. As discussed in Sect. 10.3, in a superlattice with all layers equal, and assuming that the nucleation of a soliton occurs via a surface spin-flop transition at the bottom surface, if we require a soliton to propagate upwards, the only option is to apply fields H such as $H_{SSF} < H \lesssim H_{BSF}$, resulting into the soliton moving beyond the center of the stack. However, with H approaching to H_{BSF} is the soliton width substantially increases and the exact configuration of the superlattice near a bulk-flopped state is very sensitive to small variations in layer properties. On the contrary, the process shown here for region R1 is much more robust: the soliton extension is well controlled, since the maximum field needed is H_{SSF}, the propagation direction is well defined, and it does not depend on the number of layers of the superlattice, as long as it is even.

10.6 Experimental Realization for the Injection and Propagation of Solitons in CoFeB/Ru Ferrimagnetic Superlattices

After showing the general behavior of N = 6 ferrimagnetic superlattices using macrospin Monte Carlo simulations, in this section we reproduce experimentally the behavior described for the area R1 of the diagram, the part with soliton propagation. For that, we carried out experiments in CoFeB/Ru superlattices as the one previously described in Fig. 10.10, subjected to the same type of field cycle explained in the previous section. These superlattices have N = 6, $H_J = 37.6$ Oe, t = 5 nm, $H_u = 20$ Oe, $t_1 = 15$ nm and $H_{u1} = 36$ Oe, i.e. their properties are such as to be in the R1 region.

Figure 10.13 shows the Kerr signal obtained experimentally (a) and using simulations (b) for this superlattice. The black line is the signal along the negative-to-positive major field cycle, and the blue line is the signal when a minor loop as the

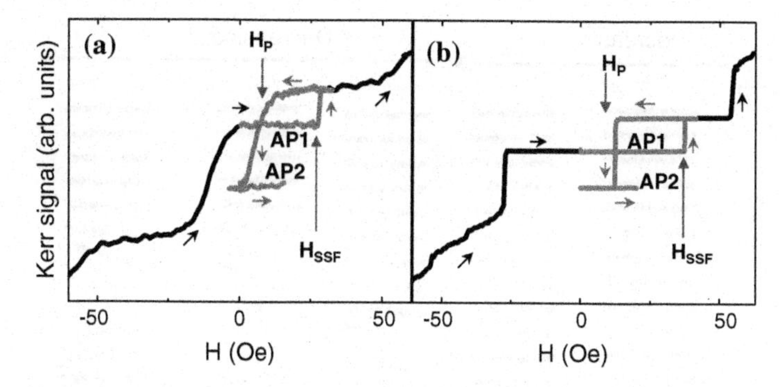

Fig. 10.13 Experimental (**a**) and simulated (**b**) Kerr signal in an N = 6 CoFeB/Ru superlattice under minor loops. The system, starting from negative saturation, goes from AP1 at first remanence, to AP2 at the second, after injecting a soliton at H_{SSF}. Simulations show how this occurs via the propagation of a soliton upwards the system. Adapted from [55]

one previously explained is followed. A good agreement between experiments and simulations is found: as shown in Fig. 10.12a, starting from negative saturation, the system is brought to remanence (state I). As the bottom layer is 3 times thicker than the others, AP1, of the two possible antiparallel states, is the one with the bottom layer pointing left ($\Phi = -1$). As the field is increased further, becoming positive, the bottom layer switches at $H_{SSF} \approx 25$ Oe, creating a soliton at the bottom of the superlattice. As the field starts to decrease, the thick bottom layer blocks the downwards expulsion of the soliton, leading to upwards propagation at $H_P \approx +12$ Oe. As a result, the other antiparallel configuration AP2 ($\Phi = +1$) is formed when the second remanent state is reached (F). These experiments show therefore the successful injection and propagation of solitons in superlattices using a modified surface-spin flop mechanism in ferrimagnetic superlattices.

10.7 Influence of the Anisotropy/Coupling Ratio and Number of Layers for Soliton Propagation

The results shown here, both experimentally and via simulations, correspond to bulk coupling/anisotropy ratios $|H_J|/H_u = 1.75$. As described in Sect. 10.4, for this ratio a wide R1 region is found for the formation of mobile solitons in the case of superlattices with N = 6. However, previous literature has mainly focused on antiferromagnetic superlattices with significantly larger ratios, typically larger than 10. It is therefore important to complement the diagram of Sect. 10.5 by performing simulations with larger ratios, to understand the importance of an appropriate anisotropy/coupling value for the injection and motion of solitons. For the same diagram shown in Fig. 10.11, and focusing on the point ($H_{u1}/H_u = 1.5$, $t_1/t = 3$), we have studied the behavior of the system with H_J/H_u progressively increasing. We observe

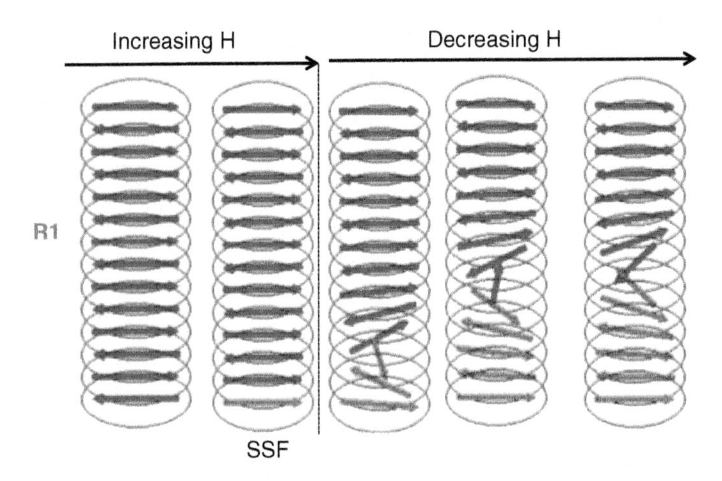

Fig. 10.14 Simulation snapshots for an N = 14 superlattice with parameters such as those which resulted in a behaviour in the R1 region for a N = 6 superlattice

that a good injection and propagation of solitons occurs as long as $H_J/H_u < 2.25$. For higher ratios, a soliton nucleated at the bottom layer is not stable within the system, being immediately expelled up the stack at H_{SSF}; this result was expected, in agreement with our initial discussion about the dependence of soliton width with H_J/H_u ratio. A similar trend is expected for other points of the diagram, which makes us conclude that the wide R2 region found for a specific coupling/anisotropy ratio in Fig. 10.11 shrinks as this ratio increases. In any case, a good operation is conserved for a significant range, as soon as it is not much larger than 1.

Another factor which influences the stability of solitons within superlattices is the number of layers N. Figure 10.14 shows the same system and field sequence as described in Fig. 10.12a, but now with N = 14 layers instead of 6. We can observe how, after the injection of a soliton at the bottom of the superlattice and propagation when the field is decreased, the system ends up in a non-antiparallel remanent state (F) where the soliton, instead of being fully ejected from the superlattice, resulting in the second antiparallel state, stays in the middle. The complex energy potential created by the edges of the superlattice is responsible of the difference observed as a function of N. This result suggests that it should be possible to inject solitons in superlattices which would be stable at remanence, just by using the mechanisms here described. However, the evolution in properties of the superlattices, affecting coupling and anisotropy as they are grown is major issue which prevents the creation of extended systems. Furthermore, as the stack height increases Kerr signals become very complex and so probing the state of the system is challenging, requiring more complex techniques such as polarized neutron reflection [57].

10.8 Conclusions and Outlook

In summary, we present here a new concept to move binary information along the vertical direction, based on magnetic solitons in antiferromagnetic superlattices. Solitons are walls separating the two possible antiparallel states in a superlattice formed by ferromagnetic layers coupled antiferromagnetically between them via RKKY interactions. The main advantage of using these excitations is their intrinsic mobility along the vertical (perpendicular to the thin film plane), in contrast to conventional domain walls in planar nanowires, as well as the possibility to control their width by tuning anisotropy of layers and exchange indirect coupling between them.

Moreover, we have shown how it is possible to inject and move solitons in superlattices in a controlled manner using the surface spin-flop transition followed by minor field cycles. For that, we have extended previous works, investigating ferrimagnetic superlattices formed by an edge layer with different properties from the rest. Via macrospin simulations, we have generalized the surface spin-flop transition to this type of asymmetric superlattices, showing how the degree of ferrimagnetism of the superlattice has a great importance for the injection and propagation of solitons. In particular, we have identified a large area of the parameter space where solitons can be injected and moved unidirectionally. The importance of a correct coupling/anisotropy ratio for the controlled motion of solitons has also been discussed. Finally, we have shown the first experiments for controlled injection and propagation of solitons in superlattices.

The work presented in this chapter, part of it previously published in [55], is part of the activity of our group on solitons. Additionally, and complementing this work, we have shown how the injection mechanism presented here can be generalized to a non-macrospin situation, where the real micromagnetic configuration of each layer is considered, and the presence of collective effects during switching is present [57]. Also, we have managed to block the expulsion of broad solitons using appropriate boundary conditions, creating helical states at remanence [57], which could have applications in energy-storage spintronic devices [58]. Furthermore, we have shown how it is possible to move solitons in superlattices synchronously with external magnetic fields. In the case of broad solitons such as those shown here, this is achieved by using rotating magnetic fields and exploiting how the sense of rotation of the magnetic field is coupled to the intrinsic chirality of the solitons, resulting into bi-directional vertical shift registers [59]. In the case of sharp solitons, this is achieved instead of by creating a ratchet energy profile for soliton propagation, by periodically alternating the properties of the layers forming the superlattice; this results into systems behaving as unidirectional vertical shift registers for data storage [60–62] and logic [63] applications.

Acknowledgments A. Fernández-Pacheco acknowledges support by EPSRC and Winton Program for the Physics of Sustainability. R. Lavrijsen acknowledges support from the Netherlands Organization for Scientific Research and Marie Curie Cofund Action. We acknowledge research funding from the European Community under the Seventh Framework Programme Contracts No. 247368: 3SPIN and No. 309589: M3d.

References

1. T. Hesjedal, T. Phung, Magnetic logic element based on an S-shaped Permalloy structure. Appl. Phys. Lett. **96**, 072501 (2010)
2. D.A. Allwood, Characterization of submicrometer ferromagnetic NOT gates. J. Appl. Phys. **95**, 8264 (2004)
3. L. O'Brien et al., Bidirectional magnetic nanowire shift register. Appl. Phys. Lett. **95**, 232502 (2009)
4. D.A. Allwood et al., Magnetic domain-wall logic. Science **309**, 1688–1692 (2005)
5. J.H. Franken, H.J.M. Swagten, B. Koopmans, Shift registers based on magnetic domain wall ratchets with perpendicular anisotropy. Nat. Nanotechnol. **7**, 499–503 (2012)
6. S.S.P. Parkin, M. Hayashi, L. Thomas, Magnetic domain-wall racetrack memory. Science **320**, 190–194 (2008)
7. M. Hayashi, L. Thomas, R. Moriya, C. Rettner, S.S.P. Parkin, Current-controlled magnetic domain-wall nanowire shift register. Science **320**, 209–211 (2008)
8. M.A. Basith, S. McVitie, D. McGrouther, J.N. Chapman, J.M.R. Weaver, Direct comparison of domain wall behavior in permalloy nanowires patterned by electron beam lithography and focused ion beam milling. J. Appl. Phys. **110**, 083904 (2011)
9. J. Garcia, A. Thiaville, J. Miltat, MFM imaging of nanowires and elongated patterned elements. J. Magn. Magn. Mater. **249**, 163–169 (2002)
10. Y. Jang, S.R. Bowden, M. Mascaro, J. Unguris, C.A. Ross, Formation and structure of 360 and 540 degree domain walls in thin magnetic stripes. Appl. Phys. Lett. **100**, 062407 (2012)
11. M. Laufenberg et al., Observation of thermally activated domain wall transformations. Appl. Phys. Lett. **88**, 052507 (2006)
12. M. Klaäui et al., Head-to-head domain-wall phase diagram in mesoscopic ring magnets. Appl. Phys. Lett. **85**, 5637 (2004)
13. M. Kläui, Head-to-head domain walls in magnetic nanostructures. J. Phys. Condens. Matter **20**, 313001 (2008)
14. P. Roy et al., Antivortex domain walls observed in permalloy rings via magnetic force microscopy. Phys. Rev. B **79**, 060407 (2009)
15. D. Petit et al., Magnetic imaging of the pinning mechanism of asymmetric transverse domain walls in ferromagnetic nanowires. Appl. Phys. Lett. **97**, 233102 (2010)
16. R.P. Cowburn, D.A. Allwood, G. Xiong, M.D. Cooke, Domain wall injection and propagation in planar Permalloy nanowires. J. Appl. Phys. **91**, 6949 (2002)
17. T. Ono, Propagation of a magnetic domain wall in a submicrometer magnetic wire. Science **80**(284), 468–470 (1999)
18. K. Shigeto, T. Shinjo, T. Ono, Injection of a magnetic domain wall into a submicron magnetic wire. Appl. Phys. Lett. **75**, 2815 (1999)
19. J. Akerman, M. Muñoz, M. Maicas, J.L. Prieto, Stochastic nature of the domain wall depinning in permalloy magnetic nanowires. Phys. Rev. B **82**, 064426 (2010)
20. M.T. Bryan, D. Atkinson, D.A. Allwood, Multimode switching induced by a transverse field in planar magnetic nanowires. Appl. Phys. Lett. **88**, 032505 (2006)
21. M.-Y. Im, L. Bocklage, P. Fischer, G. Meier, Direct observation of stochastic domain-wall depinning in magnetic nanowires. Phys. Rev. Lett. **102**, 147204 (2009)
22. M.-Y. Im, L. Bocklage, G. Meier, P. Fischer, Magnetic soft X-ray microscopy of the domain wall depinning process in permalloy magnetic nanowires. J. Phys. Condens. Matter **24**, 024203 (2012)
23. D. Petit, A.-V. Jausovec, D. Read, R.P. Cowburn, Domain wall pinning and potential landscapes created by constrictions and protrusions in ferromagnetic nanowires. J. Appl. Phys. **103**, 114307 (2008)
24. A. Beguivin, L.A. O'Brien, A.V. Jausovec, D. Petit, R.P. Cowburn, Magnetisation reversal in permalloy nanowires controlled by near-field charge interactions. Appl. Phys. Lett. **99**, 142506 (2011)

25. T.J. Hayward et al., Pinning induced by inter-domain wall interactions in planar magnetic nanowires. Appl. Phys. Lett. **96**, 052502 (2010)
26. L. O'Brien et al., Tunable remote pinning of domain walls in magnetic nanowires. Phys. Rev. Lett. **106**, 087204 (2011)
27. L. O'Brien et al., Near-field interaction between domain walls in adjacent permalloy nanowires. Phys. Rev. Lett. **103**, 077206 (2009)
28. Y. Nakatani, A. Thiaville, J. Miltat, Head-to-head domain walls in soft nano-strips: a refined phase diagram. J. Magn. Magn. Mater. **290–291**, 750–753 (2005)
29. M. Eltschka et al., Nonadiabatic spin torque investigated using thermally activated magnetic domain wall dynamics. Phys. Rev. Lett. **105**, 056601 (2010)
30. M. Hayashi, L. Thomas, C. Rettner, R. Moriya, S.S.P. Parkin, Direct observation of the coherent precession of magnetic domain walls propagating along permalloy nanowires. Nat. Phys. **3**, 21–25 (2006)
31. S. Lepadatu et al., Domain-wall spin-torque resonators for frequency-selective operation. Phys. Rev. B **81**, 060402 (2010)
32. G.S.D. Beach, M. Tsoi, J.L. Erskine, Current-induced domain wall motion. J. Magn. Magn. Mater. **320**, 1272–1281 (2008)
33. M. Donolato et al., On-chip manipulation of protein-coated magnetic beads via domain-wall conduits. Adv. Mater. **22**, 2706–2710 (2010)
34. A. Beguivin et al., Simultaneous magnetoresistance and magneto-optical measurements of domain wall properties in nanodevices. J. Appl. Phys. **115**, 17C718 (2014)
35. R. Mattheis, S. Glathe, M. Diegel, U. Hübner, Concepts and steps for the realization of a new domain wall based giant magnetoresistance nanowire device: from the available 24 multiturn counter to a 212 turn counter. J. Appl. Phys. **111**, 113920 (2012)
36. Patent-Cowburn-US20070047156.pdf
37. A. Fernández-Pacheco et al., Three dimensional magnetic nanowires grown by focused electron-beam induced deposition. Sci. Rep. **3**, 1492 (2013)
38. Magnetic Data Storage (2010). https://www.google.com/patents/US20100128510?dq=cowburn +US+20100128510&hl=en&sa=X&ei=XlVrVOKPENKvacy3gogM&ved=0CB8Q6AEwAA
39. S. Parkin, Systematic variation of the strength and oscillation period of indirect magnetic exchange coupling through the 3d, 4d, and 5d transition metals. Phys. Rev. Lett. **67**, 3598–3601 (1991)
40. E.L. Starostin, G.H.M. van der Heijden, The shape of a Möbius strip. Nat. Mater. **6**, 563–567 (2007)
41. D. Mills, Surface spin-flop state in a simple antiferromagnet. Phys. Rev. Lett. **20**, 18–21 (1968)
42. D. Mills, W. Saslow, Surface effects in the Heisenberg antiferromagnet. Phys. Rev. **171**, 488–506 (1968)
43. D. Elefant, R. Schäfer, J. Thomas, H. Vinzelberg, C. Schneider, Competition of spin-flip and spin-flop dominated processes in magnetic multilayers: magnetization reversal, magnetotransport, and domain structure in the NiFe/Cu system. Phys. Rev. B **77**, 014426 (2008)
44. J. Meersschaut et al., Hard-axis magnetization behavior and the surface spin-flop transition in antiferromagnetic Fe/Cr(100) superlattices. Phys. Rev. B **73**, 144428 (2006)
45. C. Micheletti, R. Griffiths, J. Yeomans, Surface spin-flop and discommensuration transitions in antiferromagnets. Phys. Rev. B **59**, 6239–6249 (1999)
46. S. Te Velthuis, J. Jiang, S. Bader, G. Felcher, Spin flop transition in a finite antiferromagnetic superlattice: evolution of the magnetic structure. Phys. Rev. Lett. **89**, 127203 (2002)
47. U.K. Rößler, A.N. Bogdanov, Magnetic phase diagrams for models of synthetic antiferromagnets. J. Appl. Phys. **101**, 09D105 (2007)
48. U. Rößler, A. Bogdanov, Magnetic states and reorientation transitions in antiferromagnetic superlattices. Phys. Rev. B **69**, 094405 (2004)
49. J.-P. Nguenang, A.J. Kenfack, T.C. Kofané, Soliton-like excitations in a deformable spin model. J. Phys. Condens. Matter **16**, 373–403 (2004)
50. M.G. Pini et al., Surface spin-flop transition in a uniaxial antiferromagnetic Fe/Cr superlattice induced by a magnetic field of arbitrary direction. J. Phys. Condens. Matter **19**, 136001 (2007)

51. R. Wang, D. Mills, E. Fullerton, J. Mattson, S. Bader, Surface spin-flop transition in Fe/Cr(211) superlattices: Experiment and theory. Phys. Rev. Lett. **72**, 920–923 (1994)
52. S. Rakhmanova, D. Mills, E. Fullerton, Low-frequency dynamic response and hysteresis in magnetic superlattices. Phys. Rev. B **57**, 476–484 (1998)
53. R. Lavrijsen et al., Magnetism in Co[sub 80-x]Fe[sub x]B[sub 20]: effect of crystallization. J. Appl. Phys. **109**, 093905 (2011)
54. E. Fullerton, M. Conover, J. Mattson, C. Sowers, S. Bader, Oscillatory interlayer coupling and giant magnetoresistance in epitaxial Fe/Cr(211) and (100) superlattices. Phys. Rev. B **48**, 15755–15763 (1993)
55. A. Fernández-Pacheco et al., Controllable nucleation and propagation of topological magnetic solitons in CoFeB/Ru ferrimagnetic superlattices. Phys. Rev. B—Condens. Matter Mater. Phys. **86**, (2012)
56. B. Dieny, J.P. Gavigan, J.P. Rebouillat, Magnetisation processes, hysteresis and finite-size effects in model multilayer systems of cubic or uniaxial anisotropy with antiferromagnetic coupling between adjacent ferromagnetic layers. J. Phys. Condens. Matter **2**, 159–185 (1990)
57. A. Fernández-Pacheco, No Title. *to be Publ.*
58. E.Y. Vedmedenko, D. Altwein, Topologically protected magnetic helix for all-spin-based applications. Phys. Rev. Lett. **112**, 017206 (2014)
59. D. Petit, R. Mansell, A. Fernández-Pacheco, J.H. Lee, R.P. Cowburn, in *VLSI: Circuits for Emerging Applications*, ed. by T. Wojcicki (CRC Press, Boca Raton, 2014)
60. R. Lavrijsen et al., Magnetic ratchet for three-dimensional spintronic memory and logic. Nature **493**, 647–650 (2013)
61. J.H. Lee et al., Soliton propagation in micron-sized magnetic ratchet elements. Appl. Phys. Lett. **104**, 232404 (2014)
62. J.-H. Lee et al., Domain imaging during soliton propagation in a 3D magnetic ratchet. SPIN **03**, 1340013 (2013)
63. R. Lavrijsen et al., Multi-bit operations in vertical spintronic shift registers. Nanotechnology **25**, 105201 (2014)

Index

J. Seidel (ed.), *Topological Structures in Ferroic Materials*, Springer Series in Materials Science 228, DOI 10.1007/978-3-319-25301-5

CPSIA information can be obtained at www.ICGtesting.com
Printed in the USA
BVOW10*0824030316

438912BV00007B/2/P

9 783319 252995